Handbook of Antioxidants: Chemical, Biological and Synthetic Aspects

Handbook of Antioxidants: Chemical, Biological and Synthetic Aspects

Edited by **Nick Gilmour**

SYRAWOOD
PUBLISHING HOUSE
New York

Published by Syrawood Publishing House,
750 Third Avenue, 9th Floor,
New York, NY 10017, USA
www.syrawoodpublishinghouse.com

Handbook of Antioxidants: Chemical, Biological and Synthetic Aspects
Edited by Nick Gilmour

International Standard Book Number: 978-1-68286-324-4(Hardback)

Printed in China.

Contents

Preface

Antioxidants are primarily used as reducing agents. The applications of antioxidants range from food preservatives to stabilizers in lubricants. This book contains some path-breaking studies on antioxidants, their properties and applications. Included in this book are detailed discussions on topics such as antioxidant metabolism, industrial uses of antioxidants, synthesis of antioxidants and their impacts on health, etc. The book presents researches and studies performed by experts across the globe which will help both undergraduate and post graduate students. It will also prove beneficial for professionals and anyone who wants to delve deeper into this field.

This book is a result of research of several months to collate the most relevant data in the field.

When I was approached with the idea of this book and the proposal to edit it, I was overwhelmed. It gave me an opportunity to reach out to all those who share a common interest with me in this field. I had 3 main parameters for editing this text:

1. Accuracy - The data and information provided in this book should be up-to-date and valuable to the readers.
2. Structure - The data must be presented in a structured format for easy understanding and better grasping of the readers.
3. Universal Approach - This book not only targets students but also experts and innovators in the field, thus my aim was to present topics which are of use to all.

Thus, it took me a couple of months to finish the editing of this book.

I would like to make a special mention of my publisher who considered me worthy of this opportunity and also supported me throughout the editing process. I would also like to thank the editing team at the back-end who extended their help whenever required.

Editor

Phytochemicals and Antioxidative Properties of Borneo Indigenous Liposu (*Baccaurea lanceolata*) and Tampoi (*Baccaurea macrocarpa*) Fruits

Mohd Fadzelly Abu Bakar [1,2,*], Nor Ezani Ahmad [1], Fifilyana Abdul Karim [1] and Syazlina Saib [1]

[1] Institute for Tropical Biology and Conservation, Universiti Malaysia Sabah, Jalan UMS, Kota Kinabalu, Sabah 88400, Malaysia; E-Mails: misznarney@gmail.com (N.E.A.); fifilyana1111@gmail.com (F.A.K.); syazlina1991@gmail.com (S.S.)

[2] Faculty of Science, Technology and Human Development, Universiti Tun Hussein Onn Malaysia (UTHM), Parit Raja, Batu Pahat, Johor 86400, Malaysia

* Author to whom correspondence should be addressed; E-Mail: mofadz@ums.edu.my

---

Abstract: Two underutilized indigenous fruits of Borneo, Liposu (*Baccaurea lanceolata*) and Tampoi (*Baccaurea macrocarpa*) were investigated for their total phenolic (TPC), flavonoid (TFC), anthocyanin (TAC) and carotenoid (TCC) contents as well as antioxidant properties *in vitro*. The fruits were separated into three different parts (*i.e.*, pericarp, flesh and seed) and extracted using 80% methanol. Antioxidant activity was determined using DPPH (2,2-diphenyl-1-picrylhydrazyl) free radical scavenging, ABTS decolorization and FRAP (Ferric Reducing Antioxidant Power) assays. The results showed that *B. macrocarpa* pericarp contained the highest amount of total phenolics, total flavonoid, total anthocyanin and total carotenoid with the values of 60.04 ± 0.53 mg GAE/g, 44.68 ± 0.67 mg CE/g, 1.23 ± 0.20 mg c-3-gE/100 g and 0.81 ± 0.14 mg BCE/g. Results from DPPH, ABTS and FRAP assays also showed that the pericarp of *B. macrocarpa* displayed the highest antioxidant capacity. The antioxidant activity of the extract was significantly correlated with the total phenolic and flavonoid contents, but not with the carotenoid contents. In conclusion, *B. macrocarpa* displayed high potential as natural source of phytochemicals with antioxidant properties.

Keywords: *Baccaurea lanceolata*; *Baccaurea macrocarpa*; phenolic; flavonoid; anthocyanin; carotenoid; antioxidant activity

1. Introduction

Exogenous antioxidants are essential in the human body due to depletion of natural antioxidants [1]. They prevent or delay oxidation when exposed to free radicals and reactive oxygen species which are generated continuously inside the human body [2]. Reactive oxygen species might cause oxidative damage that leads to chronic diseases such as cancer, neurodegenerative disease, arthritis and diabetes mellitus [3]. Niki [4] reported that antioxidants can be evaluated by many methods but often give inconclusive and conflicting results. Therefore, assessment of antioxidants using more than one method is suggested to give more consistent results [5].

Previous studies have reported that high consumption of fresh fruits and vegetables may have protective effects against a wide range of chronic diseases caused by oxidative stress in the body [6]. This is mainly due to the presence of antioxidants constituents and bioactive compounds such as vitamin C, carotenoid, phenolics, flavonoids, tannins and anthocyanidins that are able to scavenge free radicals and inhibit lipid peroxidation [7–9]. A comprehensive analysis conducted based on the epidemiological studies available in the literature shows that there is undoubted evidence that higher consumption of fruits and vegetables reduces the risk of cardiovascular disease as well as cancer [10–12].

Despite many research on commonly consumed fruits, little information is available for underutilized indigenous fruits. Underutilized fruit is defined as fruit that is less popular, with under-exploited potential and not ready for commercialization [13]. Borneo (Malaysia—Sabah and Sarawak; Indonesia—Kalimantan and Brunei) has more than hundred types of indigenous fruits that can be found in backyards, small-scale orchards and also tropical rainforests. However, many of the indigenous fruits in Borneo are considered underutilized since they are only available locally and not much research has been conducted on these fruits. These underutilized fruits may contain a significant amount of phytochemicals or other unique compound which might have health-promoting properties. Their antioxidant capacity may be comparable or even superior to that of the more extensively studied fruits [14].

Baccaurea macrocarpa and *Baccaurea lanceolata* come from the genera of *Baccaurea* (family: Euphorbiaceae). Most of the *Baccaurea* species are endemic to Borneo. Both *B. macrocarpa* and *B. lanceolata* fruits are green to purple when young, and turn into yellow to orange when ripe. Both species are also dioecious, having the male and female reproductive organs borne on separate individuals of the same species, and can be propagate by seed. Although both species almost resembles each other, unlike *B. lanceolata*, *B. macrocarpa* trees have buttresses. *B. lanceolata* usually have one to four seeded berries with globose to ellipsoid fruit shape while the *B. macrocarpa* only have two to three seed berries with oblongoid to globose fruit shape [15]. For *B. lanceolata*, the pericarp and flesh are edible while its seed is discarded. Meanwhile, only the flesh of *B. macrocarpa* is edible while the other part, pericarp and seed are discarded [16]. *B. lanceolata* bears thick skin, white arils, with a tasty, sour flavor which is common in the wild and unknown in cultivation while the *B. macrocarpa* have tangy sweet taste, white arils around the seeds is uncommon in the wild and already introduced in garden cultivation [17].

Plants of *Baccaurea* species have been shown to contain diverse phytochemicals and pharmacological properties [16,18–20]. The present study was conducted to investigate the potential of *B. lanceolata* and *B. macrocarpa* fruits as a new source of natural bioactive phytochemicals and nutraceutical.

2. Materials and Methods

2.1. Plant Material and Sample Preparation

The whole fruits of *B. macrocarpa* and *B. lanceolata* were collected from Beaufort, Sabah, Malaysia. The fruits were cleaned and separated into pericarp, flesh and seed. Authentication of the fruits was done by Mr Johnny Gisil from Institute for Tropical Biology and Conservation of Universiti Malaysia Sabah, Malaysia. The pericarp, flesh and seed of the fruits were freeze-dried and ground to a fine powder. The samples were then kept in air-tight container and stored in freezer with temperature of −20 °C.

2.2. Extraction

All samples were extracted using 80% methanol (at a ratio of 1:20) at room temperature on an orbital shaker for 12 h at 200 rpm [21]. The resulting slurry was filtered through a Whatman No. 1 filter paper and subsequently used for determination of total phenolic, flavonoid, anthocyanin and carotenoid contents as well as total antioxidant activity.

2.3. Determination of Total Phenolic Content (TPC)

Total phenolic content was determined spectrophotometrically using Folin-Ciocalteu's reagent as described by Velioglu *et al.* [21]. One hundred micro liters of sample extract was mixed with 750 μL of Folin-Ciocalteu reagent solution. The solution was mixed well using vortex and then allowed to stand at room temperature for 5 min. About 750 μL of sodium bicarbonate solution was then added to the mixture. The solution again allowed to stand at room temperature for 90 min. After 90 min, absorbance was measured at 725 nm using spectrophotometer. Gallic acid was used as standard. Total phenolic contents of the extracts were determined from the standard graph. The results were expressed as mg gallic acid equivalents in 1 g of sample (mg GAE/g).

2.4. Determination of Total Flavonoid Content (TFC)

Total flavonoid content was determined using aluminum chloride spectrophotometry assay as described by Zhishen *et al.* [22]. About 1 mL of sample extract was added to a beaker. Four (4) mL of distilled water were then added to the mixture followed by 0.3 mL of 5% sodium nitrite. After 6 min, 0.6 mL 10% $AlCl_3$ was added and allowed to stand for 5 min. Then, 2 mL 1M NaOH and 2.1 mL of distilled water were added to the mixture and mixed using vortex. Absorbance of the mixture was determined at 510 nm. Catechin was used as standard. The total flavonoid contents of the extracts were determined from the standard graph. The total flavonoid content of the extracts was expressed as mg catechin equivalents in 1 g of sample (mg CE/g).

2.5. Determination of Total Anthocyanin Content (TAC)

The total anthocyanin content was determined by the pH-differential method as described by Giusti and Wrolstad [23] where the anthocyanin pigments undergo reversible structural transformations with a change in pH manifested by strikingly different absorbance spectra, which permits accurate and rapid measurement of the total anthocyanins. The colored oxonium form predominates at pH 1.0 and the colorless hemiketal form at pH 4.5. 0.5 mL of sample extract was mixed with 3.5 mL of 0.025 M potassium chloride buffer pH 1. The mixture were then allowed to mix properly using a vortex and allowed to stand for 15 min. After 15 min, absorbance at 515 and 700 nm was measured using spectrophotometer. The same mixture was then combined with 3.5 mL of 0.025 M sodium acetate buffer pH 4.5 and allowed to stand for 15 min. After 15 min, absorbance again was measured at 515 and 700 nm using spectrophotometer. The total anthocyanin content was expressed as mg of cyanidin-3-glucoside equivalents in 100 g of dried sample (mg c-3-gE/100 g dried sample) and was calculated using the formula as follows:

$$\text{The total anthocyanin content (mg/100 g of dried sample)} = A \times MW \times DF \times 1000/(e \times C) \quad (1)$$

where A is absorbance = $(A515 - A700)_{pH1.0} - (A515 - A700)_{pH4.5}$; MW is a molecular weight + cyanidin-3-glucoside = 449.2; DF is the dilution factor of the samples, e is the molar absorbity of cyanidin-3-glucoside = 26,900; and C is the concentration of the buffer in mg/mL.

2.6. Determination of Total Carotenoid Content (TCC)

Total carotenoid content was determined according to Hess *et al.* [24]. Three hundred micro liters of sample extract was added to 300 μL of distilled water and 600 μL of 80% methanol in a test tube. The mixture was then mixed with 1200 μL of *n*-hexane solution. The final solution was then centrifuged at 2000 rpm at 4 °C for 5 min. After that, the solution should form two layers of solvents. The upper layer (contained carotenoid) was transferred to a cuvette and the absorbance was measured at 350 nm. The results were expressed as mg beta carotene equivalents in 1 g of sample (mg BCE/g).

2.7. Antioxidant Assessment

2.7.1. DPPH (2,2-Diphenyl-1-pycryl-hydrazyl) Free Radical Scavenging Assay

The free radical scavenging ability of the extracts was determined according to Mensor *et al.* [25]. One mililiter of 0.3 mM DPPH solution was added to a test tube. Then, 2.5 mL of sample extracts or standards was added to the mixture. The mixture was then allowed to stand for 30 min in room temperature. The mixture was transferred to a cuvette and the absorbance value was then measured using spectrophotometer at 518 nm. The Antioxidant Activity (AA) was calculated as:

$$AA\% = 100 - [(\text{Abs sample} - \text{Abs empty sample})/(\text{Abs control})] \times 100 \quad (2)$$

where Abs is absorbance.

The result obtained is expressed as ascorbic acid equivalent antioxidant capacity (AEAC) in 1 g of dry sample using the following equations:

$$AEAC = (IC_{50\,(AA)}/IC_{50(sample)}) \times 10^5 \quad (3)$$

where AA is ascorbic acid.

2.7.2. ABTS Decolourization Assay

This assay was conducted by using ABTS as a model of free radical [26]. Briefly, an equal amount of 7 mM ABTS solution and 2.45 mM potassium persulfate was mixed and allowed to stand for 15 h in the dark under room temperature. Later, the solution was diluted with methanol to obtain absorbance of 0.7 ± 0.2 units at 734 nm. The plants extracts were separately dissolved in methanol to yield a concentration of 1 mg/mL. About 200 μL of methanolic test solution of each sample was added to 2 mL of ABTS free radical cation solution. The solution was allowed to mix well using vortex for 45 s. The solution was then transferred into a cuvette and measured at 734 nm. The result obtained was expressed as the ascorbic acid equivalent antioxidant capacity (AEAC) in 1 g of dry sample.

2.7.3. FRAP (Ferric Reducing/Antioxidant Power) Assay

This procedure was done according to Benzie and Strain [27] with slight modification. The working FRAP reagent was produced by mixing 20 mL of TPTZ, 200 mL acetate buffer, and 20 mL ferric chloride and 24 mL distilled water in a 10:1:1 ratio prior to use and heated to 37 °C in a water bath. A total of 0.3 mL FRAP reagent was added to a cuvette and blank reading was then taken at 593 nm using spectrophotometer. Then, a total of 100 μL of sample extracts and 300 μL distilled water was then transferred to a cuvette and measured at 593 nm using spectrophotometer. The sample extracts were then added to the FRAP reagent. After 4 min, second reading at 593 nm was performed. The change in absorbance after 4 min from the initial blank reading was then compared with the standard curve. A standard of known Fe(II) concentration were run using several concentrations from 100 to 1000 μM. A standard curve was prepared by plotting the FRAP value of each standard *versus* its concentration. The final result was expressed as the concentration of antioxidant having ferric reducing ability.

2.8. Statistical Analysis

All experiments were carried out in triplicates and results are expressed as average of three independent experiments (mean ± standard deviation). Correlation analysis between antioxidant activity and phytochemical contents was analyzed using Pearson's correlation in SPSS version 11.0 program (SPSS Inc., Chicago, IL, USA).

3. Results and Discussion

Determination of total phenolic (TPC), flavonoid (TFC), anthocyanin (TAC) and carotenoid (TCC) contents of pericarp, flesh and seed extracts of *B. lanceolata* and *B. macrocarpa* were determined spectrophotometrically using standard methods. TPC of the fruits extracts ranged from 3.29 to 4.81 mg GAE/g for *B. lanceolata*, while it ranged from 2.74 to 60.04 mg GAE/g for *B. macrocarpa* (Table 1).

Pericarp of *B. macrocarpa* contained the highest TPC while the lowest TPC was observed in the seed of *B. macrocarpa*. The same trend of TFC was observed in *B. macrocarpa* extracts. This study was supported by Khir Jauhari *et al.* [18] who reported that peel of Belimbing Dayak (*B. angulata*) contained higher TPC and TFC as compared to the other parts of the fruits.

Table 1. Phytochemicals contents of *B. lanceolata* and *B. macrocarpa*.

Samples	TPC [a]	TFC [b]	TAC [c]	TCC [d]
B. lanceolata				
Pericarp	3.31 ± 0.48	2.29 ± 0.01	0.50 ± 0.13	0.75 ± 0.00
Flesh	4.81 ± 0.14	4.73 ± 0.27	0.37 ± 0.08	0.67 ± 0.15
Seed	3.29 ± 0.33	1.97 ± 0.19	0.07 ± 0.00	0.64 ± 0.28
B. macrocarpa				
Pericarp	60.04 ± 0.53	44.68 ± 0.67	1.23 ± 0.20	0.81 ± 0.14
Flesh	4.60 ± 0.10	1.51 ± 0.11	0.01 ± 0.01	0.69 ± 0.22
Seed	2.74 ± 0.24	2.14 ± 0.17	0.04 ± 0.00	0.47 ± 0.17

[a] TPC was expressed as mg gallic acid equivalent (mg GAE) in 1 g of dry sample; [b] TFC was expressed as mg catechin equivalent (mg CE) in 1 g of dry sample; [c] TAC was expressed as mg of cyanidin-3-glucoside equivalents (mg c-3-gE) in 100 g of dry sample; [d] TCC was expressed as mg beta carotene equivalents (mg BCE) in 1 g of dry sample.

Meanwhile, the amount of total anthocyanin content was relatively low in all samples tested (Table 1). In *B. lanceolata*, anthocyanin content ranged from 0.07 to 0.50 mg c-3-gE/100g. For *B. macrocarpa*, TAC were in the ranged only from 0.01 to 1.23 mg c-3-gE/100g. Pericarp of *B. macrocarpa* contained the highest TAC while the flesh of *B. macrocarpa* displayed the lowest TAC. According to Vangdal and Slimestad [28], TAC usually higher in dark-red color fruit (*i.e.*, cherry) as compared to pale-yellow color fruit (*i.e.*, plum).

On the other hand, a very low total carotenoid content was observed in all sample extracts. The TCC of the samples were in the ranged of 0.47 to 0.81 mg BCE/g (Table 1). Pericarp of *B. macrocarpa* contained the highest amount of TCC with the value of 0.81 mg BCE/g while seed of *B. macrocarpa* contained the lowest TCC with the value of 0.47 mg BCE/g.

Three methods (DPPH, ABTS and FRAP) were used for antioxidant estimation. In DPPH assay, extract of *B. macrocarpa* (pericarp) showed the highest scavenging effects; followed by the flesh of *B. macrocarpa* and the flesh of *B. lanceolata* (Table 2). In ABTS assay, the antioxidant capacity ranged from 2.16 to 9.97 mg/g; where *B. macrocarpa* (pericarp) displayed the highest scavenging activity, while pericarp of *B. lanceolata* displayed the lowest scavenging effect in this assay. In FRAP assay, the pericarp of *B. macrocarpa* displayed superior antioxidant (reducing effects) as compared to other samples.

Correlation analysis revealed that the DPPH assay was moderately correlated with the total phenolic (r^2 = 0.794, $p < 0.05$) and the flavonoid content (r^2 = 0.796, $p < 0.05$). Previous studies on the antioxidant activity of nectarines and peach showed a strong correlation between phenolic content and antioxidant capacity [29]. However, no correlation was observed between DPPH with TAC and TCC. These results were supported by previous literature which reported that anthocyanins could be easily degraded under different conditions (*i.e.*, high temperature and humidity) [14,29,30].

Table 2. Antioxidant activity of *B. lanceolata* and *B. macrocarpa*.

Samples	DPPH [a]	ABTS [b]	FRAP [c]
B. lanceolata			
Pericarp	48.93 ± 0.02	2.16 ± 0.11	2.29 ± 0.12
Flesh	94.36 ± 0.02	2.99 ± 0.12	2.81 ± 0.23
Seed	85.06 ± 3.15	3.03 ± 0.11	1.93 ± 0.17
B. macrocarpa			
Pericarp	196.94 ± 0.43	9.97 ± 0.30	102.88 ± 0.57
Flesh	97.35 ± 0.21	3.04 ± 0.11	1.90 ± 0.12
Seed	42.99 ± 0.17	2.90 ± 0.09	2.31 ± 0.15

[a] DPPH free radical scavenging was expressed as mg ascorbic acid equivalent antioxidant capacity (AEAC) in 1 g of dry sample; [b] ABTS decolorization assay was expressed as mg ascorbic acid equivalent antioxidant capacity (AEAC) in 1 g of dry sample; [c] FRAP was expressed as mM ferric reduction to ferrous in 1 g of dry sample.

Unlike DPPH, analysis of correlation revealed that the ABTS was strongly correlated with the total phenolic ($r^2 = 0.994$, $p < 0.01$) and the flavonoid content ($r^2 = 0.992$, $p < 0.01$). The same trend of correlation was also found by Gorenstein *et al.* [31] whereby the correlation analyzed between the polyphenols and antioxidant activity as assessed by two different methods showed that phenolics and flavonoid may contribute maximally to the ABTS assay and to a lesser extent to DPPH assay. There is also strong positive correlation between ABTS and TAC ($r^2 = 0.862$, $p < 0.05$). However, there is no correlation between ABTS and TCC ($p > 0.05$).

A strong positive correlation ($r^2 = 0.999$, $p < 0.01$) between TPC and reducing power of extracts suggested that phenolic compounds in *Baccaurea* species might have acted as powerful reducing agent. Li *et al.* [32] reported the existence of similar linear relationships between reducing power and TPC. Similarly, a strong correlation ($r^2 = 0.998$, $p < 0.01$) was determined between TFC and FRAP assay. This result was supported by Gil *et al.* [29] and Sabli *et al.* [33]. Similarly with ABTS assay, there is a strong correlation between TAC and FRAP assay, despite their low concentration in both *Baccaurea* fruits ($r^2 = 0.907$, $p < 0.05$). However, TCC was not correlate with FRAP and DPPH assays.

Based on the results obtained, the edible part of *B. lanceolata* has the highest phytochemical content as compared to non-edible parts whereas the flesh of *B. macrocarpa* contained considerable amount of TPC and TFC. The pericarp and seed of *B. macrocarpa* showed the highest phytochemical contents compared to the flesh part. The consumption of edible part of these fruits was considered safer and suitable for human direct consumption as it may reduce the risk of chronic diseases that are caused by the oxidation of free radicals and reactive oxygen species.

4. Conclusions

In conclusion, the pericarp of *B. macrocarpa* has the highest phytochemicals and displayed strong antioxidant activity *in vitro*. The results of this study also suggested that non-edible part of fruits can be used as natural antioxidants that might be used as phytomedicine. This is due to its function as secondary metabolites in plants defense system.

Acknowledgments

This research was financially supported by Ministry of Higher Education of Malaysia (MOHE) under Experimental Research Grant Scheme, ERGS (Project No. ERGS0028-STWN-1/2012). Special thanks to Monica Suleiman (Universiti Malaysia Sabah) for proof-reading this manuscript. The authors would also like to acknowledge Institute for Tropical Biology and Conservation as well as School of Science and Technology of Universiti Malaysia Sabah, Malaysia for the use of the laboratory facilities and technical assistance.

Author Contributions

Mohd Fadzelly Abu Bakar and Syazlina Saib have done the experimental design, data analysis and data interpretation. Mohd Fadzelly Abu Bakar, Nor Ezani Ahmad and Fifilyana Abdul Karim are responsible for preparing the manuscript. All authors read and approved the final manuscript.

Conflicts of Interest

The authors declare no conflict of interest.

References

1. Pourmorad, F.; Hosseinimehr, S.J.; Shahabimajd, N. Antioxidant activity, phenol and flavonoid contents of some selected Iranian medicinal plants. *Afr. J. Biotechnol.* **2006**, *5*, 1142–1145.
2. Akinmoladun, A.C.; Ibukun, E.O.; Obuotor, E.M.; Farombi, E.O. Phytochemical constituent and antioxidant activity of extract from the leaves of *Ocimum gratissimum*. *Sci. Res. Essay* **2007**, *2*, 163–166.
3. Ebrahimzadeh, M.A.; Nabavi, S.M.; Nabavi, S.F.; Bahramian, F.; Bekhradnia, A.R. Antioxidant and free radical scavenging activity of *H. officinalis* L. var. *angustifolius*, *V. odorata*, *B. hyrcana* and *C. speciosum*. *Pak. J. Pharm. Sci.* **2010**, *2*, 611–619.
4. Niki, E. Antioxidant capacity: Which capacity and how to assess it? *J. Berry Res.* **2011**, *1*, 169–176.
5. Takashima, M.; Horie, M.; Shichiri, M.; Hagihara, Y.; Yoshida, Y.; Niki, E. Assessment of antioxidant capacity for scavenging free radicals *in vitro*: A rational basis and practical application. *Free Radic. Biol. Med.* **2012**, *52*, 1242–1252.
6. Chiva-Blanch, G.; Visioli, F. Polyphenols and health: Moving beyond antioxidants. *J. Berry Res.* **2012**, *2*, 63–71.
7. Battino, M.; Beekwilder, J.; Denoyes-Rothan, B.; Laimer, M.; McDougall, G.J.; Mezzetti, B. Bioactive compounds in berries relevant to human health. *Nutr. Rev.* **2009**, *67*, S145–S150.
8. Tenore, G.C.; Novellino, E.; Basile, A. Nutraceuticals potential and antioxidants benefits of red pitaya (*Hylocereus polyrhizus*) extracts. *J. Funct. Foods* **2012**, *4*, 129–136.
9. Scalzo, J.; Politi, A.; Pellegrini, N.; Mezzetti, B.; Battino, M. Plant genotype affects total antioxidant capacity and phenolic contents in fruit. *Nutrition* **2005**, *21*, 207–213.

10. Boeing, H.; Bechthold, A.; Bub, A.; Ellinger, S.; Haller, D.; Kroke, A.; Leschik-Bonnet, E.; Müller, M.J.; Oberritter, H.; Schulze, M.; *et al.* Critical review: Vegetables and fruit in the prevention of chronic diseases. *Eur. J. Nutr.* **2012**, *51*, 627–663.
11. Hung, H.C.; Joshipura, K.J.; Jiang, R.; Hu, F.B.; Hunter, D.; Smith-Warner, S.A.; Colditz, G.A.; Rosner, B.; Spiegelman, D.; Willett, W.C. Fruit and vegetable intake and risk of major chronic disease. *J. Natl. Cancer Inst.* **2004**, *96*, 1577–1584.
12. Alvarez-Suarez, J.M.; Giampieri, F.; Tulipani, S.; Casoli, T.; di Stefano, G.; González-Paramás, A.M.; Santos-Buelga, C.; Busco, F.; Quiles, J.L.; Cordero, M.D.; *et al.* One-month strawberry-rich anthocyanin supplementation ameliorates cardiovascular risk, oxidative stress markers and platelet activation in humans. *J. Nutr. Biochem.* **2014**, *25*, 289–294.
13. Gruère, G.P.; Giuliani, A.; Smale, M. Marketing underutilized plant species for the benefit of the poor: A conceptual framework. In *Agrobiodiversity Conservation and Economic Development*; Kontoleon, A., Pasqual, U., Smale, M., Eds.; Routledge: Abingdon, UK, 2008; pp. 73–87.
14. Abu Bakar, M.F.; Mohamed, M.; Rahmat, A.; Fry, J. Phytochemicals and antioxidant activity of different parts of bambangan (*Mangifera pajang*) and tarap (*Artocarpus odoratissimus*). *Food Chem.* **2009**, *113*, 479–483.
15. Verheij, E.W.M.; Coronel, R.E. Edible fruits and nuts. In *Plant Resources of South-East Asia 2*; Prosea Foundation: Bogor, Indonesia, 1992; pp. 186–190.
16. Kulip, J. An ethnobotanical survey of medicinal and other useful plants of Muruts in Sabah, Malaysia. *Telopea* **2003**, *10*, 81–98.
17. Sosef, M.S.M.; Hong, L.T.; Prawirohatmodjo, S. Timber trees: Lesser-known timbers. In *Plant Resources of South-East Asia*; Backhuys Publishers: Leiden, The Netherlands, 1998; Volume 5, p. 97.
18. Khir Jauhari, N.; Ibrahim, D.; Ibrahim, M.; Nor Adrosyahya, M.; Md Nor, N.; Md Isa, K.A.; Ayob, M.K.; Nor Omar, M.; Hazali, N. Proximate composition and antioxidant activity of dried belimbing dayak (*Baccaurea angulata*) fruits. *Sains Malays.* **2013**, *42*, 129–134.
19. Voon, B.H.; Kueh, H.S. The nutritional value of indigenous fruits and vegetables in Sarawak. *Asia Pac. J. Clin. Nutr.* **1999**, *8*, 24–31.
20. Wan-Yaacob, A.; Tong, S.W.; Din, L.B.; Yusoff, Y.N.I.; Latiff, A.; Mat Salleh, K. An alkaloid phytochemical survey of plants in the Gunung Danum area, Sabah, Malaysia. *Malay. Nat. J.* **2000**, *54*, 87–94.
21. Velioglu, Y.S.; Mazza, G.; Gao, L.; Oomah, B.D. Antioxidant activity and total phenolics in selected fruits, vegetables, and grain products. *J. Agric. Food Chem.* **1998**, *46*, 4113–4117.
22. Zhishen, J.; Mengcheng, T.; Jianming, W. The determination of flavonoid contents in mulberry and their scavenging effects on superoxide radicals. *Food Chem.* **1999**, *64*, 555–559.
23. Giusti, M.M.; Wrolstad, R.E. Characterization and measurement of anthocyanins by UV-visible spectroscopy. In *Current Protocols in Food Analytical Chemistry*; Wrolstad, R.E., Ed.; John Wiley & Sons: New York, NY, USA, 2001; pp. 7–47.
24. Hess, D.; Keller, H.E.; Oberlin, B.; Bonfanti, R.; Schuep, W. Simultaneous determination of retinol, tocopherols, carotenes and lycopene in plasma by means of high performance liquid chromatography on reversed phase. *Int. J. Vitam. Nutr. Res.* **1991**, *61*, 232–238.

25. Mensor, L.L.; Menezes, F.S.; Leitão, G.G.; Reis, A.S.; dos Santos, T.C.; Coube, C.S.; Leitao, S.G. Screening of Brazillian plant extracts for antioxidant activity by the use of DPPH free radical method. *Phytother. Res.* **2001**, *15*, 127–130.
26. Re, R.; Pellegrinni, N.; Proteggente, A.; Pannala, A.; Yang, M.; Rice-Evans, C. Antioxidant activity applying an improved ABTS radical cation decolorization assay. *Free Radic. Biol. Med.* **1999**, *26*, 1231–1237.
27. Benzie, I.F.F.; Strain, J.J. The ferric reducing ability of plasma (FRAP) as a measure of "antioxidant power": The FRAP assay. *Anal. Biochem.* **1996**, *239*, 70–79.
28. Vangdal, E.; Slimestad, R. Methods to determine antioxidative capacity in fruit. *J. Fruit Ornam. Plant Res.* **2006**, *14*, 123–131.
29. Gil, M.I.; Tomas-Barberan, F.A.; Hess-Pierce, B.; Kader, A.A. Antioxidant capacities, phenolic compounds, carotenoids, and vitamin C contents of nectarine, peach, and plum cultivars from California. *J. Agric. Food Chem.* **2002**, *50*, 4976–4982.
30. Tonon, R.V.; Brabet, C.; Hubinger, M.D. Anthocyanin stability and antioxidant activity of spray-dried açai (*Euterpe oleracea* Mart.) juice produced with different carrier agents. *Food Res. Int.* **2010**, *43*, 907–914.
31. Gorenstein, S.; Martin-Belloso, O.; Park, Y.; Haruenkit, R.; Lojek, A.; Ciz, M. Comparison of some biochemical characteristics of different citrus fruits. *Food Chem.* **2001**, *74*, 309–315.
32. Li, H.; Hao, Z.; Wang, X.; Huang, L.; Li, J. Antioxidant activities of extracts and fractions from *Lysimachia foenum-graecum* Hance. *Bioresour. Technol.* **2009**, *100*, 970–974.
33. Sabli, F.; Mohamed, M.; Rahmat, A.; Ibrahim, H.; Abu Bakar, M.F. Antioxidant properties of selected *Etlingera* and *Zingiber* species (Zingiberaceae) from Borneo Island. *Int. J. Biol. Chem.* **2012**, *6*, 1–9.

2

Kinetics and Mechanistic Studies on the Reaction between Cytochrome c and Tea Catechins

Lihua Wang, Elizabeth Santos, Desiree Schenk and Montserrat Rabago-Smith *

Department of Chemistry and Biochemistry, Kettering University, 1700 University Ave, Flint, MI 48504, USA; E-Mails: lwang@kettering.edu (L.W.); santose2@chemistry.msu.edu (E.S.); desi.schenk@gmail.com (D.S.)

* Author to whom correspondence should be addressed; E-Mail: mrabagos@kettering.edu

Abstract: Green tea is characterized by the presence of an abundance of polyphenolic compounds, also known as catechins, including epicatechin (EC), epigallocatechin (EGC), epicatechin gallate (EGC) and epigallocatechin gallate (EGCG). In addition to being a popular beverage, tea consumption has been suggested as a mean of chemoprevention. However, its mode of action is unclear. It was discovered that tea catechins can react with cytochrome c. When oxidized cytochrome c was mixed with catechins commonly found in green tea under non-steady-state conditions, a reduction of cytochrome c was observed. The reaction rate of the catechins was dependent on the pH and the nature of the catechin. The pseudo-first order rate constant obtained increased in the order of EC < ECG < EGC < EGCG, which is consistent with previously reported superoxide reduction activities and Cu^{2+} reduction activities of tea catechins.

Keywords: cytochrome c; catechins; kinetics; mechanistic studies; antioxidant; chemo-preventive

1. Introduction

Tea, brewed from the plant, *Camellia sinensis*, is one of the most consumed beverages in the world, consumed by two-thirds of the world's population [1]. Green tea is produced by steaming freshly harvested leaves, yielding a dry, stable product [1]. It is characterized by the presence of an abundance of polyphenolic compounds, also known as catechins. Catechins contain a benzopyran skeleton with a "phenyl group substituted at the 2-position and a hydroxyl (or ester) function at the 3-position" [1]. The major catechins found in green tea are epicatechin (EC), epigallocatechin (EGC), epicatechin gallate (EGC) and epigallocatechin gallate (EGCG). EGCG is the most abundant and the most biologically active compound [2] (Figure 1: structures of the major catechins found in green tea). Green tea contains 30%–42% catechins by dry weight; it is the main source of natural antioxidants.

Figure 1. Different catechins found in tea.

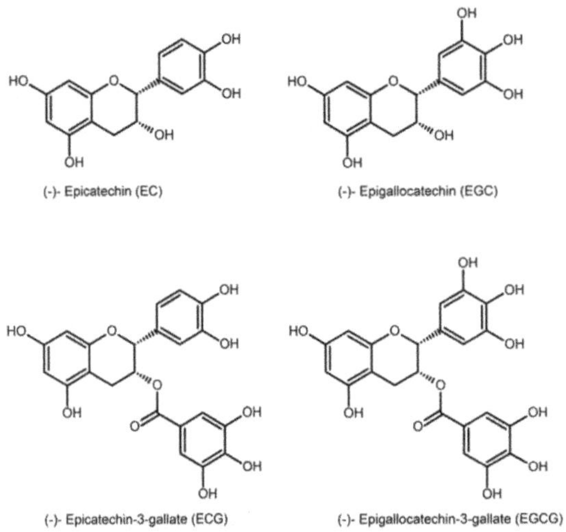

In addition to being a popular beverage, tea has been used for medicinal purposes, Tea consumption appears to decrease cancer risk and cardiovascular diseases, but its mechanism of action is uncertain [3,4]. Studies have shown that green tea and polyphenol compounds extracted from green tea have anticarcinogenic activity in animal models [5,6]. The proposed cancer prevention mechanisms include protection from oxidative stress, inhibition of enzymes, inhibition of growth factor-related cell signaling and the promotion of apoptosis in cancer cells [7,8].

The signal transduction pathways affected include the upstream of activator protein-1 (AP-1) and nuclear factor kappa B (NF-κB), activated through proline-rich kinase signaling. These transcription factors are known to be significant in tumor promoter-induced cell transformation and tumor promotion [9].

Other "studies show that many anti-cancer drugs induce the apoptosis of cancer cells and that catechin compounds induce apoptosis" [10] without affecting normal cells [1]. Additionally, it has been determined that the extent of their apoptosis-inducing effect is dependent on their polyphenolic structure.

The most commonly discussed mechanism for the biological activity of catechins involves their anti-oxidative activities. Reactive oxygen species induce oxidative stress, which has been linked to tumor promotion in many tissues. These reactive oxygen species include superoxide, hydroxyl, hydroperoxyl, peroxyl and alkoxyl radicals. Reactive nitrogen species include nitric oxide and the peroxynitrite anion. Catechins have the ability to prevent the formation of reactive oxygen and nitrogen species, which play pertinent roles in carcinogenesis [11]. Moreover, flavonoids are known to directly scavenge the superoxide anion and peroxynitrite. Recently, green tea extracts have been shown to be effective in scavenging reactive oxygen species when incorporated into plastic films for food packaging [12].

It has been reported that the main site of catechin antioxidant action depends on the structure of the catechins [13]. Computational studies predict that their activity comes from the B-rings and ortho-diphenols [14,15].

Cytochrome c is a small heme protein, important in mitochondrial respiration [16]. It is an important protein in the biological electron transport chains, assisting in the transport of electrons from organic substrates to oxygen. In addition, it has been discovered recently that it also plays an important role in the process of apoptosis. Cytochrome c is a metalloenzyme that requires a heme cofactor. In the electron transport chain, mitochondrial cytochrome c transports electrons from complex III to IV. The small, approximately 100-amino acid protein is conserved across most species. When the protein is released from the mitochondria into the cytoplasm, it has been found to initiate apoptosis in various cells with the addition of various compounds. Flavonoids were used to induce apoptosis via cytochrome c in leukemia cells. EGCG was found to do the same in bile duct cancer cells. Cytochrome c injections into the cytoplasm of neuronal cells also induced apoptosis after nerve cell growth factor deprivation.

We have discovered a new reaction between cytochrome c and catechins. Because of the importance of cytochrome c in apoptosis, understanding the interaction of tea components with cytochrome c would provide insights that may help in elucidating the mechanisms of the health benefits of green tea. This paper discusses the results of the kinetic and mechanistic study of the reaction between cytochrome c and various tea catechins.

2. Experimental Section

2.1. Materials and Equipment

Tea catechins, (−)-epicatechin (EC), (−)-epigallocatechin (ECG) and (−)-epigallocatechin gallate (EGCG), bovine heart cytochrome c, potassium phosphate and D_2O (99%) were purchased from Sigma-Aldrich (St. Louis, MO, USA and used without further purification. The kinetic studies were performed using a Cary 100 UV/Vis spectrophotometer (Palo Alto, CA, USA)).

2.2. Kinetic Study

The reactions were performed under anaerobic conditions using 10 or 20 μM of cytochrome c and an excess amount of a tea catechin (EC, ECG, EGC or EGCG). The concentration of the catechin was at least ten times that of the concentration of cytochrome c. A 0.3-M potassium phosphate buffer was used to maintain the pH at 7.0 or 7.7 at 25 °C. The reactant solutions were purged with nitrogen before being mixed by injection into a sealed cuvette. The progress of the reaction was monitored at 550 nm using a Cary 100 UV/Vis Spectrometer for a time period of four half-lives.

The kinetic studies on the presence of BHT (butylate hydroxytoluene) were performed following the same protocol; except that 0.10 or 0.25 molar equivalent of BHT *vs.* EC was mixed with EC, and then the mixture was reacted with cytochrome c. The buffer solution used for the BHT studies contained 10% ethanol, due to the low solubility of BHT in water.

A preliminary ^{1}H-NMR study was carried out by recording the proton NMR spectra every 10 min under anaerobic conditions using a 400-MHz Varian NMR. The solution contained cytochrome c (0.8 mm) and 0.5 equivalents of (–)-epicatechin in deuterated water.

3. Results and Discussion

When oxidized cytochrome c was mixed with catechins commonly found in green tea (EC, ECG and EGCG) under non-steady-state conditions, the reduction of cytochrome c was observed. Figure 2 shows the absorption spectra of cytochrome c (0.025 mm) before and after reacting with an excess amount of EGCG (1.0 mm) for an hour at room temperature. The UV-Vis absorption spectrum of the product was characteristic of the reduced cytochrome c with the calculated millimolar absorptivity of 26.9 mM^{-1} cm^{-1} at 550 nm. The millimolar absorptivity value was close to the value of 28.0 mM^{-1} cm^{-1} reported by the product specification sheet from Sigma (St. Louis, MO, USA)).

Figure 2. UV-Vis absorption spectra of cytochrome c (0.025 mm) before and after reacting with excess amount of EGCG (1.00 mm) for an hour at room temperature.

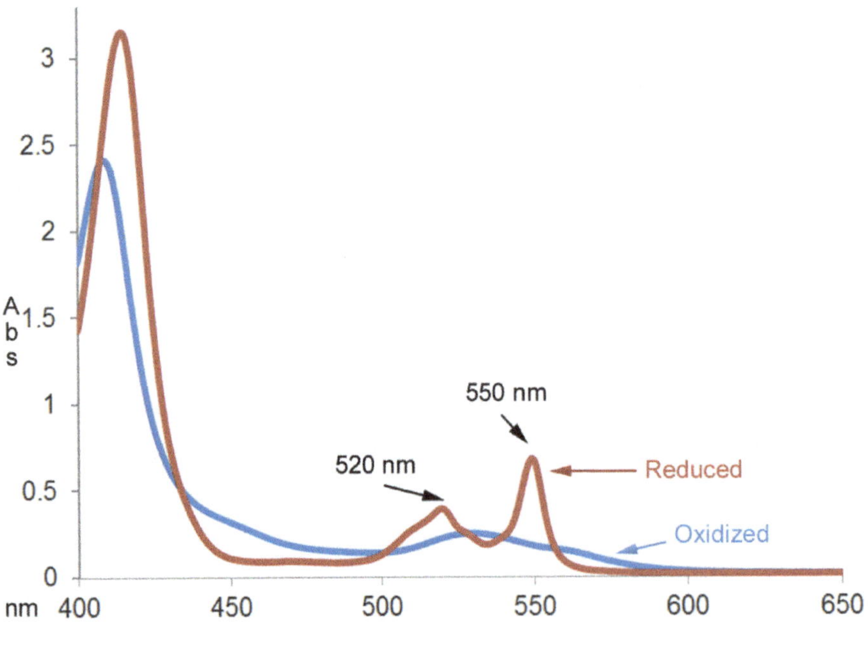

The kinetics of the reaction between cytochrome c and the tea catechins was studied as described in the Experimental Section.

The pseudo-first order rate constant (k_{obsd}) of the reaction depended on the nature and the concentration of the tea catechin used and the reaction pH. At pH 7.7, there was a linear relationship between the pseudo-first order rate constant (k_{obsd}) and the concentration of the catechin used. (Figure 3A, Equation (1)).

$$k_{obsd} = k_1 + k_2[catechin] \quad (1)$$

According to Figure 3A, the slope of the line, k_2, increased in the order of EC < EGC < ECG < EGCG. The reaction between EGCG and cytochrome c at pH 7.7 was too fast to be measured using a conventional UV-Vis spectrophotometer. There is a small *y*-intercept (k_1) that was similar for different catechins.

In order to obtain measurable rates for the reactions between EGCG and cytochrome c, the reaction was also studied at pH 7.0 for some of the catechins (Figure 3B). The reaction rate of the catechins at pH 7.0 was slower than those obtained at pH 7.7. The order of increasing k_2 values, EC < ECG < EGCG, at pH 7.0 was consistent with the order observed at pH 7.7. For ECG and EGCG, the relationship of k_{obsd} *vs.* [catechin] was consistent with Equation (1) at pH 7.0, with similar k_1 values for both catechins. However, the k_1 value for EC was determined to be zero. Because there are two terms in Equation 1 ($term_1 = k_1$ [cytochrome c] and $term_2 = k_2$ [catechin] [cytochrome c]), there should be two terms in the rate law for the reaction between cytochrome c and green tea catechins; Equation (2):

$$Rate = k_{obsd}[cyt.c] = k_1[cyt.c] + k_2[catechin][cyt.c] \quad (2)$$

Figure 3. Plot of pseudo-first order rate constant, k_{obsd}, *vs.* the concentration of the catechin for the reaction of different green tea catechins with cytochrome c at pH 7.7 (**A**) and pH 7.0 (**B**).

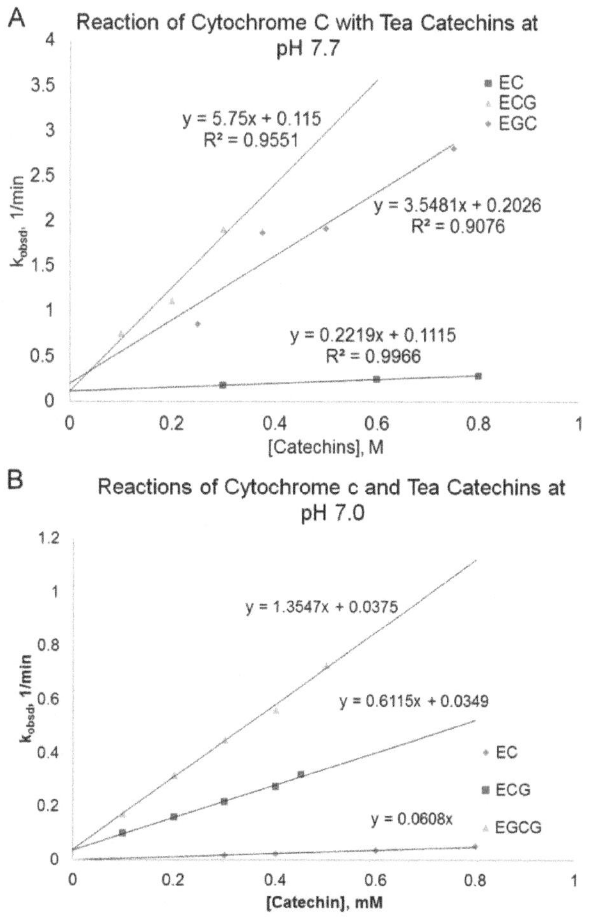

This indicates that there are two pathways for the reaction: a catechin-dependent pathway, in which the rate determining step is a second order reaction between cytochrome c and the catechins, and a catechin-independent pathway, in which the rate determining step is independent of the concentration of the catechin.

The catechin-dependent pathway can be described as follows:

Step 1 (rate determining step): electron transfer reaction between the oxidized cytochrome c and catechin to form the reduced cytochrome c (cyt.c) and a catechin free radical (catechin·):

$$cyt.c(ox) + catechin \rightarrow cyt.c(red) + catechin\cdot \tag{3}$$

Step 2: electron transfer reaction between another oxidized cytochrome c and the catechin free radical to form a reduced cytochrome c and a two-electron oxidized catechin:

$$cyt.c(ox) + catechin\cdot \rightarrow cyt.c(red) + catechin(ox) \tag{4}$$

Or two catechin free radicals react with each other to form a dimer:

$$2catechin\cdot \rightarrow (catechin_{ox})_2 \tag{5}$$

Kinetic data alone cannot differentiate the two possibilities for Step 2. These two alternatives can only been distinguished by the structural characterization of the oxidized catechins. Preliminary ^{1}H-NMR studies of EC and cytochrome c over time indicated that hydrogen 6 of EC [17] disappears (Figure 4). Characterization of the reaction products is currently underway in our laboratory.

Figure 4. Proton NMR spectra of epicatechin and cytochrome c over time.

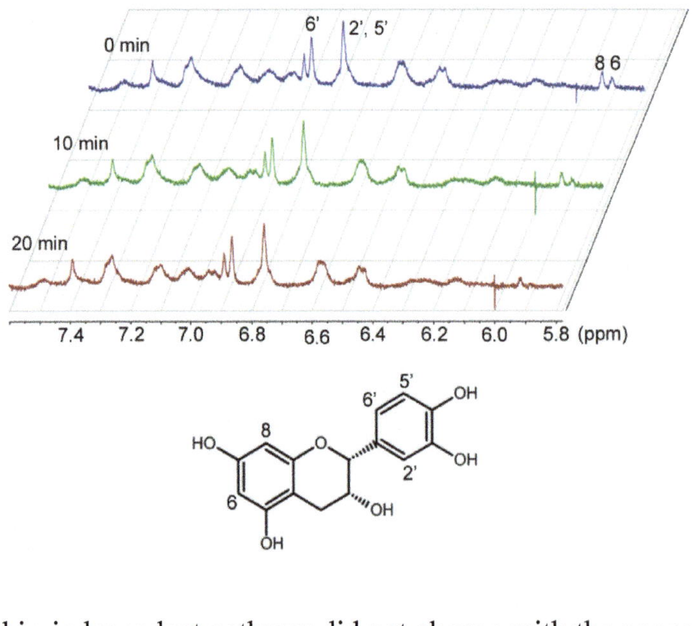

The rate of the catechin-independent pathway did not change with the concentration of the catechin. The possibility that this pathway involves the reduction of cytochrome c by a component in the solvent was ruled out, because, in the absence of the tea catechin, the reduction of cytochrome c did not occur. The proposed mechanism for the catechin-independent pathway consists of cytochrome c undergoing a slow transformation to a different form. The conformational change is the rate determining step of the pathway. The transformed cytochrome c then reacts with the catechin. Therefore, even though the catechin participates in the reaction, the rate law is independent of its concentration, because it is involved in a step after the rate-determining step. The catechin-independent pathway is described as follows:

Step 1 (rate determining step): structural transformation of cytochrome c:

$$cyt.c(ox) \rightarrow cyt.c(ox)^* \quad (6)$$

Step 2: electron transfer between the transformed cyt.c (ox)* and catechin to form the reduced cytochrome c and a catechin free radical:

$$cyt.c(ox)^* + catechin \rightarrow cyt.c(red) + catechin\cdot \quad (7)$$

The rest of the pathway should be the same as the catechin-dependent pathway, but it involves the transformed cytochrome c.

The fact that k_1 values were independent of the nature of the catechins except for EC at pH 7.0 supports the proposed mechanism (Figure 2B). The reason that EC and cytochrome c at pH 7.0 did not show the catechin-independent pathway is probably because the electron transfer reaction between EC and transformed cytochrome c at pH 7.0 is slower than the conformational change of cytochrome c. Thus, the rate determining steps in both pathways involve EC, which leads to a rate law that appears to have one pathway that is first order in EC and first order in cytochrome c.

As it can be seen from Figure 1, for reactions carried out using the same concentration of different types of tea catechins, the pseudo-first order rate constant obtained increased in the order of EC < ECG < EGC < EGCG. This is consistent with previously reported superoxide reduction activities and the Cu^{2+} reduction activities of tea catechins (Table 1) [18].

Table 1. A comparison of the relative reactivities of green tea catechins towards different oxidizing agents.

Catechins	Superoxide Reduction IC_{25} (μm) [15]	Reduction of Cu^{2+} RC_{25} (μm) [15]	Cytochrome C Reduction at pH 7.7 k_2 (mm^{-1} min)
EC	20.6	218	0.222
EGC	3.22	38.4	3.55
ECG	2.29	24.3	5.75
EGCG	1.45	12.5	Too fast to measure

Finally, the rate of the reaction was faster at pH 7.7 than at pH 7.0. This is because at pH 7.7, some of the catechins (EGCG (pK_{a1} = 7.75 [19]), ECG (pK_{a1} = 7.76 [19]), EGC (pK_{a1} = 7.73 [19]), EC (pK_{a1} = 8.72 [20])) are partially deprotonated, which results in a partial negative charge on the molecule and a stronger electron donating ability. This may facilitate the electron transfer process between the catechins and cytochrome c. The result is consistent with the pH-dependent radical scavenging capability of green tea catechins reported previously [16]. Cytochrome c is located in the intermembrane space of the mitochondria, which has a pH of 6.88 [21]. Therefore, the kinetic results at pH 7.0 are relevant to the environment of cytochrome c *in vivo*.

In the proposed mechanism, it is hypothesized that a radical catechin intermediate is formed. To test this hypothesis, the reaction was performed in the presence of BHT, a known free radical scavenger. The results obtained demonstrated a reduced reaction rate when BHT was added to the reaction

(Figure 5). Interestingly, the reduction of the rate was proportional to the concentration of BHT. These results support the existence of a radical intermediate.

Figure 5. Plot of pseudo-first order rate constant, k_{obsd}, *vs.* the concentration of epicatechin in the presence of 0.0.1 and 0.25 equivalents of BHT (butylate hydroxytoluene).

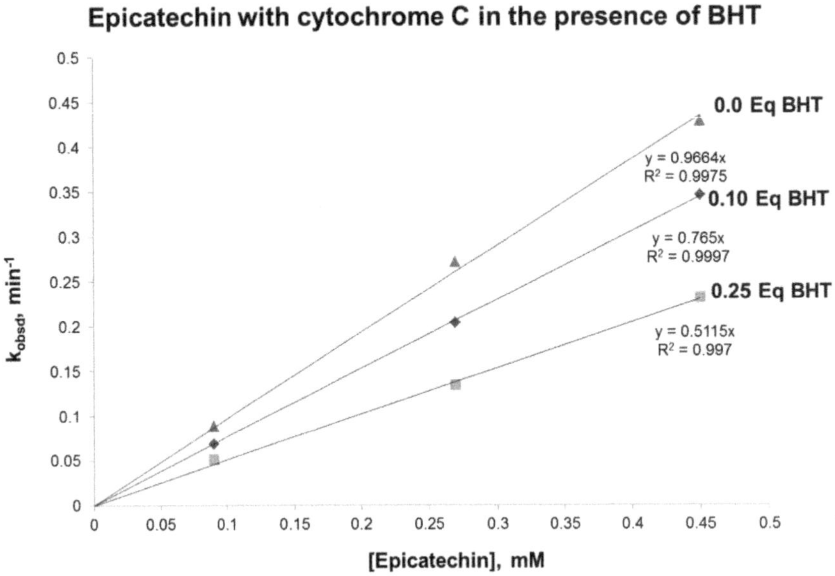

4. Conclusions

Green tea catechins can reduce cytochrome c. The reaction rate of the catechins was dependent on the pH and the nature of the catechin. The pseudo-first order rate constant obtained increases in the order of EC < ECG < EGC < EGCG, which is consistent with previously reported superoxide reduction activities and the Cu^{2+} reduction activities of tea catechins.

Author Contributions

Lihua Wang and Montserrat Rabago-Smith are faculty advisors of the research. Both Wang and Rabago-Smith were involved in the development of the project, supervising undergraduate research students who conducted the experiments, and the writing of the paper. Rabago-Smith coordinated the project and is the corresponding author of the publication. Elizabeth Santos and Desiree Schenk were undergraduate research students at the time and were involved in conducting the experiments and the writing of the paper.

Conflicts of Interest

The Authors declare no conflict of interest.

References

1. Khan, N.; Mukhtar, H. Multitargeted therapy of cancer by green tea polyphenols. *Cancer Lett.* **2008**, *269*, 269–280.
2. Hou, Z.; Lambert, J.; Chin, K.; Yang, C. Effects of tea polyphenols on signal transduction pathways related to cancer chemoprevention. *Mutat. Res.* **2004**, *55*, 3–19.
3. Dong, J.; Ye, J.; Lu, J.; Zheng, Z.; Liang, Y. Isolation of antioxidant catechins from green tea and its decaffeination. *Food Bioprod. Process*. **2011**, *89*, 62–66.
4. Moto, S.; Fujita, S.K.; Yamakazi, Y.; Kamataki, T. Inhibition by green tea catechins of metabolic activation of procarcinogens by human cytochrome P450. *Mutat. Res.* **2001**, *479*, 197–206.
5. McCarthy, S.; Caporali, A.; Enkemann, S.; Scaltriti, M.; Eschrich, S.; Davalli, P.; Corti, A.; Lee, A.; Sung, J.; Yeatman, T.; *et al*. Green tea catechins suppress the DNA synthesis marker MCM7 in the TRAMP model of prostate cancer. *Mol. Oncol*. **2007**, *1*, 196–204.
6. Huang, M.; Ferraro, T. Phenolic Compounds in Food and Cancer Prevention. *Phenol. Compd. Food Their. Eff. Health II*. **1992**, *507*, 8–34.
7. Lambert, J.; Hong, J.; Lee, M.; Sang, S.; Meng, X.; Lu, H.; Yang, C. Biotransformation and Bioavailability of Tea Polyphenols: Implications for Cancer Prevention. *Res. Phenol. Compd. Foods Nat. Health Prod.* **2005**, *1*, 212–224.
8. Lin, J.; Pan, M.; Liang, Y.; Shiau, S.; Ho, C. New Momentum on the Action Mechanisms of Black Tea Polyphenols, the Tea flavins. *Phenol. Compd. Food Nat. Health Prod.* **2005**, 197–209.
9. Bode, A.; Dong, Z. Signal Transduction Pathways: Targets for Green and Black Tea Polyphenols. *J. Biochem. Mol. Biol.* **2003**, *36*, 66–77.
10. Isemura, M.; Sacki, K.; Kiumara, T.; Hayakawa, S.; Minami, T.; Sazuka, M. Tea catechins and related polyphenols as anti-cancer agents. *BioFactors* **2000**, *13*, 81–85.
11. Kang, N.; Shin, S.; Lee, H.; Lee, K. Polyphenols as small molecular inhibitors of signaling cascades in carcinogenesis. *Pharmacol. Ther.* **2011**, *130*, 310–324.
12. Colon, M.; Nerin, C. Role of catechins in the antioxidant capacity of an active film containing green tea, green coffee, and grapefruit extracts. *J. Agric. Food Chem.* **2012**, *60*, 9842–9849.
13. Sang, S.; Tian, S.; Wang, H.; Stark, R.; Rosen, R.; Yang, C.; Ho, C. Chemical studies of the antioxidant mechanism of tea catechins: Radical reaction products of epicatechin with peroxyl radicals. *Bioorganic. Med. Chem.* **2003**, *11*, 3371–3378.
14. Leopoldini, M.; Marino, T.; Russo, N.; Toscano, M. Antioxidant properties of phenolic compounds: H-atom *versus* electron transfer mechanism, *J. Phys. Chem. A* **2004**, *108*, 4916–4922.
15. Leopoldini, M.; Russo, N.; Toscano, M. The molecular basis of working mechanism of natural polyphenolic antioxidants. *Food Chem.* **2011**, *125*, 288–306.
16. Sutin, N. *Bioinorganic Chemistry II (Advances in Chemistry Series 162)*; Raymond, K.N., Ed.; American Chemiscal Society: Washington, DC, USA, 1977; pp. 156–172.
17. Davis, A.; Cai, Y.; Davies, A.; Lewis, J.R. ^{1}H and ^{13}C NMR Assignments of Some Green Tea Polyphenols. *Magn. Reson. Chem.* **1996**, *34*, 887–890.
18. Furuno, K.; Toyokazu, K.; Sughara, N. The contribution of the pyrogallol moiety to the superoxide radical scavenging activity of flavonoids. *Biol. Pharm. Bull.* **2002**, *25*, 19–23.

19. Muzolf, M.; Szymusiak, H.; Gliszczynska-Swiglo, A.; Rietjens, I.M.; Tyrakowska, B. pH-Dependent Radical Scavenging Capacity of Green Tea Catechins. *J. Agric. Food Chem.* **2008**, *56*, 816–823.
20. Kennedy, J.A.; Munro, M.H.G.; Powell, H.K.J.; Porter, L.J.; Foo, L.Y. The protonation reactions of catechin and epicatechin and related compounds. *Aust. J. Chem.* **1984**, *37*, 885–892.
21. Porcelli, A.M.; Ghelli, A.; Zanna, C.; Pinton, P.; Rizzuto, R.; Rugolo, M. pH dependence across the outer mitochondrial measured with a green fluorescenct protein mutant. *Biochem. Biophys. Res. Commun.* **2005**, *326*, 799–804.

3

Syzyguim guineense Extracts Show Antioxidant Activities and Beneficial Activities on Oxidative Stress Induced by Ferric Chloride in the Liver Homogenate

Constant Anatole Pieme [1,*], Joseph Ngoupayo [2], Claude Herve Khou-Kouz Nkoulou [2], Bruno Moukette Moukette [1], Borgia Legrand Njinkio Nono [3], Vicky Jocelyne Ama Moor [1], Jacqueline Ze Minkande [4] and Jeanne Yonkeu Ngogang [1]

[1] Laboratory of Biochemistry, Department of Biochemistry and Physiological Sciences; Faculty of Medicine and Biomedical Sciences, University of Yaounde I, P.O. Box 1364 Yaounde, Cameroon; E-Mails: moukette2006@yahoo.fr (B.M.M.); movicky@yahoo.fr (V.J.A.M.); jngogang@yahoo.fr (J.Y.N.)

[2] Department of pharmacy, Faculty of Medicine and Biomedical Sciences, University of Yaounde I, P.O. Box 1364 Yaounde, Cameroon; E-Mails: jngoupayo@yahoo.com (J.N.); khoukouzherve@yahoo.fr (C.H.K.N.)

[3] Department of Biochemistry, Faculty of Sciences; University of Yaounde I, P.O. Box 812 Yaounde, Cameroon; E-Mail: njinoborle@yahoo.fr

[4] Department of Anesthesia and Reanimation, Faculty of Medicine and Biomedical Sciences, University of Yaounde I, P.O. Box 1364 Yaounde, Cameroon; E-Mail: minkandeze@yahoo.fr

* Author to whom correspondence should be addressed; E-Mail: apieme@yahoo.fr

---

Abstract: The aim of this study was to determine the *in vitro* antioxidant activity, free radical scavenging property and the beneficial effects of extracts of various parts of *Syzygium guineense* in reducing oxidative stress damage in the liver. The effects of extracts on free radicals were determined on radicals DPPH, ABTS, NO and OH followed by the antioxidant properties using Ferric Reducing Antioxidant Power assay (FRAP) and hosphomolybdenum (PPMB). The phytochemical screening of these extracts was performed by determination of the phenolic content. The oxidative damage inhibition in the liver was determined by measuring malondialdehyde (MDA) as well as the activity of the antioxidant enzymes superoxide dismutase (SOD), catalase (CAT) and peroxidase. Overall, the bark extract of the ethanol/water or methanol showed the highest radical scavenging activities against

DPPH, ABTS and OH radicals compared to the other extracts. This extract also contained the highest phenolic content implying the potential contribution of phenolic compounds towards the antioxidant activities. However, the methanol extract of the root demonstrated the highest protective effects of SOD and CAT against ferric chloride while the hydro-ethanol extract of the leaves exhibited the highest inhibitory effects on lipid peroxidation. These findings suggest that antioxidant properties of *S. guineense* extracts could be attributed to phenolic compounds revealed by phytochemical studies. Thus, the present results indicate clearly that the extracts of *S. guineense* possess antioxidant properties and could serve as free radical inhibitors or scavengers, acting possibly as primary antioxidants. The antioxidant properties of the bark extract may thus sustain its various biological activities.

Keywords: antioxidant activity; phenolic content; *Syzygium guineense;* radical scavenging activity; lipid peroxidation

---

Abbreviations

DPPH: 2,2-diphenyl-1-picrylhydrazyl, 1,1-diphenyl-2-picrylhydrazyl radical; FRAP: Ferric Reducing Ability power; ABTS: 2,2 -Azinobis(3-ethylbenzthiazoline)-6-sulfonic acid; BHT: Butylated hydroxytoluene; Vit C: Vitamine C; MDA: Malondialdehyde; TBA: Thiobarbituric acid; SNP: Sodium nitroprusside; SOD: Superoxide dismutase.

1. Introduction

Oxidative damage (OD) of cellular biomolecules such as lipids, proteins and DNA is thought to play a crucial role in the incidence of several chronic diseases [1–3]. It is among the major causative factors in induction of many chronic and degenerative diseases including atherosclerosis, ischemic heart disease, aging, diabetes mellitus, cancer, immunosuppression, neuro-degenerative diseases and male infertility [4]. Oxidative stress is defined as the imbalance between the pro-oxidants and antioxidants in favor of the oxidants, potentially leading to damage, while antioxidants are substances which possess free radical chain reaction breaking properties. Most living species have an efficient defense system to protect themselves against the oxidative stress induced by reactive oxygen species (ROS) [5]. These antioxidants include enzymes (superoxide dismutase, catalase, glutathione peroxidase and reductase), glutathione and vitamins (e.g. vitamin E, C), as well as many dietary components and compounds from plants [6]. Antioxidants scavenge and control the formation of free radicals thereby preventing oxidative damage to cellular components arising as a consequence of chemical reactions involving ROS [7]. Most of the antioxidants isolated from higher plants are polyphenols [8]. Evaluation of polyphenols and antioxidant activity of herbal products has become an important tool to understand the medicinal property of plants. Interest in the possible health benefits of flavonoids and other polyphenolic compounds has increased in recent years owing to their potent antioxidant and free-radical scavenging activities [9–11]. Recent investigations have shown that the antioxidant properties of plants could be correlated with oxidative stress defense and different human diseases including cancer, atherosclerosis and the aging process [2,3]. The antioxidants can interfere with the oxidation process by reacting with free radicals, chelating free catalytic metals and also by acting as oxygen scavengers.

Syzygium guineense (Myrtaceae) is an odorous species native of the wooded savannahs and tropical forests of Africa [12]. This short-trunked tree grows widely in northern Benin. Its wild, oval fruits are edible with high concentrations of Ca, Mg, Fe, K and P [13,14]. *S. guineense* is included among the African plant species that are active against malaria [15]. Its bark is used in traditional medicine to treat gasto-intestinal upsets and diarrhea [16–18].

In Cameroon, the wood of *S. guineense* is used as fuel for the household, for construction and for carpentry [19]. Twigs and leaves are used against hookworm and leaves against amenorrhea and madness. *S. guineense* sap yields a black dye used to color textiles. The antibacterial properties of the watery extract of *S. guineense* have been demonstrated on different strains of bacteria responsible for diarrhea [20,21]. Ethanol (EtOH) extracts of the stem bark of *S. guineense* showed molluscicidal activities and cardiovascular properties, mainly the reduction of blood pressure [22]. Antibacterial activity of triterpenes isolated from *S. guineense* has been demonstrated [23]. Other biological properties such as anti-inflammatory, analgesic and immunological activities of different part of *S. guineense* have been reported [24,25]. The chemical composition of essential oil from *S. guineense* was also investigated [26]. However, no studies reporting the *in vitro* antioxidant and protective effects of extracts of *S. guineense* have been demonstrated.

2. Material and Methods

2.1. Extraction and Preparation of Extract Solution

Syzygium guineense (*S. guineense*) was collected in November 2013 in the Bafia town, Mbam and Inougou Central Region of Cameroon. A sample part of the plant (leaves) was identified and deposited at the National Herbarium under the reference number 20899/SRFCam. Parts of the plant (leaves, bark and roots) of *S. guineense* were carefully cleaned with distilled water, then dried in the open air in an enclosed space away from light. The different samples were ground to a coarse powder and stored for extraction. Their powders (75 g) were soaked separately in 750 mL of solvent (water or ethanol/water v/v) for 48 h followed by filtration with Watchman No 1 filter paper (Sigma-Aldrich, Munich, Gemany). The filtrates obtained were concentrated in a rotary evaporator for hydro-alcoholic (EtOH/H_2O) extract and dried in the oven. The extracts were kept at 4 °C for testing antioxidant properties. The extraction yield was calculated. Prior to experimentation, the solutions of the plant extracts were reconstituted using the appropriate solvent and different dilutions (25, 50, 75, 150, 300 μg/mL) of each extract.

2.2. Determination of the Free Radical Scavenging Potential of the Samples

2.2.1. Scavenging Activity of DPPH Radical

This assay measures the free radical scavenging capacity of the investigated extracts [27]. DPPH is a molecule containing a stable free radical. In the presence of an antioxidant, which can donate an electron to DPPH, the purple color typical of free DPPH radical decays, and the absorbance change at $\lambda = 517$ nm is measured. The antioxidant activity of the plant extracts are examined on the basis of the scavenging effect on the stable DPPH free radical activity. Briefly, in 3 mL of each diluted extract or vitamin C used as standard, 1 mL of a methanol solution of DPPH 0.1 mM is added. The mixture is kept in the dark at room temperature for 30 min and the absorbance is measured at 517 nm against a blank. The following equation is used to determine the percentage of the radical scavenging activity of each extract.

$$\text{Scavenging effect (\%)} = 100 \times (A_o - A_s)/A_o \quad (1)$$

where A_o is the absorbance of the blank; A_s is the absorbance of the sample.

2.2.2. Scavenging Effect of the $ABTS^+$ Radical

ABTS assay was based on a previously described method with slight modifications [28]. ABTS radical cation ($ABTS^+$) was produced by the reaction of a 7 mM ABTS solution with 2.45 mM potassium persulphate. The mixture was stored in the dark at room temperature for 12 h before use. The $ABTS^+$ solution was diluted with ethanol to an absorbance of 0.70 ± 0.05 at 734 nm. After addition of 25 µL of extract sample or vitamin C (standard) to 2 mL of diluted $ABTS^+$ solution, absorbance at 734 nm was measured at exactly 6 min. The decrease in absorption was used for calculating scavenging effect values. The following equation is used to determine the percentage of the radical scavenging activity of each extract.

$$\text{Scavenging effect (\%)} = 100 \times (A_o - A_s)/A_o \quad (2)$$

where A_o is the absorbance of the blank; A_s is the absorbance of the sample.

2.2.3. Nitric Oxide Scavenging Activity

Nitric oxide scavenging activity was determined according to the Griess Illosvoy reaction [29]. The reaction mixture contained: 10 mM sodium nitroprusside (SNP) in 0.5 M phosphate buffer, pH 7.4, and various concentrations of the tested extracts to a final volume of 3 mL. After incubation for 60 min at 37 °C, Griess reagent (α-napthyl-ethylenediamine 0.1% in water and sulphanilic acid 1% in H_3PO_4 5%) was added. The pink chromophore generated during diazotization of nitrite ions with sulphanilamide and subsequent coupling with α-napthyl-ethylenediamine was measured spectrophotometrically at 540 nm. The vitamin C was used as a positive control. Nitric oxide scavenging ability (%) was calculated by using the formula:

$$\text{Percentage of NO radical scavenging activity (\%)} = 100 \times (A_o - A_s)/A_o \quad (3)$$

where A_o is the absorbance of the blank; A_s is the absorbance of the sample.

2.2.4. Hydroxyl Radical Scavenging Activity

The scavenging activity of the extracts on hydroxyl radical was measured according to a previously described method [30]. In 1.5 mL of each diluted extract, 60 µL of $FeCl_3$ (1 mM), 90 µL of 1,10-phenanthroline (1 mM), 2.4 mL of 0.2 M phosphate buffer, pH 7.8 and 150 µL of H_2O_2 (0.17 M) was added respectively. The mixture was then homogenized and incubated at room temperature for 5 min. The absorbance was read at 560 nm against the blank. The percentage of the radical scavenging activity of each extract was calculated from the equation below:

$$\text{Percentage of OH radical scavenging activity (\%)} = 100 \times (A_o - A_s)/A_o \qquad (4)$$

where A_o is the absorbance of the blank; A_s is the absorbance of the sample.

2.3. Determination of the Total Antioxidant Potential of the Different Samples

2.3.1. Total Antioxidant Activity by Ferric Reducing Antioxidant Power Assay (FRAP)

The FRAP was determined using a previously described method with slight modifications [31]. This test measures the reduction of ferric ion to the ferrous form in the presence of antioxidant compounds. The fresh FRAP reagent consists of 500 mL of acetate buffer (300 mM pH 3.6), 50 mL of 2,4,6-tri(2-pyridyl)-*s*-triazine (TPTZ) (10 mM), and 50 mL of $FeCl_3 \cdot 6H_2O$ (50 mM). The colorimetric measurement was performed at 593 nm and the reaction monitored up to 12 min on 75 µL of each extract and 2 mL of FRAP reagent. The vitamin C was used to draw a standard curve and the butylated hydroxyl toluene (BHT) was used for the comparison. The absorbance was read at 593 nm. The results were expressed as mg equivalent vitamin C/g of dried extract.

2.3.2. Phosphomolybdenum Total Antioxidant Assay (PPMB)

The total antioxidant activity of extracts was evaluated by green phosphomolybdenum complex according to a described method [32]. An aliquot of 10 µL of sample solution was mixed with 1 mL of reagent solution (0.6 M sulphuric acid, 28 mM sodium phosphate and 4 mM ammonium molybdate) in a micro centrifuge tube. Tubes were incubated in a dry thermal bath at 95 °C for 90 min. After cooling, the absorbance of the mixture was measured at 695 nm against a blank. Vitamin C was used as reference to draw the standard curve and BHT was used for the comparison. The reducing capacities of the analyzed extracts were expressed as mg of vitamin C equivalents/g of dried extract.

2.3.3. Reducing Power Assay

The reducing power of the extracts was determined by the method described by Oyaizu [33]. Different concentrations of extracts (20, 40, 80, 120, 240 µg/mL) in 1 mL of distilled water were mixed with 2.5 mL of phosphate buffer (0.2 M, pH 6.6) and 2.5 mL of potassium ferrocyanide (1%). The mixture was incubated at 50 °C for 20 min. Aliquots (2.5 mL) of trichloroacetic acid (10%) were added to the mixture and centrifuged at 3000 rpm for 10 min. The upper layer of the solution (2.5m L) was mixed with 2.5 mL of distilled water and 0.5 mL of $FeCl_3$ (0.1%). Increased absorbance was measured at 700 nm against the blank indicating increased reducing power.

2.4. Determination of the Phenolic Content of the Extracts

2.4.1. Total Phenol Determination

The total phenol was determined by the Folin-Ciocalteu method [34], the reaction mixture contained: 200 μL of diluted spice extract, 800 μL of freshly prepared diluted Folin-Ciocalteu reagent and 2 mL of 7.5% sodium carbonate. The final mixture was diluted to 7 mL with deionized water. The mixtures were kept in the dark at ambient conditions for 2 h to complete the reaction. The absorbance at 765 nm was measured. Caffeic acid was used as standard and the results were expressed as mg caffeic acid (CAE)/g dried extract.

2.4.2. Determination of Total Flavonoid Content

Total flavonoid content was determined using aluminum chloride ($AlCl_3$) according to a known method using quercetin as a standard [35]. The spice extract (0.1 mL) was added to 0.3 mL distilled water followed by 5% $NaNO_2$ (0.03 mL). After 5 min at 25 °C, $AlCl_3$ (0.03 mL, 10%) was added. After further 5 min, the reaction mixture was treated with 0.2 mL of 1 mM NaOH. Finally, the reaction mixture was diluted to 1 mL with water and the absorbance was measured at 510 nm. The results were expressed as mg quercetin (QE)/g of dried extract (QE/g dried extract).

2.4.3. Determination of Total Flavonols

Total flavonols in the plant extracts were estimated using a known method with modifications [36]. To 2.0 mL of sample (standard), 2.0 mL of 2% $AlCl_3$ ethanol and 3.0 mL (50 g/L) sodium acetate solutions were added. The absorption at 440 nm was read after 2.5 h of incubation at 20 °C. Extract samples were evaluated at a final concentration of 0.1 mg/mL. Total flavonol content was calculated as quercetin (mg/g) using the following equation $y = 5.3911\ x + 0.0313$, $R^2 = 0.9967$, based on the calibration curve, where x was the absorbance and the concentration was expressed as the quercetin equivalent (mg/g) dried extract (QE/g dried extract).

2.4.4. Phytochemical Analysis

The phytochemical analysis of the extracts was carried out to identify the presence of proteins, lipids, alkaloids, saponins, steroids, cardiac glycosides, tannins, resins and coumarins using a standard method.

2.5. Protective Properties of the Plant against Oxidative Damage

2.5.1. Preparation of Liver Homogenate

The liver was isolated from three normal albino *Wistar* rats. The organs were weighed and 10% (w/v) homogenate was prepared in phosphate buffer (0.1 M, pH 7.4 having 0.15 M KCl) using a homogenizer at 4 °C [37]. The homogenate was centrifuged at 3000 rpm for 15 min and clear cell-free supernatant was used for the study.

2.5.2. Preparation of the Pro-Oxidative Solution

The oxidant solution was prepared directly before its utilization by adding a solution of ferric chloride 100 mM to H_2O_2 0.50% prepared in phosphate buffer (0.1 M, pH 7.4). This solution was used for the investigation of the protective assays on liver enzymes.

2.5.3. Total Protein Content

The total protein content of the mixture of liver was measured according to the protein kit supplier methods (Human Kit-Hu102536, Boehringer Ingelheim, Ingelheim, Germany). This result was used to express the activities of the different enzymes per g of organs.

2.5.4. *In vitro* Lipid Peroxidation Assay

Lipid peroxidation assay was performed by a formerly described protocol [38]. Phosphate buffer 0.58 mL (0.1 M; PH 7.4), 200 µL samples, 200 µL liver homogenate and 20 µL ferric chloride (100 mM) were combined to form a mixture which was placed in a shaking water bath for 1 h at 37 °C. Reaction was terminated by adding 1 mL TCA (10%), TBA 1 mL (0.67%) to all the tubes which were placed in boiling water bath for 20 min. Then the test tubes were moved to a crushed ice bath and were centrifuged at 3000 rpm for 10 min. Absorbance of the supernatant was checked at 535 nm and was calculated as nM of MDA tissue using a molar extinction coefficient of 1.56×10^5/M.cm.

2.5.5. Determination of Peroxidase Activity

Peroxidase activity was determined by the peroxidase kit supplier with modifications. A solution containing 1 mL of the oxidant solution ($FeCl_2$, 100 mM) and extract or vitamin C (standard) for a final concentration of 100 µg/mL was incubated for 1 h a water bath at 37 °C. An aliquot of PBS (0.1 mL), hydrogen peroxide (50 µL), and pyrogallol solution (110 µL) were added to distilled water (625 µL) that had been earlier dispensed into an Eppendorf tube (Fisher Scientific, Wohlen, Switzerland). The plant extract (75 µL) from the mixture was added thereafter. The same reagents were always used, except for the extract which was replaced by distilled water (75 µL), for the blank, the oxidant solution for the control and vitamin C for the standard. The reaction was mixed and incubated for at least 10 min. The solution containing 100 mM, pH 6.0 PBS (40 µL) and 0.002% (v/v) diluted liver homogenate (40 µL) was added to the blank and test mixtures respectively. These were mixed, and the increase in absorbance at 420 nm was measured at every 10 s interval for 3 min using a spectrophotometer (BioMate 3S UV-Visible, Thermo Scientific™Manufacturer, Wohlen, Switzerland). One unit of peroxidase was defined as the change in absorbance/seconds/mg of protein at 420 nm.

2.5.6. Determination of Catalase Activity

Prior to the test, a solution containing a mixture of 1 mL of the oxidant solution and extract or vitamin C (standard) for a final concentration of 100 µg/mL was incubated for 1h in a water bath at 37 °C. Catalase activity of plant extracts on homogenate was assayed as previously described with modifications [38]. Briefly, an aliquot of hydrogen peroxide (0.8 mL) was dispensed into an Eppendorf tube. Phosphate buffer (1.0 mL), extracted sample/vitamin C/oxidant solution (75 μL) and (0.002% v/v) diluted homogenate (125 μL) were added. The reaction mixture (0.5 mL) was dispensed into 5% dichromate reagent (1.0 mL) and vigorously shaken. The mixture was heated in a Clifton water bath for 10 min, and allowed to cool. The absorbance at 570 nm was measured spectrophotometrically. The absorbance obtained was extrapolated from a prepared standard graph. The catalase activity was thereafter expressed as Unit/min/mg of protein:

$$\text{CAT}((\text{unit})/\text{mg protein}) = (Abs/\text{min} \times 30{,}000 \text{ units})/(40 \text{ cm/M} \times \text{mgprotein}) \times df \quad (5)$$

where df = dilution factor, Abs = absorbance.

2.5.7. Superoxide Dismutase (SOD) Activity

The measurement of total SOD activity was performed according to the Misra and Fridovich method with modifications similar to those on catalase assay [39]. The principle of this method is based on the inhibition of epinephrine autoxidation. A solution of 0.2 mL distilled water and 2.5 mL sodium carbonate buffer 0.05 M, pH 10.2.was added to 0.3 mL buffered epinephrine to initiate the reaction. The absorbance at 480 nm was read for 150 s at 30 s intervals against a blank made up of 2.5 mL buffer, 0.3 mL epinephrine and 0.2 mL distilled water. The following equation allowed the calculation of the SOD activity:

$$\text{SOD (unit/mg protein)} = \text{SOD (units/mL/min)/protein (mg/mL)} \times df \quad (6)$$

where df = dilution factor.

2.6. Statistical Analysis

The results were presented as mean ± SEM of triplicate assays. Analyses of data were conducted using the Kruskal Wallis test and Dunnett's multiple test (SPSS program version 18.0 for Windows, IBM Corporation, New York, NY, USA). The Log probit was used to determinate the IC_{50} and the software XLstat version 7 (Addinsoft, New York, NY, USA) was used to achieve the Pearson Correlation Analysis (PCA). The differences were considered as significant at $p < 0.05$.

3. Results and Discussion

3.1. Extraction

Table 1 shows the yield of extraction of the leaves, bark and roots of *S. guineense*. The yield varies from 10.4% (roots) to 16% (bark) for EtOH/H_2O and 5% (roots) to 6.67% (leaves or bark) for MeOH.

Table 1. Yield extract of *S. guineense*.

	Extracts	Yield of Extraction	
		EtOH/H_2O	**MeOH**
	Leaves	15.6%	6.67%
S. guineense	Bark	16%	6.67%
	Roots	10.4%	5%

3.2. Phytochemical Extracts

The results of the phytochemical composition of *S. guineense* extracts are given in Table 2. These results show that all samples contain sugars, proteins, lipids, polyphenols, alkaloids, saponins, steroids, cardiac glycosides, flavonoids, tannins; except for the EtOH/H_2O extract of the bark, which contains more coumarins. The results of the polyphenolic content of the extracts show the presence of total polyphenols, flavonoids and flavonols in the different parts of the *S. guineense* extracts studied (Table 3). The total polyphenol content of the extracts varies between 0.38 and 4.6 mg/g of dried material (DM) depending on type of solvent and the part of the plant. This level of polyphenol varied significantly ($p < 0.05$) between the different parts of the plant between the solvents. The result indicates that the bark has the highest level of polyphenols and that EtOH/H_2O was the best solvent.

Table 2. Phytochemical composition of extracts of *S. guineense*.

Phytochemical	EtOH/H_2O			MeOH		
Compounds	Leaves	Bark	Roots	Leaves	Bark	Roots
Sugars	+	+	+	+	+	+
Proteins	+	+	+	+	+	+
	+	+	+	+	+	+
Lipids	+	+	+	+	+	+
Alkaloids	+	+	+	+	+	+
Saponins	+	+	+	+	+	+
Stéroids/terpenoids	+	+	+	+	+	+
Cardiac glycosides	+	+	+	+	+	+
Tannins	+	+	+	+	+	+
Resins	−	−	−	−	−	−
Acids	−	−	−	−	−	−
Coumarins	−	+	−	−	−	−

(+) presence, (−) absence.

Regarding total flavonoids, the level varied from 0.05 to 0.79 mg/g of DM. This level of flavonoids was significantly higher ($p < 0.05$) in the bark extract in both solvents. In EtOH/H_2O extracts, the flavonoid content ranged from 0.79 ± 0.04 mg/g DM (bark) to 0.25 ± 0.02 mg/g DM (root) representing 16.92% and 8.62% respectively of the total polyphenol content. While in the MeOH extracts the flavonoids content ranges between 0.58 ± 0.04 mg/g DM (bark) and 0.05 ± 0.01 mg EQ/g DM (root) representing 13.84% (bark) and 13.16% (root) of the total polyphenol content. The extract of the bark in the EtOH/H_2O solvent possesses the highest flavonoid content. The flanovol level represents 9.8% and 5.34% of the total flavonoid content of the root and leaf extracts respectively with EtOH/H_2O. Concerning the MeOH extracts, the flavonol content ranged from 0.93% to 16.2% of the flavonoid content respectively for bark and roots. The EtOH/H_2O extract of the bark has the highest content of flavonols (Table 3).

3.3. Results of the Antioxidant Capacity of the Extracts of S. guineense

Given the complexity of the oxidation process and the diverse nature of antioxidants, two methods (FRAP and PPMB) were used to evaluate the antioxidant activity of extracts from different parts of *S. guineense.* The results of an antioxidant capacity are shown in Table 4. Results show that the total antioxidant capacity of the extract varies 1859.26 to 9172.22 mg AAE/g DM with FRAP and 1114.29 to 8042.86 mg of vitamin C equivalent (AAE)/g DM with PPMB. The MeOH extract of the bark has greater antioxidant capacity with the FRAP method as well as with the PPMB. These values are significantly higher compared to vitamin C used as a control (Table 4). For each solvent, the extract of the bark has the highest antioxidant capacity for both methods. Correlations between the polyphenol content and antioxidant capacity showed a significant positive correlation between total polyphenol and antioxidant capacity with FRAP and PPMB methods for the MeOH extracts of the root and bark respectively (Table 5). Similarly, a strong positive correlation was also observed for the EtOH/H_2O extracts of leaves and roots between the polyphenols and antioxidant capacity using PPMB.

Table 3. Phenolic content of *S. guineensis.*

Solvents	Extracts	Total polyphenols (mg of QE/gDM)	Flavonoids (mg of QE/gDM)	Flavonols (mg of QE/gDM)
	Leaves	$2.59 \pm 0.03^{a,*}$	$0.35 \pm 0.04^{a,*}$	0.0187 ± 0.003 *
EtOH/H_2O	Bark	$4.67 \pm 0.11^{b,*}$	$0.79 \pm 0.01^{b,*}$	0.0200 ± 0.003 *
	Roots	$2.9 \pm 0.17^{a,*}$	$0.25 \pm 0.02^{c,*}$	0.0245 ± 0.001 *
	Leaves	1.6 ± 0.05^{b}	0.06 ± 0.01^{a}	0.0069 ± 0.002
MeOH	Bark	4.19 ± 0.05^{a}	0.58 ± 0.04^{b}	0.0054 ± 0.001
	Roots	0.38 ± 0.04^{c}	0.05 ± 0.01^{a}	0.0081 ± 0.002

The results are expressed as mean ± SD ($n = 3$); the values with different superscripts in the same solvent are significantly different ($p < 0.05$); * significant difference ($p < 0.05$) for different solvents of the same portion of the plant.

Table 4. Variation antioxidant capacity of extracts of *S. guineense.*

Solvents	Extracts	FRAP (mg AAE/gDM)	PPMB (mg AAE/gDM)
	Leaves	8907.4 ± 30.6^{a}	$4000 \pm 157.14^{a,*,\alpha}$
EtOH/H_2O	Bark	9131.48 ± 84.13^{a}	$8150 \pm 135.71^{b,\alpha}$
	Roots	$9009.26 \pm 87.9^{b,*}$	$5664.29 \pm 64.29^{c,*,\alpha}$
	Leaves	$8850 \pm 28.87^{a,\alpha}$	$1442.86 \pm 71.43^{a,\alpha}$
MeOH	Bark	$9172.22 \pm 78.37^{b,\alpha}$	$8042.86 \pm 42.86^{b,\alpha}$
	Roots	$1859.26 \pm 50.1^{c,\alpha}$	$1114.29 \pm 103.02^{a,\alpha}$
	Vit C	9009.26 ± 42.43	5964.29 ± 78.57

The results are expressed as mean ± SD ($n = 3$); the values with different superscripts in the same solvent are significantly different ($p < 0.05$); * significant difference ($p < 0.05$) for different solvents of the same portion of the plant; $^{\alpha}$ significant difference ($p < 0.05$) between the extracts and vitamin C.

3.4. Antioxidant Activities of the Extracts of S. guieneense by Radical Scavenging

Results of the antioxidant activity by scavenging free radicals are shown in Table 6. From these results, the ability of extracts to trap radicals expressed as the IC_{50} varies, depending on the radical tested, the portion of the plant and the solvent extraction used. These results show that all IC_{50} values obtained are higher than that of vitamin C used as a control except for the hydroxyl radical. Among the extracts obtained from EtOH/H_2O the bark has lower IC_{50} of 5.52 g/mL, 16.25 mg/mL, 126.35 g/mL respectively for the DPPH, ABTS and OH radical. These results are similar to the extracts obtained with MeOH solvent. A significant and negative correlation is found between the polyphenol content and nitrite radical with the EtOH/H_2O extract obtained from leaves of *S. guineense* (Table 5).

Table 5. Pearson correlation analysis between polyphenol content and antioxidant properties.

Solvents	Extracts	Correlation Coefficient (R^2)					
		FRAP	PPMB	DPPH	ABTS	OH	NO
	Leaves	−0.846	0.999 *	−0.839	0.696	−0.371	−0.031 *
EtOH/H_2O	Bark	−0.068	−0.318	0.133	0.972	−0.445	0.869
	Roots	0.694	0.999 *	0.280	−0.567	0.863	−0.313
	Leaves	−0.938	0.638	−0.638	0.973	0.681	0.213
MeOH	Bark	0.425	0.999 *	0.747	−0.488	−0.981	−0.695
	Roots	0.998 *	−0.592	−0.515	−0.796	−0.750	−0.412

*$p < 0.05$ statistical significant.

Table 6. Variation of fifty percent inhibitory concentration (IC_{50}) of different extracts of *S. guineense.*

Solvents		Fifty Percent Inhibitory Concentraion (μg/mL)			
		DPPH	ABTS	OH	NO
	Leaves	$5.67 \pm 0^{a,*,\alpha}$	$92.8 \pm 2.01^{a,*,\alpha}$	$274.4 \pm 5.94^{a,\alpha}$	$276.19 \pm 21^{a,*}$
EtOH/H_2O	Bark	$5.52 \pm 0.02^{b,\alpha}$	$16.25 \pm 0.12^{b,*,\alpha}$	$126.35 \pm 1.79^{b,\alpha}$	$364.56 \pm 10.79^{b,\alpha}$
	Roots	$5.66 \pm 0.04^{a,*,\alpha}$	$116.69 \pm 3.61^{c,*,\alpha}$	$131.67 \pm 2.03^{c,*,\alpha}$	$370.34 \pm 5.96^{b,*,\alpha}$
	Leaves	$5.86 \pm 0.01^{a,\alpha}$	$64.07 \pm 1.25^{a,\alpha}$	$264.85 \pm 11.83^{a,\alpha}$	$527.62 \pm 2.25^{a,\alpha}$
MeOH	Bark	$5.55 \pm 0.02^{b,\alpha}$	$14.67 \pm 0.12^{b,\alpha}$	126.61 ± 1.23^{b}	$368.35 \pm 6.21^{b,\alpha}$
	Roots	$6.01 \pm 0.03^{c,\alpha}$	$393.98 \pm 22.06^{c,\alpha}$	$240.4 \pm 4.29^{a,\alpha}$	$397.16 \pm 8.52^{b,\alpha}$
	Vit C	5.35 ± 0.01	12.73 ± 0.3	112.74 ± 1.73	257.74 ± 1.73

The results are expressed as mean ± SD ($n = 3$); the values with different superscripts in the same solvent are significantly different ($p < 0.05$); * significant difference ($p < 0.05$) for different solvents of the same portion of the plant; α significant difference ($p < 0.05$) between the extracts and vitamin C.

3.5. Protective Effect of Extracts of S. guineensis on Some Markers Involved on Oxidative Stress

The results of the protective activity of the extracts of *S. guineensis* on some biochemical markers involved in oxidative stress are shown in Table 7. These results show for both solvents that the extract of the leaves has the best protective activity against lipid peroxidation compared to the other samples and vitamin C used as a positive control. The inhibition of lipid peroxidation by the extract of leaves EtOH/H_2O (50.82 μmol/mg of protein) is significantly higher ($p < 0.05$) than that of MeOH extract (67.67 μmol/mg of protein) of the same part of the plant, 40 fold higher than the negative control (Table 7). Extracts of *S. guineense* act differently on some enzyme markers of stress, but these values are still lower than the normal (group receiving no treatment). These results show that the extracts of MeOH and the root bark significantly protect the CAT (1.42, 1.14) and SOD (0.0193, 0.0174) against the $FeCl_3$ compared to vitamin C (0.66, 0.0174).

Table 7. Variation of markers involved in oxidative stress.

Solvents	Parts of Plant	SOD (µmol/min/mg of Protein)	CAT (µmol/min/mg of Protein)	Peroxidase (µmol/min/mg of Protein)	MDA (µmol/mg of Protein)
	Leaves	0.0522 ± 0.012 *	0.65 ± 0.08 *	52 ± 5.2	50.82 ± 0.66 ^{*,a}
$EtOH/H_2O$	Bark	0.0291 ± 0.0020 *	0.61 ± 0.05 *	67 ± 4.5	73.71 ± 4.33 ^{*,b}
	Roots	0.0097 ± 0.0005	0.71 ± 0.11 *	61 ± 5.7	53.7 ± 1.52 ^{*,a}
	Leaves	0.0135 ± 0.0034	0.98 ± 0.02 ^{*,a}	62 ± 5.5	67.67 ± 3.67 ^{*,a}
MeOH	Bark	0.0174 ± 0.0002	1.14 ± 0.08 ^{*,b}	83 ± 5.5 *	74.43 ± 2.64 ^{*,b}
	Roots	0.0193 ± 0.0001	1.42 ± 0.12 ^{*,b,c}	58 ± 0.6	73.43 ± 0.1 ^{*,b}
	Nor	0.1082 ± 0.0005	1.59 ± 0.15	423 ± 49.99	74.72 ± 4.75
	Ctrl Neg	0.0251 ± 0.0041	0.32 ± 0.02	52 ± 11.4	122.95 ± 1.09
	VitC	0.0174 ± 0.0002	0.66 ± 0.06 *	54 ± 0.1	70.98 ± 1.8 *

The results are expressed as mean ± SD ($n = 3$); the values with different superscripts in the same solvent are significantly different ($p < 0.05$); * significant difference ($p < 0.05$) for different solvents of the same portion of the plant; α significant difference ($p < 0.05$) between the extracts and vitamin C.

Natural products continue to be one of the most important sources of lead compounds for the pharmaceutical industry. Plants have been used for a long time as an alternative source of antioxidant, medicines and remedies for treating human diseases. Our study describes the phytochemical characterization of various *S. guineense* leaves, root and bark and antioxidant effects of extracts including the protective activity. The extraction yields and the content of phenols vary with the extraction solvent used. Differences in the chemical constitution were observed in phytochemical screening of the tested samples. All samples contain sugars, proteins, lipids, polyphenols, alkaloids, saponins, steroids, cardiac glycosides, flavonoids, tannins. Quantitative determination of phenols in the extracts demonstrated that the $EtOH/H_2O$ bark extract possesses a comparatively higher amount of phenols (Table 3). Our results confirmed that the extraction of polyphenol can be carried out with acidified methanol or ethanol solvents [40]. Studies on anthocyanin extraction demonstrated that methanol extraction was 20% more efficient than that of ethanol and 73% less efficient than water [40]. The mixture $EtOH/H_2O$, more polar than the MeOH, can easily solubilize phenolic Compositae. The polarity of a solvent can be increased by adding polar co-solvents to improve the affinity and the relative power of the polar solvent to molecules [41]. Our study showed that among the two solvents used ($EtOH/H_2O$, MeOH), the highest concentrations of polyphenols, flavonoids and total flavonols were obtained with an $EtOH/H_2O$ mixture. Phenolic compounds are important phyto-constituents and have potential effects against different diseases because of their antioxidant properties [42]. The antioxidant activity of the phenolic compounds is due primarily to their redox properties and chemical structure, which can play an important role in the chelation of transition metals and free radicals [43]. Flavonoids containing hydroxyl functional groups, are responsible for the antioxidant effects and act by scavenging or chelating. Several methods can be used to determine the *in vitro* antioxidant capacity of natural substances due to the complex composition of phytochemicals and oxidative process [44]. In our study we used two methods, the first is based on the reduction of ferric tripyridyltriazine to ferrous complex at low pH and the second on the reduction of ammonium molybdate (PPMB). All tested samples showed good antioxidant capacities. The EtOH/H_2O extract of the bark showed a significant antioxidant activity compared to vitamin C and other extracts. This activity of the extract is related to its high phenolic content (phenols, flavonoids and flavonols) found in this part of the plant compared to other extracts. Antioxidant activity of polyphenols from natural origin occurs in various mechanisms such as the prevention of chain initiation, peroxide decomposition, the prevention of further hydrogen abstraction, trapping free radicals, reducing ability, and the binding of transition metal ions [45]. Phenolic compounds have the ability to transfer hydrogen atoms which are generally associated with the presence of reducing agents [45,46]. In addition, the number and position of the hydroxyl groups of phenolic compounds also regulate their antioxidant activity [45]. The determination of antioxidant properties of the extract through radical scavenging (DPPH, ABTS, OH and NO) is considered to be a good indication of the ability of antioxidants to divest hydrogen even though they are not biologically relevant [47]. EtOH/H_2O and MeOH extracts of the bark showed scavenging activity significantly higher than other extracts for radical DPPH, ABTS and OH. The free radical scavenging activity of the extracts is completed by transfering electrons or hydrogen atoms through the redox properties of polyphenols [48,49].

The production of oxidants and the protection against them is intrinsic to every living cell [50,51]. To minimize oxidative damage, organisms have developed antioxidative mechanisms thought to be triggered by increased ROS production [50]. Oxidants such as ROS are taken care of by an antioxidative defense system that consists of enzymes and metabolites in all subcellular compartments [51–53]. In stress conditions, normal capacities of these mechanisms are insufficient; triggering cells to increase and expand, their antioxidative network is reduced. In this context, superoxide dismutase (SOD) and catalase (CAT) together constitute an important defense against reactive oxygen species (ROS). Catalase catalyzes the conversion of hydrogen peroxide to water and oxygen. Thus, catalase reduces the tissue injury by removing the H_2O_2. Superoxide dismutase (SOD) catalyzes the conversion of superoxide anion free radical to H_2O_2 through a dismutation reaction [51,54]. In our study, enzymes activities of SOD, CAT and peroxidase varied depending on the part of the plants. All activities of the CAT and peroxidase in the groups treated with the oxidant and extracts of *S. guineense* were higher than the negative group (only with oxidant) and positive group (oxidant and vitamin C) (Table 7). The exposure of liver homogenate to ferric chloride reduces the activity of these enzymes through several mechanisms through oxidative stress while the presence of extracts restored these activities through the antioxidant property. This antioxidant activity might be exerted through modulation of antioxidant enzymes in the form of oxidative stress although the possibility of the extract acting as a ROS scavenger cannot be excluded. This property is attributed to phenolic compounds such as polyphenols, flavonoids and flavonols. Studies have shown that polyphenols are also associated with the CAT activity [55].

Lipid per-oxidation is a natural metabolic process under normal aerobic conditions and it is one of the most investigated consequences of ROS action on membrane structure and function. Polyunsaturated fatty acids (PUFA), the main components of membrane lipids, are susceptible to per-oxidation. The final stable end products of per-oxidation are aldehydes which react with TBA to form thiobarbituric acid malonaldehyde adduct with an absorbance maximum at 532 nm. In our studies the presence of various *S. guineense* extracts *in vitro* led to a reduction of adducts formation indicating their lipoprotective potential. This activity could be attributed to the hydroxyl radical scavenging by phytochemicals present in the potent extracts. Moreover reducing power and chelation of metal ion could also be responsible for providing membrane protective efficacy in extracts. Phenolics are a group of non-essential dietary components and their hydrogen donating property is responsible for the inhibition of free radical induced lipid peroxidation [56,57]. Phenolic and other non-phenolic phytochemicals present in extracts might also be involved in imparting some degree of membrane protection. Different variation observed for each of the markers of oxidative stress determined can also be influenced by various factors. The results might be due to: (a) the availability of substrates for peroxidation, (b) the presence/amount of peroxidation inducers, such as ascorbate, Fe^{2+}, oxygen, initiators of free radical reactions and the functioning of the electron transport chain, which serve as a source of ROS, (c) the levels of the antioxidant defense, and (d) the differences in the membrane lipid, such as fluidity and surface charge [55,58]. The beneficial effects of *S. guineense* extracts found in this study could probably be of importance for formulating dietary supplements, as well as for developing new ingredients with improved antioxidant properties from other plant sources.

4. Conclusions

S. guineense extracts exhibited significant scavenging and antioxidant properties. The protective effects identified in this study are attributed to the phenolic compounds. However, before suggesting the potential therapeutic use of *S. guineense* extracts in preventing oxidative stress, further studies will be necessary to determine the biological benefits on kidney, heart and to obtain a better understanding of the mechanism of this important protective effect.

Acknowledgements

We are grateful Nana (National Herbarium) and SENE Olivier Valéry for helping in collecting and identifying the plant material.

Author Contributions

Moukette Moukette Bruno and Khou-Kouz Nkoulou Claude Herve conducted the study; Njinkio Nono Borgia Legrand perfomed the phytochemical analysis; Ama Moor Vicky Jocelyne helped in the writing and correction of the manuscript; Pieme Constant Anatole and Ngoupayo Joseph designed the research co-directed the research work under the supervision of Ze Minkande Jacqueline and Ngogang Jeanne Yonkeu, Constant Anatole Pieme also provided reagents. All the authors read and approved the final manuscript.

Conflict of Interests

The authors declare that no conflict of interests.

References

1. Dhalla, N.S.; Temsah, R.M.; Netticadan, T. Role of oxidative stress in cardiovascular diseases. *J. Hypertens.* **2000**, *18*, 655–673.
2. Bokov, A.; Chaudhuri, A.; Richardson, A. The role of oxidative damage and stress in aging. *Mech. Ageing Dev.* **2004**, *125*, 811–826.
3. Madamanchi, N.R.; Vendrov, A.; Runge, M.S. Oxidative stress and vascular disease. *Arterioscler. Thromb. Vasc. Biol.* **2005**, *25*, 29–38.
4. Pham-Huy, A.L.; He, H.; Pham-Huy, C. Free radicals, antioxidants in disease and health. *Int. J. Biomed. Sci.* **2008**, *4*, 89–96.
5. Habibur, R.; Manjula, K.; Anoosha, T.; Nagaveni, K.; Chinna Eswaraiah, M.; Dipankar, B. *In vitro* antioxidant activity of *Citrullus lanatus* seed extracts. *Asian J. Pharm. Clin. Res.* **2013**, *6*, 152–157.
6. Devasagayam, A.P.T.; Tilak, C.J.; Boloor, K.K.; Sane, S.K.; Ghaskadbi, S.S.; Lele, D.R. Free radicals and antioxidants in human health: Current status and future prospects. *J. Assoc. Physicians India* **2004**, *52*, 794–804.
7. Pandey, A.K.; Mishra, A.K.; Mishra, A. Antifungal and antioxidative potential of oil and extracts derived from leaves of Indian spice plant *Cinnamomum tamala*. *Cell. Mol. Biol.* **2012**, *58*, 142–147.
8. Vinson, J.A.; Liang, X.Q.; Proch, J.; Hontz, B.A.; Dancel, J.; Sandone, N. Polyphenols antioxidants in citrus juices *in vitro* and *in vivo* studies relevant to heart diseases. *Adv. Exp. Med. Biol.* **2002**, *50*, 5113–5122.
9. Pier-Giorgio, P. Flavonoids as antioxidants. *J. Nat. Prod.* **2000**, *63*, 1035–1042.
10. Heim, K.E.; Tagliaferro, A.R.; Bobilya, D.J. Flavonoid antioxidants: Chemistry, metabolism and structure-activity relationships. *J. Nutr. Biochem.* **2002**, *13*, 572–584.
11. Sealbert, A.; Johnson, J.; Saltmarsh, M. Polyphenols: Antioxidants and beyond. *Am. J. Clin. Nutr.* **2005**, *81*, 2155–2175.
12. Tchobsala, D.; Mbolo, M. Characterization and impact of wood logging on plant formations in Ngaoundéré District, Adamawa Region, Cameroon. *J. Ecol. Nat. Env.* **2013**, *5*, 265–277.
13. Saka, J.D.K.; Msonthi, J.D. Nutritional value of edible fruits of indigenous wild stress in Malawi. *For. Ecol. Manag.* **1994**, *64*, 245–248.
14. Ambé, G.A. Wild edible fruits from Guinean savannas of Ivory Coast: State of knowledge by local Malinke population. *Biotechnol. Agron. Soc. Environ.* **2001**, *5*, 43–58.
15. Segawa, P.S.; Kasenene, J.M. Plants for malaria treatment in Southern Uganda: traditional use, preference and ecological viability. *J. Ethnobiol.* **2007**, *27*, 110–131.
16. Hamil, F.A.; Apio, S.; Mubiru, N.K.; Mosango, M.; Bukenya-Ziraba, R.; Maganyi, O.W.; Soejarto, D.D. Traditional herbal drugs of southern Uganda, I. *J. Ethnopharmacol.* **2000**, *70*, 281–300.
17. Oluwolé, O.G.A.; Pricilla, S.D; Jerome, D.M.; Lydia, P.M. Some herbal remedies from Manzini region of Swaziland. *J. Ethnopharmacol.* **2002**, *79*, 109–112.
18. Tsakala, T.M.; Penge, O.; John, K. Screening of *in vitro* antibacterial activity from *Syzygium guineense* (Willd) hydrosoluble dry extract. *Ann. Pharm. Fr.* **1996**, *54*, 276–279.

19. Matig, O.E.; Ndoye, O.; Kengue, J.; Awono, A. Edible forest fruit of Cameroon. In *Books IRAD, IPGRI and CIFOR Collaboration*; IPGRI: Yaounde, Cameroon, 2006.
20. Mukherjee, P.K.; Saha, K.; Murugesan, T.; Mandal, S.C.; Pal, M.; Saha, B.P. Screening of anti-diarrhoeal profile of some plant extracts of a specific region of West Bengal, India. *J. Ethnopharmacol.* **1998**, *60*, 85–89.
21. Ashebır, M.; Ashenatı, M. Assessment of the antibacterial activity of some traditional medicinal plants on some food-borne pathogens. *Ethiop. J. Health Dev.* **1999**, *13*, 211–216.
22. Oketch-Rabah, H.A.; Dossaji, S.F. Molluscicides of plant origin: Molluscicidal activity of some Kenyan medicinal plants. *South Afr. J. Sci.* **1998**, *94*, 299–301.
23. Djoukeng, J.D.; Abou-Mansour, E.; Tabacchi, R.; Tapondjou, A.L.; Boud, H.; Lontsi, D. Antibacterial triterpenes from *Syzygium guineense* (Myrtaceae). *J. Ethnopharmacol.* **2005**, *101*, 283–286.
24. Parakashtha, G.; Tom Erik, G.; Anders, R.; Mona, S.; Bent, R.; Drissa, D.; Terje Einar, M.; Marit, I.; Berit, S.P. Chemical composition and immunological activities of polysaccharides isolated from the Malian medicinal plant *Syzygium guineense*. *J. Pharmacog. Phytoth.* **2010**, *2*, 76–85.
25. Ior, L.D.; Otimenyin, S.O.; Umar, M. Anti-inflammatory and analgesic activities of the ethanolic extract of the leaf of *Syzygium guineense* in rats and mice. *IOSR J. Pharm.* **2012**, 33–36.
26. Noudogbessi, J.P.; Yédomonhan, P.; Sohounhloué, D.C.K.; Chalchat, J.C.; Figuérédo, G. Chemical composition of essential oil of *Syzygium guineense* (Willd.) DC. var. *guineense* (Myrtaceae) from Benin. *Rec. Nat. Prod.* **2008**, *2*, 33–38.
27. Molyneux, P. The use of the stable free radical diphenylpicrylhydrazyl (DPPH) for estimating antioxydant activity. *J. Sci. Tecnol.* **2004**, *26*, 211–219.
28. Re, R.; Pellegrini, N.; Proteggente, A.; Pannala, A., Yang, M.; Rice-Evans, C. Antioxidant activity applying an improved ABTS radical cation decolorization assay. *Free Rad. Biol. Med.* **1999**, *26*, 1231–1237.
29. Garratt, D. *The Quantitative Analysis of Drugs*. Chapman and Hall Ltd.: Tokyo, Japan, 1964, pp. 456–458.
30. Yu, W.; Zhao, Y.; Shu, B. The radical scavenging activities of *radix puerariae* isoflavonoids: A chemiluminescence study. *Food Chem.* **2004**, *86*, 525–529.
31. Benzie, F.; Strain, J. The Ferric Reducing Ability of Plasma (FRAP) as a Measure of "Antioxidant Power": The FRAP Assay. *Anal. Biochem.* **1996**, *239*, 70–76.
32. Prieto, P.; Pineda, M.; Aguilar, M. Spectophotometric quantitative of antioxidant capacity through the formation of a phosphomolybdenum complex: Specific application to the determination of vitamin E. *Anal. Biochem.* **1999**, *269*, 337–341.
33. Oyaizu, M. Studies on products of browning reaction. Antioxidative activities of products of browning reaction prepared from glucosamine. *Japn. J. Nutr. Dietetics* **1986**, *44*, 307–315.
34. Singleton, V.; Draper, D. The transfer of polyphenolic compuonds from grape seeds into wine. *Am. J. Enol. Vitic.* **1964**, *15*, 34–40.
35. Zhishen, J.; Mengcheng, T.; Jianming, W. The determination of flavonoid contents in mulberry and their scavenging effects on superoxide radicals. *Food Chem.* **1999**, *64*, 555–559.

36. Kumaran, A.; Karunakaran, R. *In vitro* antioxidant activities of methanol extracts of *Phyllanthus* species from India. *LWT-Food Sci. Technol.* **2007**, *40*, 344–352.
37. Kumar, S.; Mishra, A.; Pandey, A.K. Antioxidant mediated protective effect of *Parthenium. hysterophorus* against oxidative damage using *in vitro* models. *BMC Compl. Altern. Med.* **2013**, *13*, 120.
38. Biapa, P.; Matei, H.; Bâlici, S.; Oben, J.; Ngogang, J. Protective effects of stem bark of *Harungana. madgascariensis* on the red blood cell membrane. *BMC Compl. Altern. Med.* **2013**, *13*, 1–9.
39. Misra, H.; Fridovich, I. Estimation of superoxide dismutase. *J. Biochem.* **1972**, *247*, 3170–3178.
40. Antolovich, M.; Prenzler, P.; Patsalides, E.; McDonald, S.; Robards, K. Methods for testing antioxidant activity. *Analyst* **2001**, *127*, 183–198.
41. Ignat, I.; Volf, I.; Popa, V.I. A critical review of methods for characterisation of polyphenolic compounds in fruits and vegetables. *Food Chem.* **2011**, *126*, 1821–1835.
42. Penchev, P.I. Study Methods of Extraction and Purification of Bioactive Products from Plants by Coupling of Separation Techniques for Low and High Pressures. Ph.D. Thesis, University of Toulouse, Toulouse, France, July 2010.
43. Rice-Evans, C. Flavonoids antioxidants. *Curr. Med. Chem.* **2001**, *8*, 797–807.
44. Mohamed, A.A.; Ali, S.I.; El-Baz, F.K. Antioxidant and antibacterial activities of crude extracts and essential oils of *Syzygium. cumini* Leaves. *PLoS One* **2013**, *8*, e60269.
45. Inchuen, S.; Narkrugsa, W.; Pornchaloempong, P. Effect of drying methods on chemical composition, color and antioxidant properties of Thai red curry powder. *Kasetsart J.* **2010**, *44*, 142–151.
46. Hamrouni-Sellami, I.; Rahali, F.Z.; Rebey, I.B.; Bourgou, S.; Limam, F.; Marzouk, B. Total phenolics, flavonoids, and antioxidant activity of sage (*Salvia officinalis* L.) plants as affected by different drying methods. *Food Bioprocess Technol.* **2013**, *6*, 806–817.
47. Shimada, K.; Fujikawa, K.; Yahara, K.; Nakamura, T. Antioxidative properties of xanthan on the autoxidation of soybean oil in cyclodextrin emulsion. *J. Agric. Food Chem.* **1992**, *40*, 945–948.
48. Sunayana, S.; Verma, H.N. Antioxidant activity of *Benincasa Hispida* Seeds: Antioxidant activity of *Benincasa hispida*. *J. Nat. Prod. Plant Resour.* **2013**, *3*, 34–37.
49. Sepici-Dincel, A.; Açıkgöz, Ş.; Çevik, C.; Sengelen, M.; Yeşilada, E. Effects of *in vivo* antioxidant enzyme activities of myrtle oil in normoglycaemic and alloxan diabetic rabbits. *J. Ethnopharmacol.* **2007**, *110*, 498–503.
50. Adedapo, A.A. Jimoh, F.O.; Afolayan, A.J.; Masika, P.J. Antioxidant activities and phenolic contents of the methanol extracts of the stems of *Acokanthera. oppositifolia* and *Adenia. gummifera. BMC Compl. Altern. Med.* **2008**, *8*, 1–7.
51. Debosree, G.; Syed, B.F.; Elina, M.; Monalisa, D.; Debasish, B. Protective effect of aqueous leaf extract of *Murraya koenigi* against lead induced oxidative stress in rat liver, heart and kidney: A dose response study. *Asian J. Pharm. Clin. Res.* **2012**, *5*, 54–58.
52. Flora, S.J.S.; Pande, M.; Kannan, G.M.; Mehta, A. Lead induced oxidative stress and its recovery following co-administration of melatonin or *N*-acetycysteine during chelation with succimer in male rats. *Cell. Mol. Biol.* **2004**, *50*, 543–551.

53. Chattopadhyay, A.; Biswas, S.; Bandyopadhyay, D.; Sarkar, C.; Datta, A.G. Effect of isoproterenol on lipid peroxidation and antioxidant enzymes of myocardial tissue of mice and protection by quinidine. *Mol. Cell. Biochem.* **2003**, *245*, 43–49.
54. Henry, W.L.; Chao-Lin, K.; Wen-Hu, Y.; Chia-Hsien, L.; Hong-Zin, L. Antioxidant enzymes activity involvement in luteolin-induced human lung squamous carcinoma CH27 cell apoptosis. *Eur. J. Pharmacol.* **2006**, *534*, 12–18.
55. Mukherjee, D.; Roy, S.G.; Bandyopadhyay, A.; Chatyopadhyay, A.; Basu, A.; Mitra, E.; Ghosh, A.K.; Reiter, R.J.; Bandyopadhyay, D. Melatonin protects against isoproterenol-induced myocardial injury in the rat: Antioxidative mechanisms. *J. Pineal. Res.* **2010**, *48*, 251–262.
56. Lívia, S.O.; Caroline, D.; Cláudia, F.; Henriques, J.O.A.; Mirian, S. Hepatoprotective, cardioprotective, and renal-protective effects of organic and conventional grapevine leaf extracts (*Vitis. labrusca* var. Bordo) on Wistar rat tissues. *An. Acad. Brasil. Ciênc.* **2011**, *83*, 1403–1411.
57. Mishra, A.; Sharma, A.K.; Kumar, S.; Saxena, A.K.; Pandey, A.K. *Bauhinia variegata* leaf extracts exhibit considerable antibacterial, antioxidant and anticancer activities. *BioMed. Res. Int.* **2013**, *2013*, doi:10.1155/2013/915436.
58. Sheena, N.; Ajith, T.A.; Janardhanan, K.K. Anti-inflammatory and antinociceptive activities of *Ganoderma lucidum* occurring in South India. *Pharm. Biol.* **2003**, *41*, 301–304.

4

Flavonoids Affect Host-Microbiota Crosstalk through TLR Modulation

Francisco J. Pérez-Cano [1,2,*], Malen Massot-Cladera [1,2], Maria J. Rodríguez-Lagunas [2,3] and Margarida Castell [1,2]

1 Departament de Fisiologia, Facultat de Farmàcia, Universitat de Barcelona; Av. Joan XXIII s/n, Edifici B, 3ª planta, E-08028 Barcelona, Spain; E-Mails: malen.massot@ub.edu (M.M.-C.); margaridacastell@ub.edu (M.C.)

2 Institut de Recerca en Nutrició i Seguretat Alimentària (INSA-UB), E-08028 Barcelona, Spain; E-Mail: mjrodriguez@ub.edu

3 Departament de Ciències Fisiològiques II, Facultat de Medicina, Universitat de Barcelona, Feixa Llarga s/n, L'Hospitalet de Llobregat, E-08907 Barcelona, Spain

* Author to whom correspondence should be addressed; E-Mail: franciscoperez@ub.edu

External Editors: Jeffrey B. Blumberg and Oliver Chen

Abstract: Interaction between host cells and microbes is known as crosstalk. Among other mechanisms, this takes place when certain molecules of the micro-organisms are recognized by the toll-like receptors (TLRs) in the body cells, mainly in the intestinal epithelial cells and in the immune cells. TLRs belong to the pattern-recognition receptors and represent the first line of defense against pathogens, playing a pivotal role in both innate and adaptive immunity. Dysregulation in the activity of such receptors can lead to the development of chronic and severe inflammation as well as immunological disorders. Among components present in the diet, flavonoids have been suggested as antioxidant dietary factors able to modulate TLR-mediated signaling pathways. This review focuses on the molecular targets involved in the modulatory action of flavonoids on TLR-mediated signaling pathways, providing an overview of the mechanisms involved in such action. Particular flavonoids have been able to modify the composition of the microbiota, to modulate TLR gene and protein expression, and to regulate the downstream signaling molecules involved in the TLR pathway. These synergistic mechanisms suggest the role of some flavonoids in the preventive effect on certain chronic diseases.

Keywords: TLR; polyphenols; inflammation; vegetables; intracellular signaling

---

1. Host-Microbiota Crosstalk and Dietary Factors

1.1. Interaction between Microbiota and the Host

The intestinal mucosa is constantly interacting with a high load of antigens, which comprises, besides those from food, those derived from microbes, both commensal microbiota and invading pathogens. For this reason, the gastrointestinal tract contains the intestinal epithelial cells (IEC) that act as a physical barrier, and the largest part of the immune system: the gut-associated lymphoid tissue (GALT), with approximately 70% of the body immune cells [1]. The main function of the IEC and immune cells located in the intestine is to discriminate precisely between the potential harmful antigens, and then mounting an effective immune response, and those which are innocuous and therefore do not require counteraction from the host, a process known as oral tolerance [2]. The key and ineludible first action in this discrimination consists of the recognition of microbial components in the intestine.

The microbial interaction with host cells (both immune and non-immune cells) is known as crosstalk. Certain microbiota components can be recognized through the pattern-recognition receptors (PRRs). These are host innate immune receptors involved in the detection of pathogens that recognize conserved molecular structures known as pathogen-associated molecular patterns (PAMPs) as well as microbe-associated molecular patterns (MAMPs) (host-commensal interactions) and induce the production of innate effector molecules. These can be located on the cell membrane or in the cytosol joining to bacterial-surface-exposed structures or nucleic acids, respectively. The cytosolic receptors are composed of two major families: the nucleotide oligomerization domain (NOD)—like receptors (NLRs) such as NOD1, NOD2 and NLRP1, and the retinoic acid inducible gene I (RIG-1)—like receptors (RLRs) which recognizes viral RNAs. The cell membranes' PRRs include toll-like receptors (TLRs) that may reside outside the cell or in specialized vesicular compartments (endosomes) where they detect DNA and RNA derived from viruses and bacteria [3].

PRRs represent the first line of host defense against pathogens, play a pivotal role in both innate and adaptive immunity and can lead to the subsequent inflammatory process. Their activation results not only in local mucosal immunity but also in systemic immune responses. In this sense, small amounts of MAMPS and exogenous antigens from microbes are handled in a homeostatic manner when the PRRs are operating in a healthy manner, leading then to an appropriate pathway activation and balanced mediators secretion. However, dysregulation in the activity of such receptors can lead to the development of chronic and severe inflammation as well as immunological disorders, including septic shock, atherosclerosis, diabetes and cancer [4,5].

It can be of importance to study the influence of diet in such initial activation. Among components present in the diet, polyphenols, mainly flavonoids, have been suggested as dietary factors able to modulate TLR-mediated signaling pathways [5].

1.2. Toll-Like Receptors

Each host TLR recognizes distinct PAMPs derived from various microbial pathogens including viruses, bacteria, protozoa and fungi. However, besides their microbial-sensing activity, TLRs also play a role in the recognition of endogenous intracellular molecules released by activated or necrotic cells and extracellular matrix, called damage-associated molecular patterns (DAMPs), which also lead to an inflammatory response which is needed to promote tissue repair [6].

In addition, TLR activation induces the expression of many immune and inflammatory genes via the stimulation of nuclear factor κB (NF-κB) and mitogen-activated protein kinases (MAPKs), such as extracellular signal-regulated kinase (ERK), p38 and c-jun N-terminal kinase (JNK) [7]. Generally, NF-κB and MAPKs, involving p38MAPK or JNK, contribute in activating some proinflammatory cytokines [8,9], whereas activation of ERK promotes cell survival [10].

There are 11 human TLRs identified so far, divided into three groups according to the target in which they interact: lipids and lipopeptides (TLR1, 2, 4 and 6), proteins (TLR5) and nucleic acids (TLR3, 7, 8 and 9). TLR1 forms heterodimers with TLR2 (TLR1/2) and recognizes triacyl lipopeptides. TLR2 in concert with TLR1 or TLR6 recognizes a wide variety of PAMPs, including peptidoglycan, lipopeptides and lipoproteins of Gram-negative bacteria, mycoplasma lipopeptides and fungal zymosan. TLR6 in association with TLR2 (TLR2/6) recognizes diacyl lipopeptides. TLR4 interacts with bacterial lipopolysaccharide (LPS). TLR5 recognizes bacterial flagellin. TLR3 recognizes double-strand RNA, and TLR7 and TLR8 interact with the single-strand RNA found during viral replication. TLR9 recognizes unmethylated deoxycytidyl-phosphate-deoxyguanosine (CpG) motifs commonly present in bacterial and viral genomes. Finally, human TLR11 has been reported to be non-functional because of the presence of a stop codon in the gene and TLR10 is able to homodimerize or heterodimerize with TLR1 and TLR2, but its ligand remains unknown [11].

Recognition of PAMPs by host TLRs triggers intracellular signaling cascades through a set of toll/interleukin-1 receptor (TIR) -domain-containing adaptors, including myeloid differentiation primary response protein 88 (MyD88), TIR domain-containing adaptor protein (TIRAP/Mal), TIR domain-containing adaptor inducing interferon (IFN) (TRIF/TICAM1) and TRIF-related adaptor molecule (TRAM/TICAM2) [12]. Each TLR recruits a specific combination of adaptors to activate different transcription factors, giving rise to appropriate inflammatory responses. All TLRs, with the exception of TLR3, share the adaptor MyD88 which recruits IL-1 receptor-associated kinase-4 (IRAK-4) and leads to its phosphorylation. The phosphorylated IRAK-4 then induces the phosphorylation of IRAK-1 [13], which, in turn, activates the tumor necrosis factor receptor-associated factor 6 (TRAF6). TRAF6 induces phosphorylation of the IκB kinase (IKK) complex, resulting in activation of the transcription factor NF-κB [14].

TIRAP mediates the activation of the MyD88-dependent pathway downstream of TLR2 and TLR4 [15]. TRIF is another adaptor molecule for TLR3 and TLR4 that produces a MyD88-independent [16,17] signaling pathway (TRIF-dependent). This pathway activates both NF-κB and interferon regulatory factor 3 (IRF3) to induce the activation of type I interferons (IFNs), and also causes delayed NF-κB activation mediated through the receptor-interacting protein-1 (RIP1) [17]. TRAM selectively mediates the TRIF-dependent pathway downstream of TLR4, but not TLR3.

1.3. Flavonoids

Flavonoids are vegetal compounds that contribute to the brilliant shades of blue, scarlet and orange colors in leaves, flowers and fruits of plants. In addition, they are found in seeds, nuts, grains and spices, and in some beverages such as wine, tea and beer [18]. Chemically, flavonoids are polyphenols which are commonly found conjugated to sugars (as a glycosylated form) although some of them can exist as free aglycones [19].

The basic flavonoid structure is the flavan nucleus consisting of 15-carbon skeleton arranged in two phenyl rings bound by a three-carbon bridge commonly cyclized with oxygen, then forming three rings labeled A, B, C (Figure 1) [18]. The various classes of flavonoids differ in the level of oxidation and pattern of substitution on the C ring while individual compounds within a class differ in the pattern on substitution of the A and B rings. The major classes of flavonoids are flavonols, flavones, flavanones, isoflavones, flavanols and anthocyanidins (Figure 1).

Figure 1. Chemical structure and representative compounds from the main families of flavonoids.

Basic structure

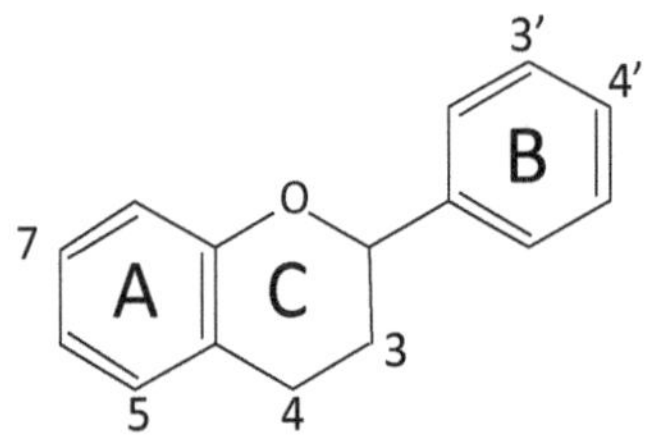

Chemical structure	Family name: representative compounds
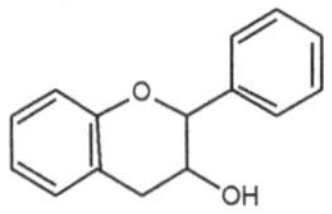	**Flavanols:** epicatechin, epicatechin-3-gallate, epigallocatechin-3-gallate
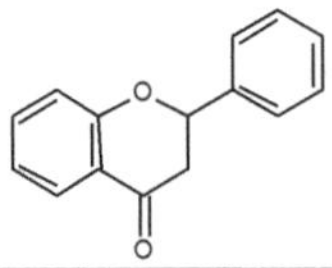	**Flavanones:** eriodictyol, hesperetin, naringenin
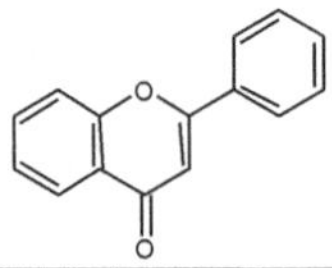	**Flavones:** baicalein, chrysin, eupalin, luteolin
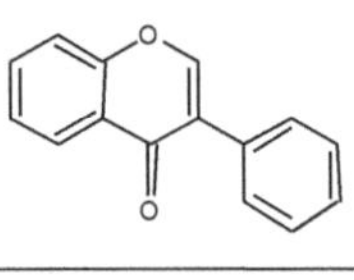	**Isoflavones:** daidzein, equol, genistein
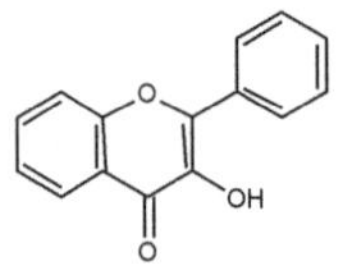	**Flavonols:** quercetin, morin, kaempferol, myricetin
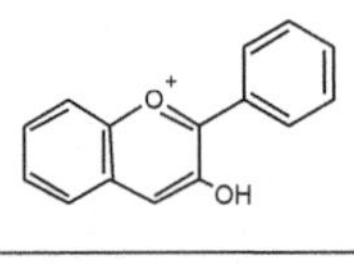	**Anthocyanidins:** cyanidin, delphinidin, malvidin

Flavonols are ubiquitous in plants and are found mainly in fruit skin, being the yellow and red onions one of the richest sources of them [20]. The flavonol class includes kaempferol, quercetin, and myricetin typically found as glycosides. The flavones, such as apigenin, luteolin, and baicalin, are less widely distributed, although substantial amounts of apigenin and luteolin have been detected in cereal grain and some aromatic herbs (parsley, rosemary, thyme) [18,20]. The most abundant *flavanone* is naringenin, followed by hesperidin, eriodictyol and hesperetin. They are present in especially high amounts in citrus fruits and usually occur as rutinosides and neohesperidosides, which differ in taste [19]. Isoflavones are found most often in leguminous plants with substantial quantities of daidzein and genistein occurring in soybean [20]. *Flavanols* (flavan-3-ols) include monomeric and polymeric forms called procyanidins (condensed tannins). Monomeric forms are catechins and gallic acid esters of catechins and epicatechins [19]. They are widely distributed in plants and fruits and also in tea leaves [18]. *Anthocyanidins* have a fully aromatized C ring and as a result they are positively charged. The most common of these natural water-soluble compounds are pelargonidin, cyaniding, delphinidin, and malvidin, which form conjugates with sugars and organic acids to generate a multitude of anthocyanins that confer a wide range of colors, ranging from orange and red to blue and purple, on fruits (specially in berries like strawberries, blueberries, blackberries and blackcurrants), vegetables and plants [21,22].

2. Objective

There is increasing scientific evidence of a relationship between diets rich in natural antioxidants and the prevention of several chronic diseases. The inverse association between dietary antioxidants and inflammatory-like disorders is supported by several epidemiological and interventional studies [23]. Polyphenols, including flavonoids, are the most commonly consumed dietary antioxidants and therefore they may be one of the major responsible for the healthy effect of fruit-and-vegetable-enriched diets. As the activation of the TLR pathway constitutes the first step in the inflammatory cascade activation, and its deregulation can lead to severe chronic inflammation and immune disorders, it is plausible to hypothesize a downmodulatory action of flavonoids in the TLR-induced pathways.

This review focuses on the molecular targets and mechanisms involved in the modulatory action of flavonoids on TLR-mediated signaling pathways. In this context, it aims to provide an overview of the three major levels involved in such action: by varying the composition of the microbiota, by modulating the expression and activation of TLRs and, finally, by modifying the downstream signaling pathways involved. The synergy of all these mechanisms may explain the high potential of flavonoids for preventing certain types of disorders (Figure 2). The suppression of TLR activation by flavonoids offers new potential interventions for ameliorating inflammatory diseases and therefore constitutes a new alternative to current approaches.

3. Mechanisms of Modifying "Cross-Talk" by Flavonoids

3.1. Influence on Growth and Composition of Microbiota

Bacterial composition (*i.e.*, types of MAMP and PAMP) is one of the major factors influencing TLR activation. In turn, dietary habits are known to modulate the composition of the gut microbiota [24]. For this reason, the earliest level involved in modifying the host-bacteria crosstalk by dietary flavonoids involves the modification of the microbial inhabitants in the gut (Figure 2). There is evidence showing that several kinds of flavonoids, including flavanones, flavonols, flavanols, and isoflavones, modify the microbiota composition. This has been recently reviewed by Etxeberria *et al.* [22] who extensively compile the latest investigations regarding the impact of phenolic compounds, as pure chemicals or in the food context, on microbial communities in both *in vitro* and *in vivo* studies (in animal and human interventions).

Figure 2. Overview of the mechanisms involved in the regulation of microbiota-host crosstalk by flavonoids. They can act at three different levels by modulating: (**1**) microbiota composition, by means of directly (flavonoid) or indirectly (metabolite) affecting the growth; (**2**) Toll-like receptor (TLR) activation, by means of acting on the receptor and its adaptor proteins; (**3**) signal transduction, by means of interfering with upstream and downstream kinases as well as the transcription factors involved in the inflammatory and immune response activation.

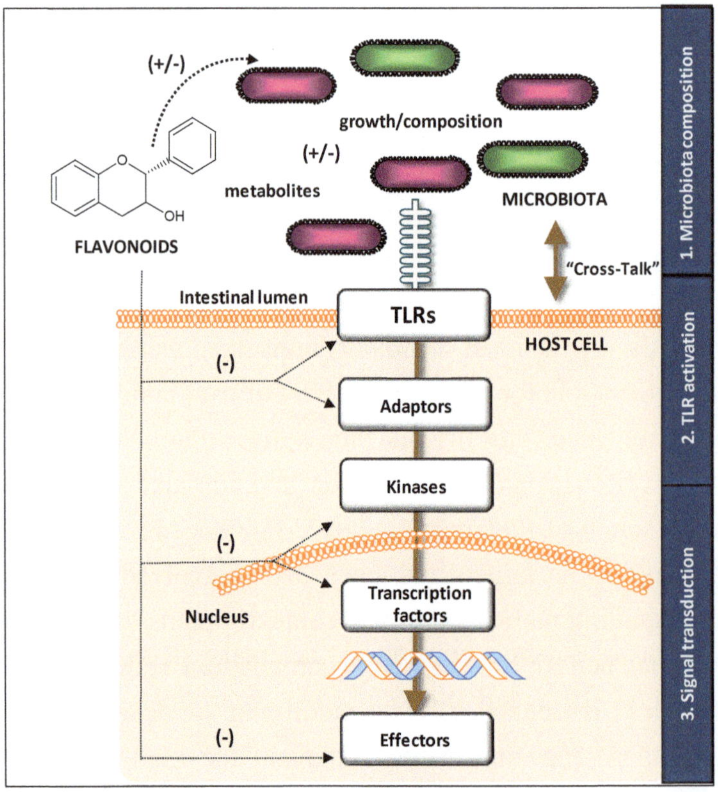

With regard to pure flavonoids, an *in vitro* study tested the impact of particular flavonols (rutin, quercetin), flavanols (catechin) and flavanones (naringin, naringenin, hesperidin and hesperetin) on the growth of pure cultures of six bacteria species [25]. The author showed the dose-dependent inhibitory effect of the aglycones naringenin and quercetin on the growth of all bacteria species studied, which was stronger than that of their glycosides [25]. On the other hand, both quercetin (aglycone) and rutin (glycoside) significantly increased the growth of the major intestinal *Bacteroidetes* and *Firmicutes* phyla, with a greater increase in the former than in the latter *in vitro* [26]. However hesperetin and catechin (both aglycones) showed no or weak antibacterial activity against most of the analysed bacteria [25], which was later corroborated by an *in vitro* study evaluating the antimicrobial activity of five aglycones on the growth of the probiotic *Bifidobacterium adolescentis* [27]. Similarly, the repressing and/or stimulating effect induced by *flavanols*, such as catechin and epicatechin, on the growth of certain bacteria has been reported in *in vitro* and batch cultures studies [28,29].

Besides those studies of particular flavonoids on bacteria *in vitro*, there are a number of studies showing the changes induced by flavonoid-rich foods on microbiota composition. In order to synthesize the available data, and based on the recompilation performed by Etxeberria *et al.* [22], the Figure 3 below classifies the number of studies or treatments with polyphenols, mainly flavonoids, into the antimicrobial activity or the growth-promoting effect on the more prevalent bacterial groups.

Figure 3. Summary of the number of treatments using beverages (cocoa, tea and wine), fruits or vegetables (soy, pomegranate, grapes, berries and apples) rich in flavonoids and their impact on bacterial growth. There is a high number of studies with flavonoids (*in vitro* and *in vivo*) showing both antimicrobial activity (**left**) and growth-promoting effects (**right**) on the more prevalent bacterial groups.

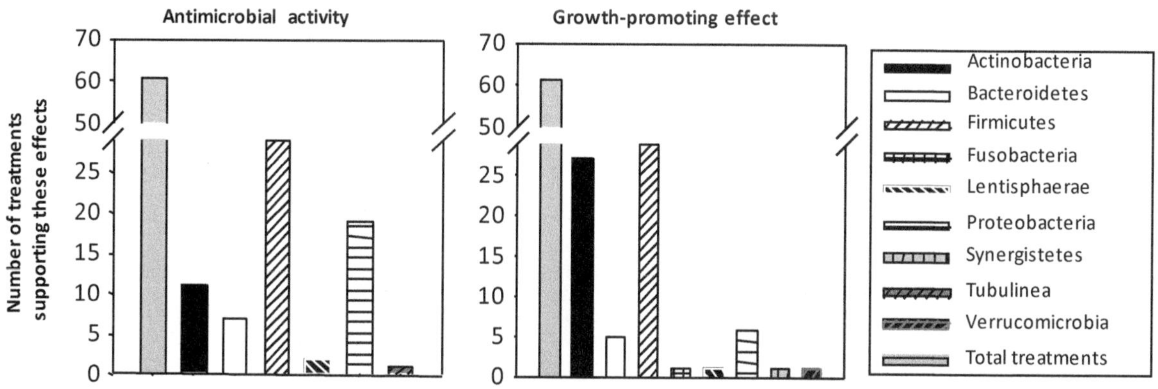

Globally, more than 100 experiments have shown that bacterial groups in gut microbiota are affected by beverages (cocoa, tea and wine) and fruits/vegetables rich in flavonoids, either showing antimicrobial activity or, on the contrary, promoting the growth of certain bacteria. These effects have been shown in all types of flavonoids.

Concerning the effect of flavonoid-rich beverages, the intake of cocoa, which mainly contains epicatechin and catechin, and their polymeric forms, was able to inhibit the growth of *Clostridium histolyticum/C. perfringens, Staphylococcus* and *Streptococcus* genera in preclinical and clinical studies [28,30]. In the same way, tea flavonoids exerted an antimicrobial activity against selective bacterial species [31], whereas bifidogenic properties have also been attributed to them [32,33]. Differences in the proportion of the four major bacteria phyla in humans (*Proteobacteria, Fusobacteria, Firmicutes*, and *Bacteroidetes*) have been observed after a nutritional intervention with red wine which contains a complex mixture of polyphenols, mainly including flavanols and anthocyanins [34]. Similar effects have been described in rats treated for 16 weeks with wine polyphenol, which showed a predominant proportion of *Lactobacillus* and *Bifidobacterium* spp. [35], while controversial results were obtained in an *in vitro* batch culture [33,36]. Nevertheless, these effects were not so pronounced or even disappeared in humans when consuming dealcoholized red wine [34].

Changes have also been observed in the *Firmicutes*, *Actinobacteria* and/or *Bacteroidetes* proportions induced by soy products, which mainly contain isoflavones, in *in vitro*, preclinical and clinical studies [37–40]. In addition, the potential modulation of gut microbiota composition by fruits has been thoroughly reviewed by Etxeberria *et al.* [22]. To date, a large amount of evidence suggests that fruits and their derivatives are able to significantly boost the growth of colonic-friendly bacteria, such as *Bifidobacterium* and *Lactobacillus* [41–44].

Overall, it can be summarized that the intake of flavonoids is able to modulate the composition of the gut microbiota, potentiating the growth of specific beneficial bacteria strains (*Lactobacillus and Bifidobacterium* in >70% of the studies), and also inhibiting the growth of certain pathogenic bacteria (*i.e., Clostridium*). It must be added that, despite the evidence described above, natural flavonoids are not the only responsible for this effect on gut microbiota but their metabolites are also involved. Flavonoids and related compounds can reach the colon where they become substrate for the microbial metabolism [45,46]. Indeed, intestinal bacteria have the ability to metabolize flavonoids, more or less efficiently, depending on their chemical structure, to simpler and more bioavailable metabolites [20,47]. This transformation modulates the biological activities of these compounds, which, in general, are more beneficial after colonic metabolism than in their original forms [45,46,48–50]. Besides the known biological activity of most small phenolic metabolites, several microbial catabolites from epicatechin, epigallocatechin or naringenin have been described possessing important antimicrobial activity, especially against Gram-negative species [51].

3.2. Modulatory Action on TLR Gene and Protein Expression

TLR pathways can be modulated by flavonoids at different levels (Figure 2). Both TLR gene expression and cell membrane expression, which are directly related with TLR functionality, can be affected by flavonoids. Under homeostatic conditions IEC have low expression of TLR2 and TLR4, and therefore in a healthy context, TLR activation is low; however, in an inflammatory scenario, TLR expression on IEC increases and then TLR signaling is triggered [52].

Both *in vitro* studies (Table 1), using primary cells or cell lines, and *in vivo* studies (Table 2) have evidenced the influence of different classes of flavonoids on TLR gene and protein expression. The best results regarding the down-regulatory effect of flavonoids have been raised in the presence of TLR activators (*i.e.*, LPS as an *in vitro* TLR4 activator) whereas lack of effect is found under non stimulatory conditions (*i.e.*, in healthy animals/subjects).

LPS-stimulated J774 macrophages treated with the flavonols engeletin and astilbin showed inhibition of TLR4 gene expression [53]. In addition, kaempferol-3-*O*-sophoroside down-regulated the cell surface expression of TLR2 and TLR4 in endothelial cells [54] as did quercetin in PBMC after challenge [55]. Flavanones as naringenin and hesperidin have shown similar effects. In particular, naringenin down-regulates TLR2 and TLR4 protein expression and TLR2 mRNA levels in PBMC after being challenged with *Chlamyidia trachomatis* [56]. Other assays performed *in vitro* in 3T3-L1 cells during adipocyte differentiation and *in vivo* in C57Bl/6j mice after a high fat diet (HFD) agree with the inhibitory effect of naringenin on TLR2 gene expression, although no effect was found for TLR4 [57].

Table 1. Flavonoids on TLR gene and protein expression in *in vitro* studies.

Flavonoid	Dose	Duration [1]	TLR [2]	Studied Target	Cell (Challenge)	Reference
FLAVONOLS Engeletin	10/50 µM	2 h	↓ TLR4	mRNA	J774 macrophages (LPS)	[53]
Astilbin			↓ TLR4	mRNA		
Kaempferol-3-*O*-sophoroside	–	6 h	↓ TLR2 ↓ TLR4	Protein	Endothelial cells	[54]
Quercetin	25 µM	1 h	↓ TLR2 ↓ TLR4	Protein	PBMC	[55]
FLAVANONES Naringenin	1 µg/mL	48 h	↓ TLR2	Protein/mRNA	J774 macrophages (*C. trachomatis*)	[56]
			↓ TLR4	mRNA		
Naringenin	100 µM	–	↓ TLR2	mRNA	3T3L1 cells (adipocyte differentiation)	[57]
FLAVANOLS Epigallocatechin-3-gallate	1 µM	24 h	↓ TLR4	Protein/mRNA	RAW 264.7 macrophages (LPS)	[58]
Epigallocatechin-3-gallate	10 µM	24 h	↔ TLR4	mRNA	Murine bone marrow-derived DCs (unstimulated)	[59]
FLAVONES Baicalin	40/80 µM	24 + 3 h	↔ TLR1-9	mRNA	Human oral keraynocytes (LPS)	[60]
Baicalin	5/10 µg/mL	0.5–6 h	↓ TLR2	Protein/mRNA	PC12 and primary rat neurons (oxygen glucose deprivation)	[61]
			↓ TLR4	Protein/mRNA		

Notes: [1] Treatment duration is showed when detailed in the article, if it is not provided a (–) is showed; [2] Changes in gene expression are summarized by means of an increase (↑), decrease (↓) or not affected (↔) by the intervention.

Table 2. Flavonoids on TLR gene or protein expression in *in vivo* studies.

Flavonoid	Dose	Duration	TLR [1,2]	Studied Target	Cell	Reference
FLAVONES Luteolin	Daily dose of 10 and 25 mg/kg body weight	24/78 h	↓ TLR4 ↓ TLR5	Protein/ mRNA	Cerebral cortex from. p.o [3] fed SD rats	[62]
Baicalin	One dose of 50 mg/kg body weight	4 h	↓ TLR2 ↓ TLR4	Protein/ mRNA mRNA	Mice hippocampus cells (carotid arteries ligation)	[61]
FLAVANONES Naringenin	1% included in food	16 weeks	↓ TLR2	mRNA	Adipocytes C57Bl/6J mice after HFD	[57]
EXTRACTS *Achyrocline satureoides* (Quercetin and Luteolin)	One dose of 100 mg/kg	1 h	↓ TLR4	Protein	Neutrophils from p.o fed Wistar rats	[63]
Cocoa (Procyanidins)	5%/10% included in food	3 weeks	↔, ↓ TLR2 ↔, ↔ TLR4 ↔,↓ TLR7 ↔, ↑ TLR9	mRNA	Small intestine and mesenteric lymph nodes from p.o fed Wistar rats	[64]
Cocoa (Procyanidins)	10% included in food	7 weeks	↔, ↓, ↔ TLR2 ↓, ↑, ↔ TLR4 ↔, ↓, ↔ TLR7 ↔, ↑, ↑ TLR9	mRNA	Small intestine, Peyer's patches and mesenteric lymph nodes from p.o fed Wistar rats	[65]
Cocoa (Procyanidins)	10% included in food	6 weeks	↔ TLR2 ↔ TLR4 ↔ TLR7 ↓ TLR9	mRNA	Large intestine from p.o fed Wistar rats	[30]
Orange juice (Flavanones)	One dose of 300 kcal drink of orange juice	1/3/5 h	↓ TLR2 ↓ TLR4	mRNA Protein	Mononuclear cells from healthy subjects given high-fat high-carbohydrate meal	[66]
Orange juice (Flavanones)	One dose of 300 kcal drink of orange juice	1/3/5 h	↔ TLR4	mRNA Protein	Mononuclear cells from healthy subjects given high-fat high-carbohydrate meal	[67]

Notes: [1] The TLR changes are expressed as the effect on each studied tissue in the same order as mentioned on the cell/challenge column separated among them with a comma; [2] Changes in gene expression are summarized by means of an increase (↑), decrease (↓) or not affected (↔) by the intervention; [3] p.o. means *per os* or administration by oral route.

Flavones have been shown to also modulate TLR gene expression in both *in vitro* and *in vivo* studies. Isoliquiritigenin decreased, among other targets, the gene expression of TLR4 in RAW 264.7 macrophage line [68], and luteolin administrated at different doses to rats was able to suppress the overexpression of TLR4 and TLR5 induced by ischemia in the cerebral cortex at both protein and gene level [62]. Baicalin added to human oral keratynocytes culture was able to down-regulate *P. gingivalis* LPS-induced expression of downstream genes associated with TLR signaling, but almost no changes were found in the gene expression of TLR1-9 determined by the PCR array [60]. However, in another study, baicalin down-regulated protein and gene expression of TLR2 and TLR4 both *in vitro*—PC12 and primary neurons under oxygen glucose deprivation conditions—and *in vivo*—hippocampus cells from mice after carotid arteries ligation [61].

With regard to flavanols, gene and cell surface expression and total protein level of TLR4 were markedly reduced by the addition of epigallocatechin gallate (EGCG) to LPS-stimulated RAW 264.7 macrophages [58]. However, it did not affect TLR4 gene expression when murine bone marrow-derived DCs were in resting conditions [58], which is in agreement with other studies where the flavonoid is not able to modify TLRs in the absence of stimulatory conditions. On the contrary, a semi-synthetic flavonoid called flavopiridol did not alter the TLR4 cell surface expression in LPS-stimulated RAW 264.7 macrophages [69].

Some extracts rich in flavonoids have also shown an effect on TLR gene expression, mainly in *in vivo* studies. Thus, the oral treatment of Wistar rats with a hydroalcoholic extract of *Achyrocline satureoides*, a rich source of quercetin and luteolin, decreased TLR4 expression on neutrophils [63]. Wistar rats fed cocoa, a rich source of procyanidins among other flavonoids, showed changes in the expression of several TLR in different tissues of the intestinal immune system [64]: a cocoa diet induced an up-regulation of TLR4 and TLR9 gene expression and a down-regulation of TLR2 and TLR7 gene expression in Peyer's patches (PPs) and mesenteric lymph nodes (MLNs). Conversely, in small intestine (SI) tissue, cocoa-fed animals showed lower values for TLR4 and TLR9 mRNA and a higher value for TLR2 and TLR7 [65]. However, a similar design did not show a clear effect on colonic tissue, where microbiota is mainly found [30]. It should be taken into account that the effects induced by these extracts cannot be exclusively attributed to flavonoids because other components may be acting as well (*i.e.*, theobromine and fiber in cocoa).

In humans, two interventional studies using orange juice have evaluated the increase in TLR2 and TLR4 mRNA and protein expression in PBMC, induced by a high-fat high-carbohydrate meal, in volunteers [66,67].

In addition to flavonoids, the effect on TLR genes or surface expression by other polyphenols has also been reported. Thus, curcumin decreases TLR2 gene and surface expression in immune cell lines (monocytic THP-1 cells and dHL-60 promyelocytic leukaemia cells) and in primary cells (peripheral blood polymorphonuclear neutrophils, PMN) [70], and a complement containing resveratrol prevented the increase in TLR4 gene expression in PBMC induced by a high-fat, high-carbohydrate meal in healthy subjects [71].

Finally, it has recently been described that TLR gene expression is subject to circadian rhythms [72,73] which might be related to the presence of flavonoids and other exogenous regulators of TLR in the diet.

It should be also mentioned that certain flavonoids *in vivo* may not be the only direct responsible of the induced effects because indirect effects, through one or more of its microbial metabolites, might be also considered. However, in the *in vitro* studies the TLR modulation should be due to direct effects because the microbial metabolism is not present in such approach. Further studies with this aim should be developed.

3.3. Modulatory Action in TLR Activation

The activation of TLR intracellular signaling activation requires further processes that can also be targeted by flavonoids. These processes include the accumulation of lipid rafts, microdomains rich in cholesterol and sphingolipids that act as platforms for downstream molecules and seem to be a key factor in the activation of immune cells, TLR dimerization/oligomerization [74], its glycosilation and the participation of adaptor proteins.

Thus, quercetin and luteolin could interact with lipid rafts [75], whereas isoliquiritigenin was able to suppress LPS-induced dimerization of TLR4 in a dose-dependent manner in RAW 264.7 macrophage cell line [68]. Polyphenols such as curcumin and resveratrol have also shown this effect [74]. With regard to the influence on TLR glycosilation, *Helicobacter pylori* infection seems to initiate the inflammatory cascade in gastric epithelial cells by TLR4 glycosilation, fact that has been reported suppressed by EGCG [47].

There are at least five adaptor proteins of TLR and each of these has a TIR domain (*i.e.,* TRIF or MyD88), which directly or indirectly binds to the TIR domain of a TLR [76]. Flavonoids seem to modulate the pathways involving such adaptors, however from the literature it is difficult to ascertain which is the specific target affected and mainly focus on downstream molecules (Figure 4). As examples, isoliquiritigenin in RAW 264.7 macrophages was able to suppress the MyD88-dependent pathway (*i.e.*, IRAK1 degradation) [68] whereas other *in vitro* studies performed in RAW 264.7 cells and also in 293T (human embryonic kidney cell line) showed that EGCG treatment was able to modulate both MyD88- and TRIF-dependent signaling pathways [77].

On the other hand, it has to be taken into account that TLR gene expression can be self-regulated by downstream molecules participating in its signaling (*i.e.*, NF-κB) [78], which potentiates the down-modulatory effect of those flavonoids able to act on both targets.

3.4. Effect on Signal Transduction Molecules

A large number of articles can be found regarding the effect of flavonoids on the activity of different protein kinases that regulate various intracellular signaling cascades, such as MAPK [79,80]. For this reason flavonoids may also modify the downstream kinases activated after TLR triggering. Furthermore, oxidative stress accompanying inflammation, which also leads to MAPK activation, seems to be controlled by the direct action of flavonoids (Figure 4).

The activation of MAPK engages ERK 1 and 2, JNK and p38 MAPK linked to stress stimuli, such as ROS overproduction or inflammation, and among other actions, this stimulation induces the phosphorylation of inhibitor proteins such as κB inhibitor proteins (IκBs), involved in the NF-κB pathway. The modulation of the activity of NF-κB, a pro-inflammatory transcription factor, is one of the clearest effects of certain flavonoids [81,82]. NF-κB triggers the expression of over 100 genes, many of them involved in the inflammatory response [83]. NF-κB is found in the cytoplasm of non-stimulated cells bound to IκBs which are modulated by MAPK action.

To date in the TLR context, the flavanols epicatechin gallate (ECG) and EGCG in human dental pulp fibroblast were able to block the ligand-stimulated TLR2 cell signaling pathways throughout the down-regulation of downstream mechanisms such as p38MAPK, SAP/JNK, ERK1/2 and phosphorylation of IkBa and p65 subunit of NF-κB [84]. Likewise, in DC, EGCG inhibited LPS-induced MAPK and NF-κB activation [59].

Flavones, flavonols and flavanones also modulate these kinases. Luteolin inhibits p38 MAPK [62] and IκB kinase activity [85] both indicating a role for this flavonoid in MyD88-dependent signaling pathway. Baicalin may inhibit *P. gingivalis* LPS-induced activation of NF-κB, p38 MAPK and JNK [60], and same inflammatory pathway when the study was performed in isquemic neurons [86]. The flavonol flavopiridol—quercetin analogue—was also able to regulate the downstream TLR2 and TLR3 pathways through the MyD88-dependent pathway but not the independent pathway, as was demonstrated by *in vitro* studies showing inhibition of the activation of NF-κB and MAPKs but not TRAF6 [69]. Flavanones such as naringenin inhibited TLR2 pathways by the modulation of IκB degradation and JNK phosphorylation [57], and also down-regulated the induced increase of phosporylation of p38 MAPK in macrophages infected with *Chlamyidia trachomatis* [56].

Foods rich in flavonoids, such as cocoa, have also shown this potential *in vitro*, either by their scavenging action or by direct interaction with the downstream kinases [87,88]. In this context, epicatechin, catechin and dimeric B procyanidins seem to be accumulated in the cytosol and act at the early stages of NF-κB activation, regulating oxidant levels and reducing IκBs phosphorylation, whereas cocoa dimeric procyanidins seem to be able to enter the nuclei and selectively avoid NF-κB binding to DNA [89].

Overall, it is plausible that the different structures of flavonoids and their regiospecific modifications in their side chains lead to different bioactivities. In this sense, Lim *et al.* [90] studied the effect of five classes of plant-derived flavonoids (flavonols, flavones, flavanones, catechins and cyanidins) on TLR2 pathway *in vitro* and found differential effects on the targets studied, such as p38 MAPK and NF-κB.

TBK1, a representative kinase found in TRIF-dependent TLR activation, participates in IRF3 activation and type I IFNs production (Figure 4). Some flavonoids, such as EGCG, luteolin, quercetin, chrysiin and eriodictyol, showed inhibitory effects on this downstream kinase whereas others, such as naringenin and hesperitin, had no effect [77,91,92]. Moreover, isoliquiritigenin was able to suppress TLR3 and TLR4 agonist-induced IRF3 activations in RAW 264.7 macrophage cell line [93]. These results demonstrated the down-regulatory activity of certain flavonoids on the TLR3 and TLR4 TRIF-dependent pathways (Figure 4).

Another mechanism can be proposed as an alternative to the direct actions of flavonoids on specific targets of the TLR pathway. In recent years, the importance of the intracellular negative regulators of TLRs such as SOCS1, TOLLIP, IRAK or SIGIRR has been described [94]. In this regard, procyanidin dimer B2-treated macrophages up-regulate IRAK-M protein and therefore also suppress TLR4 signaling and phosphorylation of MAPKs such as ERK1/2, p38 and JNK [95]. Moreover, EGCG has been shown to up-regulate TOLLIP protein expression through a 67-kDa Laminin receptor (67LR) in macrophages [58] and in DCs [59].

Furthermore, as previously described, DAMPs can also activate TLR pathways and it seems that flavonoids can target these molecules as well. For example, a DAMP known as high mobility group box protein 1 (HMGB1) acts as an endogen ligand of TLR2 and TLR4 when it is released to the extracellular compartment [6] and this release by endothelial cells is modulated by kaempferol-3-*O*-sophoroside [54].

Figure 1. Signal transduction molecules affected by the flavonoids modulatory action on TLR activation. Both, the MyD88 dependent (**left**) and independent (**right**) pathways are modulated by the effect of several flavonoids on targets from different upstream (TLR expression and activation; adaptors modulation) and downstream levels (kinases and transcription factors). Black arrows indicate stimulation of the pathway and red arrows indicate inhibition of the pathway. In brackets there are the references cited in the text.

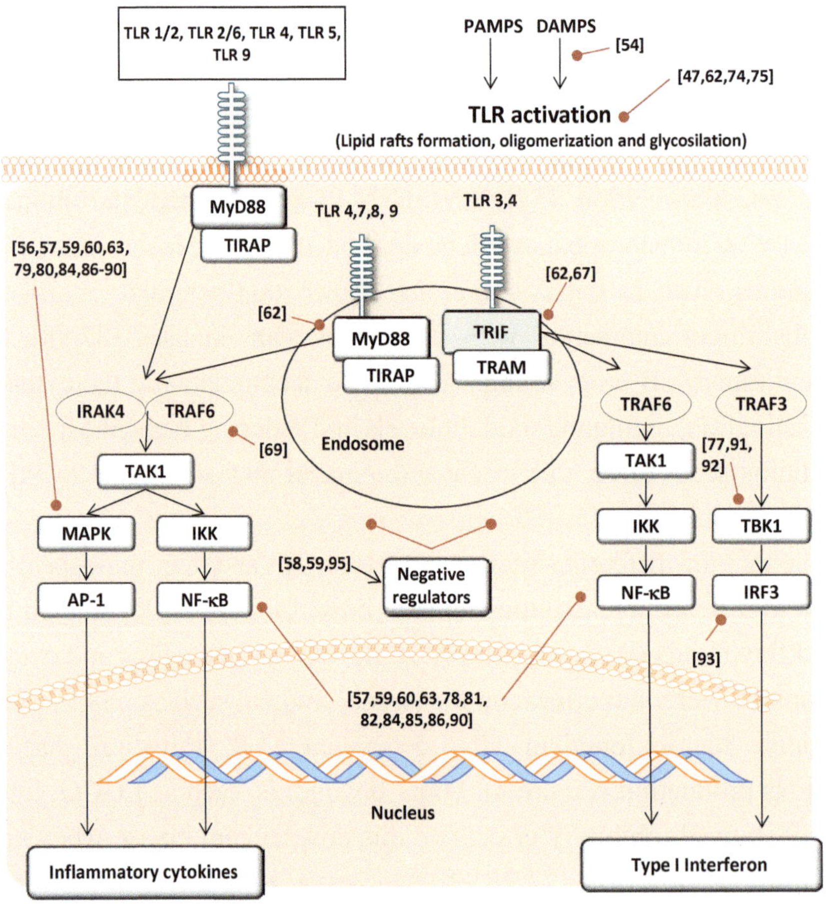

Overall, it can be suggested that changes in microbiota composition and TLR modulation by flavonoids may have also an impact on the intestinal immune activation and oral tolerance. In this sense, flavonoids by modulating intestinal microbiota (Section 3.1) may polarize the immune response towards the T-cell mediated tolerogenic pathway. At the same time the repression of the TLR signaling, which is associated to innate immune tolerance [96], can be also produced by flavonoids (Section 3.2). As certain flavonoids have been described to present immunomodulator activity by inducing changes on the GALT [97] and on systemic immune cells [98], as well as on regulatory T cells [99,100], we may suggest the possible implication of TLR signaling modulation herein described by flavonoids in such immunoregulatory effects.

4. Conclusions

Flavonoids are able to act as exogenous regulators of TLR and therefore modulate the host-microbiota crosstalk. This is possible due to their ability to modulate the TLR signaling pathways at different levels. On one hand, flavonoids can modify the composition of the microbiota, either by direct action or through their microbial catabolites, acting as antimicrobial agents on certain bacterial groups or as growth promoters for other bacteria. On the other hand, flavonoids can modulate the gene expression, the initial activation process of the receptor, and the upstream and downstream signaling molecules involved in the TLR pathway. These synergistic mechanisms make some flavonoids interesting modulatory agents on TLR-mediated inflammatory responses and therefore it is plausible to think that their preventive effect on certain chronic diseases, associated with exacerbated TLR activation, may be at least in part result from TLR regulation.

Based on these effects, we suggest that more preclinical, *in vitro* and animal model, studies, as well as human clinical trials, should be carried out with flavonoids regarding this aspect. Taking into account the lack of toxic effects and the high availability from natural sources, they should be considered as tools to counteract the increase in inflammatory-based diseases in our society.

Acknowledgements

This study was supported by grants from the Spanish Ministry of Economy and Competitiveness (AGL2011-24279).

Author Contributions

Francisco J. Pérez-Cano and Margarida Castell were mainly responsible for designing the study; Francisco J. Pérez-Cano, Maria J. Rodríguez-Lagunas, Malen Massot-Cladera and Margarida Castell carried out the scientific searches and drafted the manuscript; all authors contributed to and approved the final version of the manuscript.

Conflicts of Interest

The authors declare no conflict of interest.

References

1. Brandtzaeg, P. Function of mucosa-associated lymphoid tissue in antibody formation. *Immunol. Investing.* **2010**, *39*, 303–355.
2. Pabst, O.; Mowat, A.M. Oral tolerance to food protein. *Mucosal. Immunol.* **2012**, *5*, 232–239.
3. Marques, R.; Boneca, I.G. Expression and functional importance of innate immune receptors by intestinal epithelial cells. *Cell. Mol. Life Sci.* **2011**, *68*, 3661–3673.
4. Ospelt, C.; Gay, S. TLRs and chronic inflammation. *Int. J. Biochem. Cell. Biol.* **2010**, *42*, 495–505.
5. Jeong, E.; Lee, J.Y. Intrinsic and extrinsic regulation of innate immune receptors. *Yonsei Med. J.* **2011**, *52*, 379–392.

6. Piccinini, A.M.; Midwood, K.S. Illustrating the interplay between the extracellular matrix and microRNAs. *Int. J. Exp. Pathol.* **2014**, *95*, 158–180.
7. McGettrick, A.F.; O'Neill, L.A.J. Localisation and trafficking of toll-like receptors: An important mode of regulation. *Curr. Opin. Immunol.* **2010**, *22*, 20–27.
8. Allan, S.M.; Rothwell, N.J. Cytokines and acute neurodegeneration. *Nat. Rev. Neurosci.* **2001**, *2*, 734–744.
9. Thompson, W.L.; van Eldik, L.J. Inflammatory cytokines stimulate the chemokines CCL2/MCP-1 and CCL7/MCP-7 through NFκB and MAPK dependent pathways in rat astrocytes. *Brain Res.* **2009**, *1287*, 47–57.
10. Xia, Z.; Dickens, M.; Raingeaud, J.; Davis, R.J.; Greenberg, M.E. Opposing effects of ERK and JNK-p38 MAP kinases on apoptosis. *Science* **1995**, *270*, 1326–1331.
11. Yu, L.; Wang, L.; Chen, S. Endogenous toll-like receptor ligands and their biological significance. *J. Cell. Mol. Med.* **2010**, *14*, 2592–2603.
12. Newton, K.; Dixit, V.M. Signaling in innate immunity and inflammation. *Cold Spring Harb. Perspect. Biol.* **2012**, 4.
13. Cao, Z.; Henzel, W.J.; Gao, X. IRAK: A kinase associated with the interleukin-1 receptor. *Science* **1996**, *271*, 1128–1131.
14. Takeda, K.; Akira, S. Microbial recognition by toll-like receptors. *J. Dermatol. Sci.* **2004**, *34*, 73–82.
15. Takeda, K.; Akira, S. TLR signaling pathways. *Semin. Immunol.* **2004**, *16*, 3–9.
16. Hoebe, K.; Du, X.; Georgel, P.; Janssen, E.; Tabeta, K.; Kim, S.O.; Goode, J.; Lin, P.; Mann, N.; Mudd, S.; *et al.* Identification of Lps2 as a key transducer of MyD88-independent TIR signaling. *Nature* **2003**, *424*, 743–748.
17. Yamamoto, M.; Sato, S.; Hemmi, H.; Hoshino, K.; Kaisho, T.; Sanjo, H.; Takeuchi, O.; Sugiyama, M.; Okabe, M.; Takeda, K.; *et al.* Role of adaptor TRIF in the MyD88-independent toll-like receptor signaling pathway. *Science* **2003**, *301*, 640–643.
18. Pietta, P.G. Flavonoids as antioxidants. *J. Nat. Prod.* **2000**, *63*, 1035–1042.
19. Corradini, E.; Foglia, P.; Giansanti, P.; Gubbiotti, R.; Samperi, R.; Lagana, A. Flavonoids: Chemical properties and analytical methodologies of identification and quantitation in foods and plants. *Nat. Prod. Res.* **2011**, *25*, 469–495.
20. Del Rio, D.; Rodriguez-Mateos, A.; Spencer, J.P.E.; Tognolini, M.; Borges, G.; Crozier, A. Dietary (poly)phenolics in human health: Structures, bioavailability, and evidence of protective effects against chronic diseases. *Antioxid. Redox Signal.* **2013**, *18*, 1818–1892.
21. Crozier, A.; Del Rio, D.; Clifford, M.N. Bioavailability of dietary flavonoids and phenolic compounds. *Mol. Aspects Med.* **2010**, *31*, 446–467.
22. Etxeberria, U.; Fernández-Quintela, A.; Milagro, F.I.; Aguirre, L.; Martínez, J.A.; Portillo, M.P. Impact of polyphenols and polyphenol-rich dietary sources on gut microbiota composition. *J. Agric. Food Chem.* **2013**, *61*, 9517–9533.
23. Pérez-Jiménez, J.; Díaz-Rubio, M.E.; Saura-Calixto, F. Non-extractable polyphenols, a major dietary antioxidant: Occurrence, metabolic fate and health effects. *Nutr. Res. Rev.* **2013**, *26*, 118–129.

24. Tremaroli, V.; Bäckhed, F. Functional interactions between the gut microbiota and host metabolism. *Nature* **2012**, *489*, 242–249.
25. Duda-Chodak, A. The inhibitory effect of polyphenols on human gut microbiota. *J. Physiol. Pharmacol.* **2012**, *63*, 497–503.
26. Parkar, S.G.; Trower, T.M.; Stevenson, D.E. Fecal microbial metabolism of polyphenols and its effects on human gut microbiota. *Anaerobe* **2013**, *23*, 12–19.
27. Kawabata, K.; Sugiyama, Y.; Sakano, T.; Ohigashi, H. Flavonols enhanced production of anti-inflammatory substance(s) by *Bifidobacterium adolescentis*: Prebiotic actions of galangin, quercetin, and fisetin. *Biofactors* **2013**, *39*, 422–429.
28. Tzounis, X.; Rodriguez-Mateos, A.; Vulevic, J.; Gibson, G.R.; Kwik-Uribe, C.; Spencer, J.P.E. Prebiotic evaluation of cocoa-derived flavanols in healthy humans by using a randomized, controlled, double-blind, crossover intervention study. *Am. J. Clin. Nutr.* **2011**, *93*, 62–72.
29. Lee, H.C.; Jenner, A.M.; Low, C.S.; Lee, Y.K. Effect of tea phenolics and their aromatic fecal bacterial metabolites on intestinal microbiota. *Res. Microbiol.* **2006**, *157*, 876–884.
30. Massot-Cladera, M.; Pérez-Berezo, T.; Franch, A.; Castell, M.; Pérez-Cano, F.J. Cocoa modulatory effect on rat faecal microbiota and colonic crosstalk. *Arch. Biochem. Biophys.* **2012**, *527*, 105–112.
31. Friedman, M. Overview of antibacterial, antitoxin, antiviral, and antifungal activities of tea flavonoids and teas. *Mol. Nutr. Food Res.* **2007**, *51*, 116–134.
32. Jin, J.-S.; Touyama, M.; Hisada, T.; Benno, Y. Effects of green tea consumption on human fecal microbiota with special reference to *Bifidobacterium* species. *Microbiol. Immunol.* **2012**, *56*, 729–739.
33. Kemperman, R.A.; Gross, G.; Mondot, S.; Possemiers, S.; Marzorati, M.; van de Wiele, T.; Doré, J.; Vaughan, E.E. Impact of polyphenols from black tea and red wine/grape juice on a gut model microbiome. *Food Res. Int.* **2013**, *53*, 659–669.
34. Queipo-Ortuño, M.I.; Boto-Ordóñez, M.; Murri, M.; Gomez-Zumaquero, J.M.; Clemente-Postigo, M.; Estruch, R.; Cardona Díaz, F.; Andrés-Lacueva, C.; Tinahones, F.J. Influence of red wine polyphenols and ethanol on the gut microbiota. *Am. J. Clin. Nutr.* **2012**, *95*, 1323–1334.
35. Dolara, P.; Luceri, C.; de Filippo, C.; Femia, A.P.; Giovannelli, L.; Caderni, G.; Cecchini, C.; Silvi, S.; Orpianesi, C.; Cresci, A. Red wine polyphenols influence carcinogenesis, intestinal microflora, oxidative damage and gene expression profiles of colonic mucosa in F344 rats. *Mutat. Res.* **2005**, *591*, 237–246.
36. Sánchez-Patán, F.; Cueva, C.; Monagas, M.; Walton, G.E.; Gibson, G.R.M.; Quintanilla-López, J.E.; Lebrón-Aguilar, R.; Martín-Álvarez, P.J.; Moreno-Arribas, M.V.; Bartolomé, B. *In vitro* fermentation of a red wine extract by human gut microbiota: Changes in microbial groups and formation of phenolic metabolites. *J. Agric. Food Chem.* **2012**, *60*, 2136–2147.
37. De Boever, P.; Deplancke, B.; Verstraete, W. Fermentation by gut microbiota cultured in a simulator of the human intestinal microbial ecosystem is improved by supplementing a soygerm powder. *J. Nutr.* **2000**, *130*, 2599–2606.

38. Bedani, R.; Pauly-Silveira, N.D.; Roselino, M.N.; de Valdez, G.F.; Rossi, E.A. Effect of fermented soy product on the fecal microbiota of rats fed on a beef-based animal diet. *J. Sci. Food Agric.* **2010**, *90*, 233–238.
39. Cavallini, D.C.U.; Suzuki, J.Y.; Abdalla, D.S.P.; Vendramini, R.C.; Pauly-Silveira, N.D.; Roselino, M.N.; Pinto, R.A.; Rossi, E.A. Influence of a probiotic soy product on fecal microbiota and its association with cardiovascular risk factors in an animal model. *Lipids Health Dis.* **2011**, *10*, 126.
40. Fernandez-Raudales, D.; Hoeflinger, J.L.; Bringe, N.A.; Cox, S.B.; Dowd, S.E.; Miller, M.J.; de Mejía, E.G. Consumption of different soymilk formulations differentially affects the gut microbiomes of overweight and obese men. *Gut Microbes* **2012**, *3*, 490–500.
41. Sembries, S.; Dongowski, G.; Mehrländer, K.; Will, F.; Dietrich, H. Physiological effects of extraction juices from apple, grape, and red beet pomaces in rats. *J. Agric. Food Chem.* **2006**, *54*, 10269–10280.
42. Bialonska, D.; Kasimsetty, S.G.; Schrader, K.K.; Ferreira, D. The effect of pomegranate (*Punica granatum* L.) byproducts and ellagitannins on the growth of human gut bacteria. *J. Agric. Food Chem.* **2009**, *57*, 8344–8349.
43. Vendrame, S.; Guglielmetti, S.; Riso, P.; Arioli, S.; Klimis-Zacas, D.; Porrini, M. Six-week consumption of a wild blueberry powder drink increases bifidobacteria in the human gut. *J. Agric. Food Chem.* **2011**, *59*, 12815–12820.
44. Neyrinck, A.M.; van Hée, V.F.; Bindels, L.B.; de Backer, F.; Cani, P.D.; Delzenne, N.M. Polyphenol-rich extract of pomegranate peel alleviates tissue inflammation and hypercholesterolaemia in high-fat diet-induced obese mice: Potential implication of the gut microbiota. *Br. J. Nutr.* **2013**, *109*, 802–809.
45. Selma, M.V.; Espín, J.C.; Tomás-Barberán, F.A. Interaction between phenolics and gut microbiota: Role in human health. *J. Agric. Food Chem.* **2009**, *57*, 6485–6501.
46. Monagas, M.; Urpi-Sarda, M.; Sánchez-Patán, F.; Llorach, R.; Garrido, I.; Gómez-Cordovés, C.; Andrés-Lacueva, C.; Bartolomé, B. Insights into the metabolism and microbial biotransformation of dietary flavan-3-ols and the bioactivity of their metabolites. *Food Funct.* **2010**, *1*, 233–253.
47. Lee, K.-M.; Yeo, M.; Choue, J.-S.; Jin, J.-H.; Park, S.-J.; Cheong, J.-Y.; Lee, K.-J.; Kim, J.-H.; Hahm, K.-B. Protective mechanism of epigallocatechin-3-gallate against *Helicobacter pylori*-induced gastric epithelial cytotoxicity via the blockage of TLR-4 signaling. *Helicobacter* **2004**, *9*, 632–642.
48. Clifford, M.N. Diet-derived phenols in plasma and tissues and their implications for health. *Planta Med.* **2004**, *70*, 1103–1114.
49. Tzounis, X.; Vulevic, J.; Kuhnle, G.G.C.; George, T.; Leonczak, J.; Gibson, G.R.; Kwik-Uribe, C.; Spencer, J.P. Flavanol monomer-induced changes to the human faecal microflora. *Br. J. Nutr.* **2008**, *99*, 782–792.
50. Neilson, A.P.; Ferruzzi, M.G. Influence of formulation and processing on absorption and metabolism of flavan-3-ols from tea and cocoa. *Annu. Rev. Food Sci. Technol.* **2011**, *2*, 125–151.
51. Tuohy, K.M.; Conterno, L.; Gasperotti, M.; Viola, R. Up-regulating the human intestinal microbiome using whole plant foods, polyphenols, and/or fiber. *J. Agric. Food Chem.* **2012**, *60*, 8776–8782.

52. De Kivit, S.; Tobin, M.C.; Forsyth, C.B.; Keshavarzian, A.; Landay, A.L. Regulation of intestinal immune responses through TLR activation: Implications for pro- and prebiotics. *Front. Immunol.* **2014**, *5*, 60.
53. Huang, H.; Cheng, Z.; Shi, H.; Xin, W.; Wang, T.T.Y.; Yu, L.L. Isolation and characterization of two flavonoids, engeletin and astilbin, from the leaves of *Engelhardia roxburghiana* and their potential anti-inflammatory properties. *J. Agric. Food Chem.* **2011**, *59*, 4562–4569.
54. Kim, T.H.; Ku, S.-K.; Bae, J.-S. Inhibitory effects of kaempferol-3-*O*-sophoroside on HMGB1-mediated proinflammatory responses. *Food Chem. Toxicol.* **2012**, *50*, 1118–1123.
55. Bhaskar, S.; Shalini, V.; Helen, A. Quercetin regulates oxidized LDL induced inflammatory changes in human PBMCs by modulating the TLR-NF-κB signaling pathway. *Immunobiology* **2011**, *216*, 367–373.
56. Yilma, A.N.; Singh, S.R.; Morici, L.; Dennis, V.A. Flavonoid naringenin: A potential immunomodulator for *Chlamydia trachomatis* inflammation. *Mediators Inflamm.* **2013**, *2013*, doi:10.1155/2013/102457.
57. Yoshida, H.; Watanabe, W.; Oomagari, H.; Tsuruta, E.; Shida, M.; Kurokawa, M. Citrus flavonoid naringenin inhibits TLR2 expression in adipocytes. *J. Nutr. Biochem.* **2013**, *24*, 1276–1284.
58. Byun, E.H.; Fujimura, Y.; Yamada, K.; Tachibana, H. TLR4 signaling inhibitory pathway induced by green tea polyphenol epigallocatechin-3-gallate through 67-kDa laminin receptor. *J. Immunol.* **2010**, *185*, 33–45.
59. Byun, E.-B.; Choi, H.-G.; Sung, N.-Y.; Byun, E.-H. Green tea polyphenol epigallocatechin-3-gallate inhibits TLR4 signaling through the 67-kDa laminin receptor on lipopolysaccharide-stimulated dendritic cells. *Biochem. Biophys. Res. Commun.* **2012**, *426*, 480–485.
60. Luo, W.; Wang, C.-Y.; Jin, L. Baicalin downregulates Porphyromonas gingivalis lipopolysaccharide-upregulated IL-6 and IL-8 expression in human oral keratinocytes by negative regulation of TLR signaling. *PLoS One* **2012**, *7*, e51008.
61. Li, H.-Y.; Yuan, Z.-Y.; Wang, Y.-G.; Wan, H.-J.; Hu, J.; Chai, Y.-S.; Lei, F.; Xing, D.-M.; Du, L.-J. Role of baicalin in regulating toll-like receptor 2/4 after ischemic neuronal injury. *Chin. Med. J.* **2012**, *125*, 1586–1593.
62. Qiao, H.; Zhang, X.; Zhu, C.; Dong, L.; Wang, L.; Zhang, X.; Xing, Y.; Wang, C.; Ji, Y.; Cao, X. Luteolin downregulates TLR4, TLR5, NF-κB and p-p38MAPK expression, upregulates the p-ERK expression, and protects rat brains against focal ischemia. *Brain Res.* **2012**, *1448*, 71–81.
63. Barioni, E.D.; Santin, J.R.; Machado, I.D.; de Paula Rodrigues, S.F.; Ferraz-de-Paula, V.; Wagner, T.M.; Cogliati, B.; Corrêa Dos Santos, M.; Machado Mda, S.; de Andrade, S.F.; *et al.* *Achyrocline satureioides* (Lam.) D.C. hydroalcoholic extract inhibits neutrophil functions related to innate host defense. *Evid. Based Complement. Alternat. Med.* **2013**, *2013*, doi:10.1155/2013/787916.
64. Pérez-Berezo, T.; Franch, A.; Ramos-Romero, S.; Castellote, C.; Pérez-Cano, F.J.; Castell, M. Cocoa-enriched diets modulate intestinal and systemic humoral immune response in young adult rats. *Mol. Nutr. Food Res.* **2011**, *55*, S56–S66.

65. Pérez-Berezo, T.; Franch, A.; Castellote, C.; Castell, M.; Pérez-Cano, F.J. Mechanisms involved in down-regulation of intestinal IgA in rats by high cocoa intake. *J. Nutr. Biochem.* **2012**, *23*, 838–844.
66. Ghanim, H.; Sia, C.L.; Upadhyay, M.; Korzeniewski, K.; Viswanathan, P.; Abuaysheh, S.; Mohanty, P.; Dandona, P. Orange juice neutralizes the proinflammatory effect of a high-fat, high-carbohydrate meal and prevents endotoxin increase and toll-like receptor expression. *Am. J. Clin. Nutr.* **2010**, *91*, 940–949.
67. Deopurkar, R.; Ghanim, H.; Friedman, J.; Abuaysheh, S.; Sia, C.L.; Mohanty, P.; Viswanathan, P.; Chaudhuri, A.; Dandona, P. Differential effects of cream, glucose, and orange juice on inflammation, endotoxin, and the expression of toll-like receptor-4 and suppressor of cytokine signaling-3. *Diabetes Care* **2010**, *33*, 991–997.
68. Park, S.-J.; Youn, H.-S. Suppression of homodimerization of toll-like receptor 4 by isoliquiritigenin. *Phytochemistry* **2010**, *71*, 1736–1740.
69. Haque, A.; Koide, N.; Iftakhar-E-Khuda, I.; Noman, A.S.M.; Odkhuu, E.; Badamtseren, B.; Naiki, Y.; Komatsu, T.; Yoshida, T.; Yokochi, T. Flavopiridol inhibits lipopolysaccharide-induced TNF-α production through inactivation of nuclear factor-κB and mitogen-activated protein kinases in the MyD88-dependent pathway. *Microbiol. Immunol.* **2011**, *55*, 160–167.
70. Shuto, T.; Ono, T.; Ohira, Y.; Shimasaki, S.; Mizunoe, S.; Watanabe, K.; Suico, M.A.; Koga, T.; Sato, T.; Morino, S.; *et al.* Curcumin decreases toll-like receptor-2 gene expression and function in human monocytes and neutrophils. *Biochem. Biophys. Res. Commun.* **2010**, *398*, 647–652.
71. Ghanim, H.; Sia, C.L.; Korzeniewski, K.; Lohano, T.; Abuaysheh, S.; Marumganti, A.; Chaudhuri, A.; Dandona, P. A resveratrol and polyphenol preparation suppresses oxidative and inflammatory stress response to a high-fat, high-carbohydrate meal. *J. Clin. Endocrinol. Metab.* **2011**, *96*, 1409–1414.
72. Silver, A.C.; Arjona, A.; Walker, W.E.; Fikrig, E. The circadian clock controls toll-like receptor 9-mediated innate and adaptive immunity. *Immunity* **2012**, *36*, 251–261.
73. Froy, O.; Chapnik, N. Circadian oscillation of innate immunity components in mouse small intestine. *Mol. Immunol.* **2007**, *44*, 1954–1960.
74. Youn, H.S.; Saitoh, S.I.; Miyake, K.; Hwang, D.H. Inhibition of homodimerization of toll-like receptor 4 by curcumin. *Biochem. Pharmacol.* **2006**, *72*, 62–69.
75. Kaneko, M.; Takimoto, H.; Sugiyama, T.; Seki, Y.; Kawaguchi, K.; Kumazawa, Y. Suppressive effects of the flavonoids quercetin and luteolin on the accumulation of lipid rafts after signal transduction via receptors. *Immunopharmacol. Immunotoxicol.* **2008**, *30*, 867–882.
76. Capelluto, D.G.S. Tollip: A multitasking protein in innate immunity and protein trafficking. *Microbes Infect.* **2012**, *14*, 140–147.
77. Youn, H.S.; Lee, J.Y.; Saitoh, S.I.; Miyake, K.; Kang, K.W.; Choi, Y.J.; Hwang, D.H. Suppression of MyD88- and TRIF-dependent signaling pathways of toll-like receptor by (−)-epigallocatechin-3-gallate, a polyphenol component of green tea. *Biochem. Pharmacol.* **2006**, *72*, 850–859.

78. Wu, T.-T.; Chen, T.-L.; Loon, W.-S.; Tai, Y.-T.; Cherng, Y.-G.; Chen, R.-M. Lipopolysaccharide stimulates syntheses of toll-like receptor 2 and surfactant protein-A in human alveolar epithelial A549 cells through upregulating phosphorylation of MEK1 and ERK1/2 and sequential activation of NF-κB. *Cytokine* **2011**, *55*, 40–47.
79. Williams, R.J.; Spencer, J.P.E.; Rice-Evans, C. Flavonoids: Antioxidants or signaling molecules? *Free Radic. Biol. Med.* **2004**, *36*, 838–849.
80. Bhullar, K.S.; Rupasinghe, H.P.V. Polyphenols: Multipotent therapeutic agents in neurodegenerative diseases. *Oxid. Med. Cell. Longev.* **2013**, *2013*, doi:10.1155/2013/891748.
81. Park, Y.C.; Rimbach, G.; Saliou, C.; Valacchi, G.; Packer, L. Activity of monomeric, dimeric, and trimeric flavonoids on NO production, TNF-α secretion, and NF-κB-dependent gene expression in RAW 264.7 macrophages. *FEBS Lett.* **2000**, *465*, 93–97.
82. Comalada, M.; Camuesco, D.; Sierra, S.; Ballester, I.; Xaus, J.; Gálvez, J.; Zarzuelo, A. *In vivo* quercitrin anti-inflammatory effect involves release of quercetin, which inhibits inflammation through down-regulation of the NF-κB pathway. *Eur. J. Immunol.* **2005**, *35*, 584–592.
83. Pantano, C.; Reynaert, N.L.; van der Vliet, A.; Janssen-Heininger, Y.M.W. Redox-sensitive kinases of the nuclear factor-κB signaling pathway. *Antioxid. Redox Signal.* **2006**, *8*, 1791–1806.
84. Hirao, K.; Yumoto, H.; Nakanishi, T.; Mukai, K.; Takahashi, K.; Takegawa, D.; Matsuo, T. Tea catechins reduce inflammatory reactions via mitogen-activated protein kinase pathways in toll-like receptor 2 ligand-stimulated dental pulp cells. *Life Sci.* **2010**, *86*, 654–660.
85. Chen, C.-C.; Chow, M.-P.; Huang, W.-C.; Lin, Y.-C.; Chang, Y.-J. Flavonoids inhibit tumor necrosis factor-alpha-induced up-regulation of intercellular adhesion molecule-1 (ICAM-1) in respiratory epithelial cells through activator protein-1 and nuclear factor-kappaB: Structure-activity relationships. *Mol. Pharmacol.* **2004**, *66*, 683–693.
86. Li, H.-Y.; Hu, J.; Zhao, S.; Yuan, Z.-Y.; Wan, H.-J.; Lei, F.; Ding, Y.; Xing, D.-M.; Du, L.-J. Comparative study of the effect of baicalin and its natural analogs on neurons with oxygen and glucose deprivation involving innate immune reaction of TLR2/TNFα. *J. Biomed. Biotechnol.* **2012**, *2012*, doi:10.1155/2012/267890.
87. Kang, N.J.; Lee, K.W.; Lee, D.E.; Rogozin, E.A.; Bode, A.M.; Lee, H.J.; Dong, Z. Cocoa procyanidins suppress transformation by inhibiting mitogen-activated protein kinase kinase. *J. Biol. Chem.* **2008**, *283*, 20664–20673.
88. Ramiro-Puig, E.; Casadesús, G.; Lee, H.; Zhu, X.; McShea, A.; Perry, G.; Pérez-Cano, F.J.; Smith, M.A.; Castell, M. Neuroprotective effect of cocoa flavonoids on *in vitro* oxidative stress. *Eur. J. Nutr.* **2008**, *48*, 61.
89. Mackenzie, G.G.; Carrasquedo, F.; Delfino, J.M.; Keen, C.L.; Fraga, C.G.; Oteiza, P.I. Epicatechin, catechin, and dimeric procyanidins inhibit PMA-induced NF-κB activation at multiple steps in Jurkat T cells. *FASEB J.* **2004**, *18*, 167–169.
90. Lim, E.-K.; Mitchell, P.J.; Brown, N.; Drummond, R.A.; Brown, G.D.; Kaye, P.M.; Bowles, D.J. Regiospecific methylation of a dietary flavonoid scaffold selectively enhances IL-1β production following toll-like receptor 2 stimulation in THP-1 monocytes. *J. Biol. Chem.* **2013**, *288*, 21126–21135.

91. Lee, J.K.; Kim, S.Y.; Kim, Y.S.; Lee, W.-H.; Hwang, D.H.; Lee, J.Y. Suppression of the TRIF-dependent signaling pathway of toll-like receptors by luteolin. *Biochem. Pharmacol.* **2009**, *77*, 1391–1400.
92. Jeong, H.Y.; Sung, G.-H.; Kim, J.H.; Yoon, J.Y.; Yang, Y.; Park, J.-G.; Kim, S.-H.; Yi, Y.-S.; Yang, W.-S.; Yong, D.-H.; *et al.* Syk and Src are major pharmacological targets of a *Cerbera manghas* methanol extract with kaempferol-based anti-inflammatory activity. *J. Ethnopharmacol.* **2014**, *151*, 960–969.
93. Park, S.-J.; Song, H.-Y.; Youn, H.-S. Suppression of the TRIF-dependent signaling pathway of toll-like receptors by isoliquiritigenin in RAW264.7 macrophages. *Mol. Cells* **2009**, *28*, 365–368.
94. Liew, F.Y.; Xu, D.; Brint, E.K.; O'Neill, L.A.J. Negative regulation of toll-like receptor-mediated immune responses. *Nat. Rev. Immunol.* **2005**, *5*, 446–458.
95. Sung, N.-Y.; Yang, M.-S.; Song, D.-S.; Kim, J.-K.; Park, J.-H.; Song, B.-S.; Park, S.-H.; Lee, J.-W.; Park, H.-J.; Kim, J.-H.; *et al.* Procyanidin dimer B2-mediated IRAK-M induction negatively regulates TLR4 signaling in macrophages. *Biochem. Biophys. Res. Commun.* **2013**, *438*, 122–128.
96. Renz, H.; Brandtzaeg, P.; Hornef, M. The impact of perinatal immune development on mucosal homeostasis and chronic inflammation. *Nat. Rev. Immunol.* **2012**, *12*, 9–23.
97. Pérez-Cano, F.J.; Massot-Cladera, M.; Franch, A.; Castellote, C.; Castell, M. The effects of cocoa on the immune system. *Front. Pharmacol.* **2013**, *4*, 71.
98. Castell, M.; Pérez-Cano, F.J.; Abril-Gil, M.; Franch, À. Flavonoids on allergy. *Curr. Pharm. Des.* **2014**, *20*, 973–987.
99. Kim, W.; Lee, H. Advances in nutritional research on regulatory T-cells. *Nutrients* **2013**, *5*, 4305–4315.
100. Magrone, T.; Jirillo, E. The interplay between the gut immune system and microbiota in health and disease: Nutraceutical intervention for restoring intestinal homeostasis. *Curr. Pharm. Des.* **2013**, *19*, 1329–1342.

5

Analysis of Phenolic and Flavonoid Contents, and the Anti-Oxidative Potential and Lipid Peroxidation Inhibitory Activity of Methanolic Extract of *Carissa opaca* Roots and Its Fractions in Different Solvents

Dildar Ahmed *, Khaizran Fatima and Ramsha Saeed

Department of Chemistry, Forman Christian College (A Chartered University), Lahore-54600, Pakistan; E-Mails: khaizranfatima@yahoo.com (K.F.); ramsha.sd@gmail.com (R.S.)

* Author to whom correspondence should be addressed; E-Mail: dildarahmed@gmail.com

External Editors: Maria G. Miguel and João Rocha

Abstract: The objective of the present work was to investigate the anti-oxidative potential of methanolic extract of *Carissa opaca* roots and its fractions in solvents of different polarities. Total phenolic (TPC) and flavonoid (TFC) contents of methanolic extract were 211.95 ± 0.78 μg/mL gallic acid equivalents (GAE) and 8.35 ± 0.21 μg/mL rutin equivalents (RE), respectively. Ethyl acetate contained the highest amounts of both (TFC, 11.8 ± 0.28 RE; TPC, 342.80 ± 0.42 GAE) followed by chloroform fraction (TFC, 7.50 ± 0.14 RE; TPC, 275.85 ± 0.50 GAE). Extract and fractions displayed remarkable DPPH radical scavenging activity. EC_{50} values of methanolic extract was 0.88 mg/mL, while that of hexane, chloroform, ethyl acetate, *n*-butanolic and aqueous fractions were 0.58, 0.38, 0.29, 0.36 and 5.83 mg/mL, respectively, ethyl acetate fraction being most potent. The ethyl acetate fraction also showed the highest activity in terms of reducing power, phosphomolybdate and ABTS assays. All the fractions showed fairly good lipid peroxidation inhibitory activity, which remained almost constant over three days. Based on the results it can be concluded that roots of *Carissa opaca* contains phytochemicals with exploitable antioxidant, free radical scavenging, and lipid peroxidation inhibitory potential.

Keywords: *Carissa opaca*; roots; flavonoids; phenolics; antioxidant

1. Introduction

Reactive oxygen and nitrogen species, abbreviated as ROS and RNS, respectively, are species with odd electrons called free radicals. They are very reactive groups of atoms and have potential to damage components of living cells when produced excessively [1,2]. They are continuously produced in the human body but are simultaneously destroyed by its antioxidant defense mechanism. Continuous exposure to free radicals can contribute to fatal ailments including cancer, cardiovascular and neurodegenerative diseases, and age related disorders [3,4]. As plants constitute a virtually unending source of bioactive natural products, they are the subject of extensive research worldwide with the aim to explore novel therapeutic agents.

Carissa opaca (family Apocynaceae) is an evergreen, thorny shrub found in the Himalayan areas of Indo-Pakistan subcontinent [5–7]. The plant has a number of ethno-medicinal applications and all of its parts including leaves, fruit and roots are considered to be useful in various diseases [8,9]. Its leaves are used to cure fever, jaundice, hepatitis and asthma [10,11]. Its fruit is considered to be an aphrodisiac [12,13]. Roots of the plant are known to be purgative, and have an ability to heal wounds and injuries [14]. Ground roots are mixed with pericarp of mango in water and used as wormicide of the intestine [15]. The chemical constituents found in *C. opaca* include flavonoids, terpenoids, coumarins, glycosides, tannins and anthroquinones [16]. Antioxidant activities of both fruit and leaves of *C. opaca* have been studied under different conditions [16–18].

In continuation of our research on *C. opaca* [17,19], we investigated, in the present work, its roots for total phenolic and flavonoid contents, antioxidant and free radical scavenging properties and lipid peroxidation inhibitory activity. As far as we could explore, this is the first study of its type on the roots of *C. opaca*.

2. Experimental Section

2.1. Chemical

All the chemicals used in the present study were of analytical grade. Ammonium molybdate, sodium dihydrogen phosphate, disodium hydrogen phosphate, potassium ferricyanide, ammonium thiocyanate, potassium persulfate sodium nitrite and gallic acid were purchased from Riedel-de-Haen (Seelze, Germany). Tween 20 and ascorbic acid were bought from Fischer Chemicals (Loughborough, UK). TPTZ (2,4,6 tripyridyl-*s*-triazine), trichloroacetic acid, ABTS (2,2′-azino-bis(3-ethylbenzothiazoline-6-sulphonic acid), Trolox (6-hydroxy-2,5,7,8-tetramethylchroman-2-carboxylic acid), DPPH (2,2-diphenyl-1-picrylhydrazyl) were purchased from Sigma-Aldrich (St. Louis, MO, USA). Rutin was purchased from Alfa Aesar (Karlsruhe, Germany), Folin-Ciocalteu reagent from Merck (Darmstadt, Germany), and linoleic acid from Bio-World (Dublin, OH, USA). Solvent used for extraction and fractionation were of HPLC grade.

2.2. Collection and Preparation of Plant Material

2.2.1. Collection of Plant Material

Roots of *C. opaca* were collected from hills of Abbottabad (Pakistan) in March 2013. The identification of the plant was confirmed by the taxonomist Mr. Ajaib Khan of GC University, Lahore, Pakistan (voucher specimen: GC-Herb Bot 2271). The roots were separated from the aerial parts of the plants, washed with distilled water and allowed to air dry in shade. They were crushed and ground to obtain finely divided powder.

2.2.2. Extraction and Fractionation

A methanolic extract of the powder was obtained through cold maceration method by soaking the material in pure methanol for 15 days followed by filtration. The process was repeated thrice and the extracts were combined. The solvent was evaporated on rotary evaporator to obtain dried crude methanolic extract. The methanolic extract was then suspended in distilled water in a separatory funnel and partitioned successively with hexane, chloroform, ethyl acetate, and 1-butanol to obtained fractions in these solvents. This process left residual aqueous fraction at the end. The solvents were removed with a rotary evaporator at low pressure (about 10 mbar) to produce dried fractions.

2.3. Determination of Total Phenolic Content (TPC)

Colorimetric protocol [20] was used to determine total phenolic content of methanolic extract of *C. opaca* roots and its fractions. In a test tube, 40 μL (1 mg per 1 mL of methanol) of the plant extract (or standard gallic acid solution), 3.16 mL distilled water and 200 μL Folin-Ciocalteu reagent were put and mixed by shaking gently. After an incubation of 8 min, 600 μL sodium carbonate solution was added and mixed. The mixture was incubated at 40 °C for 30 min before recording its absorbance in a spectrophotometer at 765 nm against a blank. The blank contained 40 μL methanol in place of sample. Gallic acid was used as a standard. Its calibration curve was drawn and the total phenolic content was expressed as micrograms per milliliter of gallic acid equivalents (μg/mL of GAE).

2.4. Determination of Total Flavonoid Content (TFC)

Total flavonoid contents of plant samples were determined using aluminum chloride colorimetric method [21]. Briefly, 1 mg of plant extract was dissolved in 1 mL of methanol. In a test tube, 0.3 mL (1 mg/mL in methanol) of plant sample (or standard rutin solution) and 3.4 mL 30% aqueous methanol, 150 μL sodium nitrite solution (0.5 M) and 150 μL aluminum chloride solution (0.3 M) was mixed. After an interval of 5 min, 1 mL NaOH solution (1 M) was added. The absorbance was measured at 506 nm against a blank. The blank contained an equal amount of methanol in place of the sample. Rutin was used as a standard. Its calibration curve was drawn and total flavonoid content of a plant sample was expressed a μg/mL of rutin equivalents (RE).

2.5. DPPH Radical Scavenging Assay

Free radical scavenging activities of the methanolic extracts of *C. opaca* roots and its fractions were determined according to the DPPH methods of Ahmed *et al.* and Brand-Williams *et al.* [21,22]. The stock solution of DPPH radical (24 mg/100 mL in methanol) was prepared. By diluting this solution with methanol, a working solution was prepared to obtain an absorbance of 0.980 + 0.02 at 517 nm. Stock solution of plant extract/fraction was prepared in methanol with concentration 4 mg/mL. From this, different dilutions (0.2–2.0 mg/mL) were made. The test mixture contained 3 mL DPPH working solution and 100 μL of a sample. The mixture was incubated at 37 °C for 30 min in dark. Absorbance of each sample was recorded at 517 nm. For a negative control, 100 μL of methanol was added in place of plant sample. Ascorbic acid was used as a positive control. Inhibition curves were made and IC_{50} value were calculated for all samples. The free radical scavenging activity of each sample was calculated by using the formula:

$$\% \text{ Scavenging Activity} = 100[(Ac - As)/Ac] \quad (1)$$

where *Ac* and *As* are absorbances of negative control and sample, respectively.

2.6. ABTS Radical Scavenging Assay

ABTS (2,2′-azinobis(3-ethylbenzothiazoline)-6-sulphonic acid) assay was conducted by protocol of Re *et al.* [23]. ABTS (38.36 mg) was dissolved in distilled water and the volume was made up to 9.5 mL, and 245 μL 100 mM potassium persulfate was added. The volume was made up to 10 mL by distilled water to give the final concentration of ABTS stock solution as 7 mM. The stock solution was kept in dark for 18 h at room temperature. To prepare the ABTS working solution, ABTS stock solution was diluted with PBS buffer pH (7.4) to an absorbance of 0.7 ± 0.02). In a test tube, 2.99 mL ABTS working solution was added and mixed with 10 μL (4 mg per mL of methanol) of the plant extract solution (or standard). A negative control was prepared by adding 10 μL of methanol in place of sample. The absorbance was measured at 734 nm for eight min with the interval of 30 seconds and percentage free radical scavenging activity as per this assay was calculated by using the formula:

$$\% \text{ Activity} = 100(1 - \text{Sample Absorbance/Control Absorbance}) \quad (2)$$

Gallic acid solution (0.5 mg /mL of distilled water) and ascorbic acid solution (0.5 mg/mL of distilled water) were used as positive control. The ABTS radical scavenging activity was also determined in terms of Trolox equivalents. Trolox stock solution was prepared by dissolving 7.5 mg in 10 mL PBS. The volume was made to 20 mL to give the concentration of 1500 μM. From this stock solution, dilutions with concentrations of 1250 μM, 1000 μM, 750 μM, 500 μM, 250 μM, 100 μM and 50 μM were made by mixing 2.5 mL, 2 mL, 1.5 mL, 1 mL 0.5 mL, 0.2 mL and 0.1 mL of stock solution in distilled water to the volume of 3 mL.

2.7. Reducing Power Assay

This assay was performed as per the protocol reported by Oyaizu *et al.* [24]. In a test tube, 2.5 mL (1 mg/mL) of the plant extract (or standard gallic acid solution) was taken and mixed with 2.5 mL sodium phosphate buffer (0.2 M, pH 6.6) and 2.5 mL potassium ferricyanide solution (1%). The mixture was incubated at 50 °C for 20 min. After incubation, 2.5 mL trichloroacetic acid (10%) was added and the solution was centrifuged at 650 rpm for 10 min. Then, 5 mL supernatant was mixed with 5 mL distilled water and 1 mL ferric chloride solution (0.1%) was added. The blank had 2.5 mL methanol instead plant sample. The absorbance was measured at 700 nm.

2.8. Phosphomolybdate Assay

Total antioxidant capacity of the methanolic extracts *Carissa opaca* roots was determined as per the reported method [20,25]. In a test tube, 300 µL sample (0.3 mg/mL in methanol) was mixed with 3 mL phosphomolybdate reagent (0.6 M sulfuric acid, 28 mM sodium phosphate and 4 mM ammonium molybdate) and the test tube was capped with silver foil and incubated in water bath at 95 °C for 90 min. The mixture was cooled to room temperature. Its absorbance was measured at 765 nm against a blank. For blank, 300 µL methanol was added in place of plant sample. Ascorbic acid was used as a standard and the results were expressed as µg/mL of ascorbic acid equivalents.

2.9. Lipid peroxidation Inhibitory Assay

A reported protocol [26,27] was used to determine lipid peroxidation inhibitory activity of methanolic extract of *Carissa opaca* roots and its fractions. In a test tube, 100 µL (5 mg per mL of methanol) of plant extract (or standard BHA) was mixed with 2.4 mL potassium phosphate buffer (pH 7) and 2.5 mL linoleic acid emulsion. The mixture was incubated at 37 °C for 25 min. Then, 100 µL of this mixture was regularly taken at 24 h intervals for 120 h and dissolved in 3.7 mL ethanol. It was reacted with 100 µL of ferrous chloride (20 mM) and then 100 µL ammonium thiocyanate (30%) was added. Absorbance was measured at 500 nm. Control consisted of 2.4 mL potassium phosphate buffer, 2.5 mL of linoleic acid emulsion and 100 µL methanol. Linoleic acid emulsion was prepared by mixing 175 µL Tween 20 (emulsifier) and 155 µL of linoleic acid and the volume was made to 50 mL by adding potassium phosphate buffer (pH 7).

2.10. Statistical Analysis

All the experiments were carried out at least three times ($n = 3$) and statistical mean was calculated ± SD using Excel 2010 (Microsoft Corporation, Redmond, WA, USA). The same program was used to calculate EC_{50} values of DPPH assay.

3. Results and Discussion

Methanolic extract of *C. opaca* roots and its fractions in hexane, chloroform, ethyl acetate, 1-butanol and water were subjected to different antioxidant assays.

3.1. Total Flavonoid and Total Phenolic Contents

Total flavonoid and phenolic contents of methanolic extract of *C. opaca* roots and its fractions were determined and results are displayed in Table 1.

Table 1. Total flavonoid content (TFC) and total phenolic content (TPC) of methanolic extract of *Carissa opaca* roots and its fractions in various solvents, and their EC_{50} values in DPPH assay ($n = 3$).

Extract/Fraction	TFC, RE (μg/mL)	TPC, GAE (μg/mL)	EC_{50} (mg/mL)
Methanolic	8.35 ± 0.21	211.95 ± 0.78	0.877 ± 0.002
Hexane	nd *	111.25 ± 0.35	0.584 ± 0.004
Chloroform	7.50 ± 0.14	275.85 ± 0.50	0.380 ± 0.002
Ethyl acetate	11.8 ± 0.28	342.80 ± 0.42	0.290 ± 0.005
***n*-Butanolic**	4.85 ± 0.07	194.70 ± 0.57	0.360 ± 0.012
Aqueous	5.30 ± 0.14	73.85 ± 0.50	5.830 ± 0.007

* Mixture was turbid and there was no change in color. RE, rutin equivalents; GAE, gallic acid equivalents.

The methanolic extract of the roots was found to contain considerable quantities of polyphenols and flavonoids as evidenced by its total phenolic (TPC) and flavonoid (TFC) contents, which were 211.95 ± 0.78 μg/mL gallic acid equivalents (GAE) and 8.35 ± 0.21 μg/mL rutin equivalents (RE), respectively. Ethyl acetate contained highest amounts of both (TFC, 11.8 ± 0.28 RE; TPC, 342.80 ± 0.42 GAE) followed by chloroform fraction (TFC, 7.50 ± 0.14 RE; TPC, 275.85 ± 0.50 GAE). Order of TPC among fractions was: ethyl acetate > chloroform > *n*-butanolic > hexane > aqueous. Order of TFC was almost same except the aqueous fraction showed slightly higher value than *n*-butanolic.

3.2. DPPH Free Radical Scavenging Activity

The antioxidant or free radical scavenging activities of methanolic extract of *C. opaca* roots and its fractions were determined as a function of concentration and results are displayed in Table 1 in terms of EC_{50}.

The results revealed that roots of *C. opaca* had significant anti-oxidative potential. Ethyl acetate fraction was most active with EC_{50} of 0.29 mg/mL. Chloroform and *n*-butanolic fractions showed almost equal activity, 0.38 and 0.36 mg/mL, respectively. Aqueous fraction was least active (5.83 mg/mL). Order of activity among fractions was: ethyl acetate > chloroform > *n*-butanolic > hexane.

Although both electron transfer (ET) and hydrogen transfer (HT) pathways have been proposed for DPPH action, HT seems to predominate [28]. This assay, therefore, measures the ability of a sample to donate H radical to DPPH radical. Phenols and flavonoids have been found to be powerful free radical scavengers. The results of this assay indicated that roots of *C. opaca* have good quantities of compounds capable of free radical scavenging.

3.3. ABTS Radical Scavenging Activity

Free radical scavenging activity of methanolic extract of *C. opaca* roots and its fractions were determined according to ABTS radical assay and results are displayed in Figure 1. The activity was also determined in terms of TEAC (Trolox equivalent antioxidant capacity) values (Figure 2). TEAC measures the antioxidant capacity of a sample as compared to Trolox (6-hydroxy-2,5,7,8-tetramethylchroman-2-carboxylic acid), a known antioxidant. Methanolic extract exhibited remarkable activity in this assay. Chloroform and ethyl acetate fractions were highly active and showed comparable potency. At concentration of 1 mg/mL, their activities were (%) 94.11 ± 0.001 and 90.58 ± 0.012, respectively. Hexane fraction was least active with only about 22% activity at concentration 1 mg/mL followed by aqueous fraction. *n*-Butanolic fraction showed moderate activity. Order of activity among fractions was: chloroform > ethyl acetate > *n*-butanolic > aqueous > hexane. The TEAC values were also in the same order.

Figure 1. Percent free radical scavenging activity of methanolic extract of *Carissa opaca* roots and its fractions in ABTS assay ($n = 3$); Concentration of each sample was 1 mg/mL.

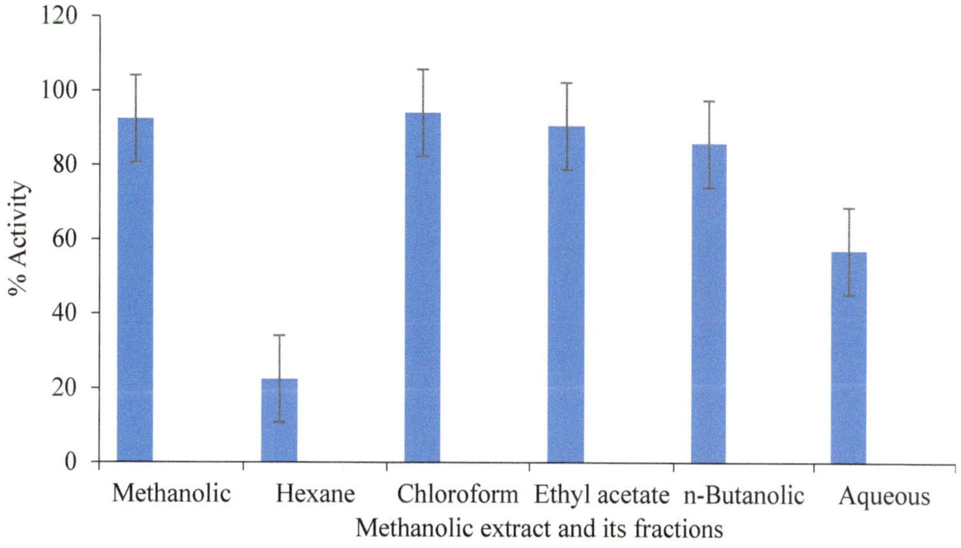

Figure 2. ABTS radical scavenging activity methanolic extract of *Carissa opaca* roots and its factions in terms of TEAC (Trolox equivalent antioxidant capacity) ($n = 3$).

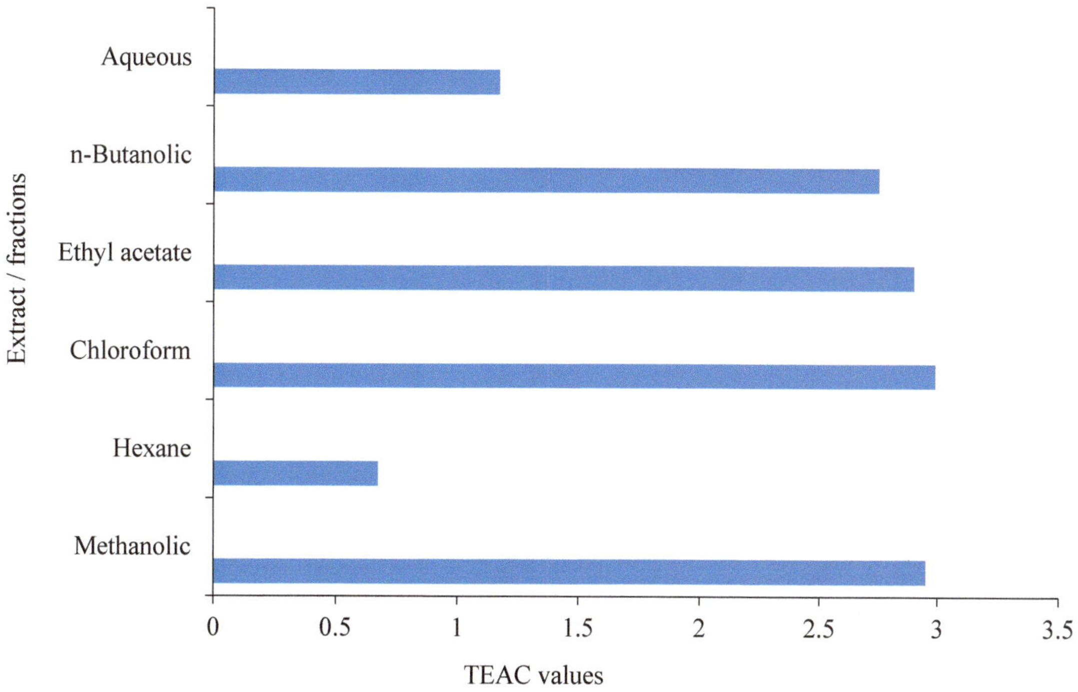

There is almost complete correspondence between DPPH and ABTS radical scavenging activities. Ethyl acetate fraction, however, showed slightly higher activity in the DPPH assay than chloroform fraction.

3.4. Antioxidant Activities as per Reducing Power Assay

Antioxidant activities of methanolic extract of *C. opaca* roots and its fractions were determined as per reducing power assay and results are shown in Figure 3. The results are in the form of mean absorbance (n = 3), which is a direct measure of antioxidant activity of a sample. Ethyl acetate and *n*-butanolic factions showed almost equal antioxidant activity in this assay. They were most potent among all the fractions. Chloroform and aqueous fractions also showed good activity. The hexane fraction, however, did not exhibit appreciable activity. This is an electron transfer (ET) assay and measures the ability of a sample to donate an electron to destroy a free radical. The order of activity among fractions was: ethyl acetate > *n*-butanolic > aqueous > chloroform > hexane.

Figure 3. Antioxidant activities as per reducing power assay (mean absorbance at 700 nm) of methanolic extract of *Carissa opaca* roots and its fractions in various solvents (n = 3).

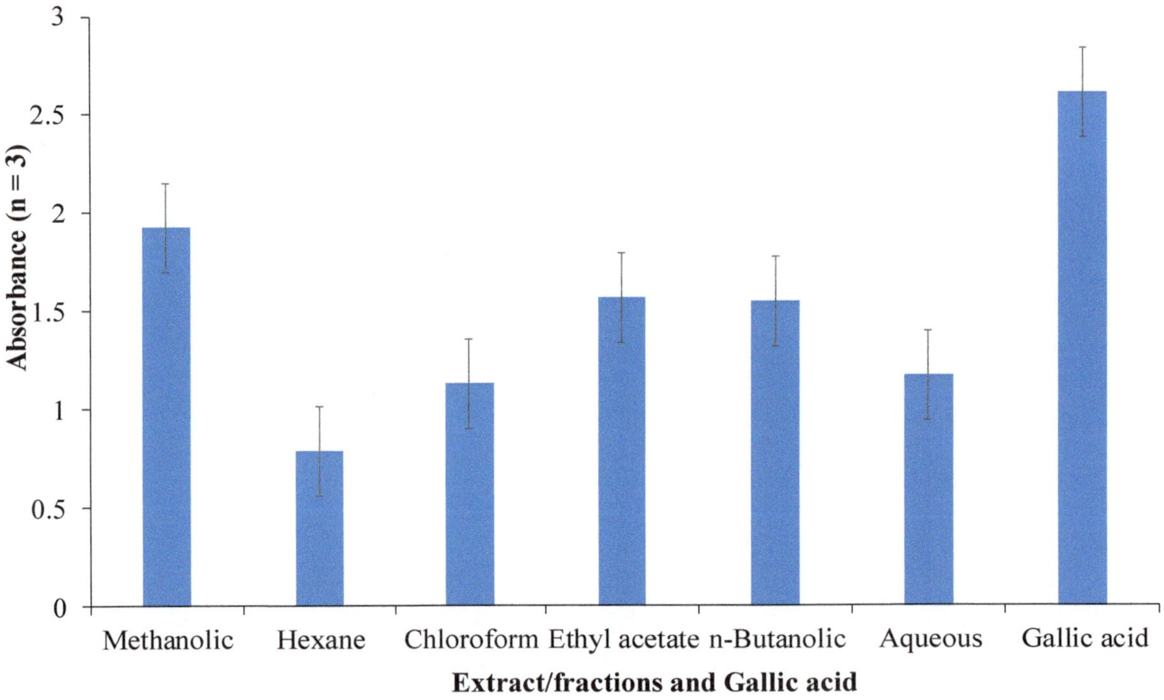

3.5. Antioxidant Capacity as per Phosphomolybdate Assay

Antioxidant capacity of methanolic extract of *C. opaca* roots and its fractions was determined as per phosphomolybdate assay and results are shown in Figure 4. The results are in terms of mg/mL of ascorbic acid equivalents (AAE). Results showed an interesting trend. While all the samples displayed remarkable activity, the most polar aqueous (271.6 ± 0.0023 mg/mL AAE) fraction exhibited the highest activity followed by the most non-polar hexane fraction (199.96 ± 0.011 mg/mL AAE). This suggested that roots of this plant have a broad range of compounds having ability to scavenge free radicals.

Figure 4. Antioxidant capacity of methanolic extract of *Carissa opaca* roots and its fractions in various solvents as per phosphomolybdate assay in units of μg/mL of ascorbic acid equivalents (AAE).

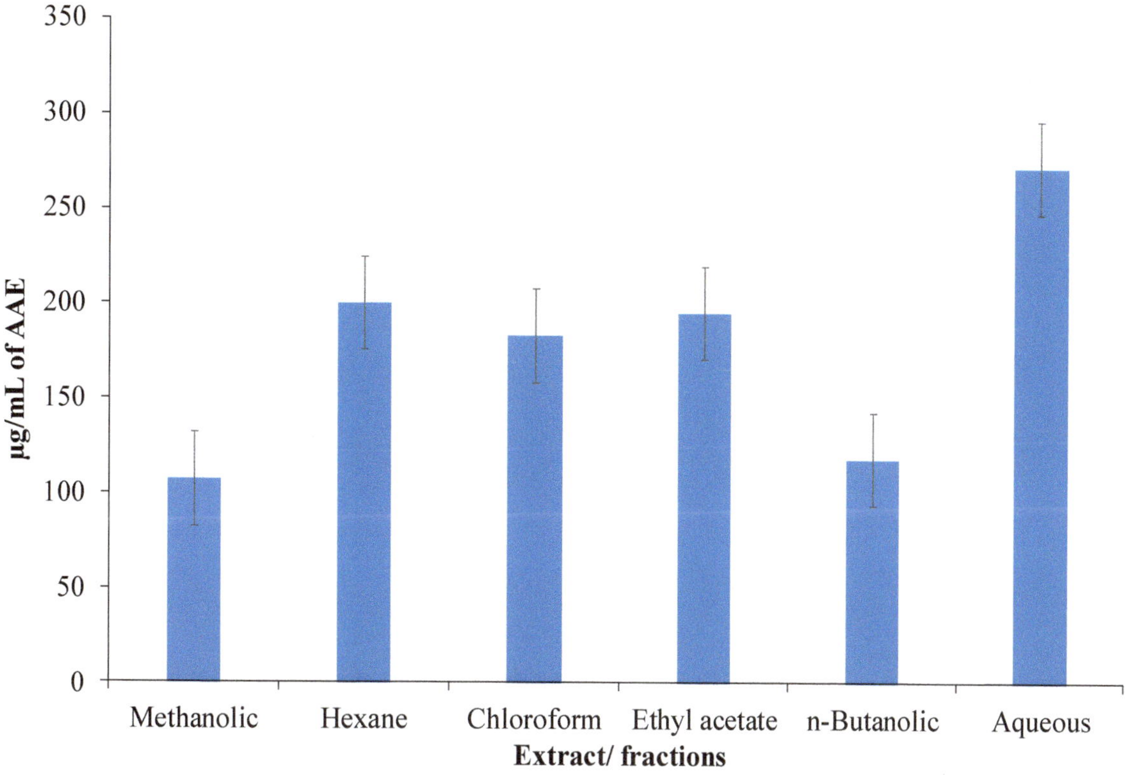

3.6. Lipid Peroxidation Inhibitory Activity

Lipid peroxidation inhibitory activity of methanolic extract of *C. opaca* roots and its fractions in different solvents were determined over a period of 120 h and results are displayed in Figure 5. The methanolic extract and its fractions showed fairly good lipid peroxidation inhibitory activity, which remained almost constant over three days. As a whole, the ethyl acetate fraction proved to be the most potent and consistent inhibitor of lipid peroxidation followed by chloroform fraction.

Roots of *C. opaca* exhibited, in assays discussed above, quite considerable antioxidant activity. There was good correlation between total phenolics and flavonoids, and between these polyphenols and antioxidant activities found through different assays. In general, moderately polar fractions, *i.e.*, ethyl acetate and chloroform, displayed the highest activity. Hexane fraction proved to be least potent, while *n*-butanolic and aqueous fraction exhibited moderate potency. The trend is in harmony with the results of many other studies. The ethyl acetate fraction, for example, has generally been noted to be the most potent antioxidant [4,29]. This trend may be attributed to the excellent ability of ethyl acetate to dissolve flavonoids and other polyphenols in it [30,31]. Phenolics constitute a class of diverse phytochemicals which are among the most potent antioxidants [32]. Investigation into ethyl acetate and chloroform fractions should yield individual chemical compounds responsible for most of their antioxidant potential.

Figure 5. Lipid peroxidation inhibitory activity of methanolic extract of *Carissa opaca* roots and its fractions in various solvents and BHA ($n = 3$).

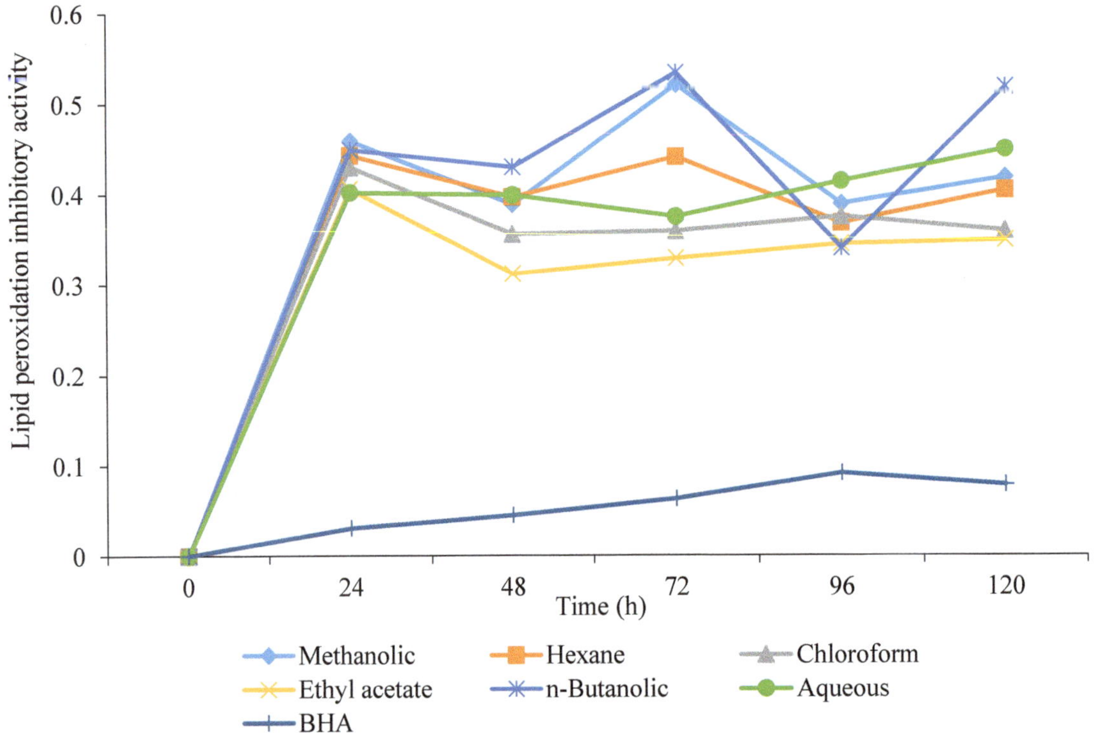

4. Conclusions

Natural antioxidants are needed for their therapeutic and "nutraceutical" applications. Roots of *Carissa opaca* possessed remarkable antioxidant activity. Use of crude extracts can be pharmacologically useful. However, isolation and characterization of phytochemicals will help to establish structure-activity relationships in the discovery of new drugs. In-depth phytochemical investigation of the highly active ethyl acetate and chloroform fractions is especially recommended.

Acknowledgements

The authors wish to thank HEC (Higher Education Commission), Islamabad, Pakistan, for providing funds for this research work as part of project No. 20-1986/R & D/11.

Author Contributions

Dildar Ahmed conceived, designed and supervised the study, and helped in paper writing. Khaizran Fatima performed the experiments, analyzed the data and wrote the paper. Ramsha Saeed conducted extraction and fractionation of the plant material.

Conflicts of Interest

The authors declare there is no conflict of interest.

References

1. Barrera, G. Oxidative Stress and Lipid Peroxidation Products in Cancer Progression and Therapy. *ISRN Oncol.* **2012**, *2012*, 1–21. Available online: http://www.hindawi.com/journals/isrn.oncology/2012/137289/ (accessed on 5 July 2014).
2. Valko, M.; Leibfritz, D.; Moncol, J.; Cronin, M.T.; Mazur, M.; Telser, J. Free radicals and antioxidants in normal physiological functions and human disease. *Int. J. Biochem. Cell. Biol.* **2007**, *39*, 44–84.
3. Fang, Y.Z.; Yang, S.; Wu, G. Free radicals, antioxidants and nutrition. *Nutrition* **2002**, *18*, 872–879.
4. Kenny, O.; Smyth, T.J.; Hewage, C.M.; Brunton, N.P. Quantitative UPLC-MS/MS analysis of chlorogenic acid derivatives in antioxidant fractionates from dandelion (*Taraxacum officinale*) root. *Int. J. Food Sci. Technol.* **2014**, doi:10.1111/ijfs.12668.
5. Rai, S.K.; Mallavarapu, G.R.; Pandey-Rai, S.; Srivastava, S.; Singh, D.; Mishra, R.; Kumar, S. Constituents of the flower oil of *Carissa opaca* growing in the Aravalli mountain range at New Delhi. *Flavour Fragr. J.* **2006**, *21*, 304–305.
6. Nazimuddin, S.; Qaiser, M. Apocynaceae. In *Flora of Pakistan*; Nasir, E., Ali, S.I., Eds.; Department of Botany, University of Karachi: Karachi, Pakistan, 1983; pp. 11–13. Available online: http://www.efloras.org/florataxon.aspx?flora_id=5&taxon_id=250077072 (accessed on 20 September 2014).
7. Handoo, S. A survey of plants used for wound healings in animals. *Vet. Scan*, **2006**, *1*. Available online: http://www.kashvet.org/vetscan (accessed on 20 September 2014).
8. Awasthi, A.K.; Kunal, K.; Bisht, G.S.; Awasthi, S. *In vitro* antibacterial and antifungal activity of *Carissa opaca* Stapf ex Haines. *Int. J. Curr. Pharm. Res.* **2013**, *5*, 15–18.
9. Ravishankar, T.; Vedavalli, L.; Nambi, A.A.; Selvam, V. *Role of Tribal People in the Conservation and Utilization of Plant Genetic Resources*; MS Swaminathan Research Foundation: Chennai, India, 1991.
10. Abbasi, A.M.; Khan, M.A.; Ahmad, M.; Zafar, M.; Khan, H.; Muhammad, N.; Sultana, S. Medicinal plants used for the treatment of jaundice and hepatitis based on socio-economic documentation. *Afr. J. Biotechnol.* **2009**, *8*, 1645.
11. Jabeen, A.; Khan, M.A.; Ahmad, M.; Zafar, M.; Ahmad, F. Indigenous uses of economically important flora of Margallah Hills National Park, Islamabad, Pakistan. *Afr. J. Biotechnol.* **2009**, *8*, 777.
12. Ahmad, S.S.; Mahmood, F.; Dogar, Z.; Khan, Z.I.; Ahmad, K.; Sher, M.; Mustafa, I.; Valeem, E.E. Prioritization of medicinal plants of Margala Hills National Park, Islamabad on the basis of available information. *Pak. J. Bot.* **2009**, *41*, 2105–2114.
13. Acharya, D.; Rai, M.K. Traditional knowledge for curing various ailments among Gonds and Bharias of Patalkot Valley, M.P., India. **2009**. Available online: http://www.motivationalmagic.com/articles4/group3/physicalfitness-575.php (accessed on 20 September 2014).

14. Dinesh, K.; Ripathi, H.C.; Tandon, S.K.; Lal, J.; Malik, J.K. Ethnoveterinary phytomedicines used in India and Nepal in the treatment of fractures, wounds and allied disorders: An update. *Ind. J. Vet. Surg.* **1997**, *18*, 65–72.
15. Ch, I.M.; Khan, M.A.; Hanif, W. Ethno-veterinary medicinal uses of plants from Samahni Valley Dist. Bhimber, (Azad Kashmir) Pakistan. *Asian J. Plant. Sci.* **2006**, *5*, 390–396.
16. Sahreen S.; Khan, M.R.; Khan, R.A. Evaluation of antioxidant activities of various solvent extracts of *Carissa Opaca* fruits. *Food Chem.* **2010**, *122*, 1205–1211.
17. Ahmed, D.; Hammad, M. Comparative study of antioxidant, antimicrobial and alpha-amylase inhibitory activities of leaves, fruits and seeds of *Carissa opaca*. *Asian J. Chem.* **2013**, *25*, 3913–3916.
18. Khan, M.A.; Shahwar, D.; Ahmad, N.; Khan, Z.; Ajaib, M. Chemical constituents of *Carissa opaca* extracts and their evaluation as antioxidants and preservatives in edible oils. *Asian J. Chem.* **2010**, *22*, 379–388.
19. Ahmed, D.; Waheed, A.; Chaudhary, M.A.; Khan, S.R.; Hannan, A.; Barkaat, M. Nutritional and antimicrobial studies on leaves and fruits of *Carissa opaca*. *Asian J. Chem.* **2011**, *23*, 2072–2076.
20. Singleton, V.L.; Rossi, J.A. Colorimetry of total phenolics with phosphomolybdic-phosphotungstic acid reagents. *Am. J. Enol. Vitic.* **1965**, *16*, 144–158.
21. Ahmed, D.; Zara, S.; Baig, H. *In vitro* analysis of antioxidant activities of *Oxalis Corniculata* Linn. fractions in various solvents. *Afr. J. Trad. Complement. Altern. Med.* **2013**, *10*, 158–165. Available online: http://www.ncbi.nlm.nih.gov/pmc/articles/PMC3746370 (accessed on 5 July 2014).
22. Brand-Williams, W.; Cuvelier, M.E.; Berset, C. Use of free radical method to evaluate antioxidant activity. *Lebensm. Wiss. Technol.* **1995**, *28*, 25–30.
23. Re, R.; Pellegrini, N.; Pannala, A.; Yang, M.; Rice-Evans, C. Antioxidant activity applying an improved ABTS radical cation decolorization assay. *Free Radic. Biol. Med.* **1999**, *26*, 1231–1237.
24. Oyaizu, M. Studies on products of browning reactions: Antioxidative activities of products of browning reaction prepared from glucosamine. *Jpn. J. Nutr.* **1986**, *44*, 307–315.
25. Prieto, P.; Pineda, M.; Aguilar, M. Spectrophotometric quantitation of antioxidant capacity through the formation of a phosphomolybdenum complex: Specific application to the determination of vitamin E. *Anal. Biochem.* **1999**, *269*, 337–341.
26. Mitsuda, H.; Yasumoto, K.; Iwami, K. Antioxidative action of indole compounds during the autooxidation of linoleic acid. *Euyo Shokuryo* **1996**, *19*, 210–214.
27. Zara, S.; Ahmed, D.; Baig, H.; Ikram, M. Evaluation of antioxidant activities of various solvent extracts of fruits and leaves of *Ehretia. serrata*. *Asian J. Chem.* **2012**, *24*, 4345–4351.
28. Marxen, K.; Vanselow, K.H.; Lippemeier, S.; Hintze, R.; Ruser, A.; Hansen, U.P. Determination of DPPH radical oxidation caused by methanolic extracts of some microalgal species by linear regression analysis of spectrophotometric measurements. *Sensors* **2007**, *7*, 2080–2095.
29. Khatoon, M.; Islam, E.; Islam, R.; Rahman, A.A.; Alam, A.H.M.K., Khondkar, P.; Rashid, M.; Parvin, S. Estimation of total phenol and *in vitro* antioxidant activity of *Albizia. procera* leaves. *BMC Res. Notes*, **2013**, *6*, 121. Available online: http://www.biomedcentral.com/1756–0500/6/121 (accessed on 6 July 2014).

30. Lee, J.H.; Johnson, J.V.; Talcott, S.T. Identification of ellagic acid conjugates and other polyphenolics in Muscadine grapes by HPLC-ESI-MS. *J. Agric. Food Chem.* **2005**, *53*, 6003–6010.
31. Maoela, M.S.; Arotiba, O.A.; Baker, P.G.L.; Mabusela, W.T.; Jahed, N.; Songa, E.A.; Iwuoha, E.I. Electroanalytical determination of catechin flavonoid in ethyl acetate extracts of medicinal plants. *Int. J. Electrochem. Sci.* **2009**, *4*, 1497–1510.
32. Brown, M.J.; Henderson, D.E.; Hunt, C. Comparison of antioxidant properties of supercritical fluid extracts of herbs and the confirmation of pinocembrin as a principle antioxidant in Mexican oregano (*Lippa gravolens*). *Electron. J. Environ. Agric. Food Chem.* **2006**, *2*, 1265–1277.

6

Antioxidant and Anti-Inflammatory Properties of an Extract Rich in Polysaccharides of the Mushroom *Polyporus dermoporus*

Celina Maria P. Guerra Dore [1], Monique Gabriela das Chagas F. Alves [1], Maria da Glória L. Santos [1], Leonardo Augusto R. de Souza [1], Iuri Goulart Baseia [2] and Edda Lisboa Leite [1,*]

1 Laboratory of Glycobiology, Department of Biochemistry, Federal University of Rio Grande do Norte, Av. Salgado Filho, 3000, Bairro L. Nova, CEP 59072-940, Natal, Brazil; E-Mails: cmpguerra@hotmail.com (C.M.P.G.D.); moniquegabi@gmail.com (M.G.C.F.A.); gloria_bios@hotmail.com (M.G.L.S.); leo_augustopdf@hotmail.com (L.A.R.S.)

2 Laboratory of Mycology, Department of Botany, Ecology and Zoology, Federal University of Rio Grande do Norte, Av. Salgado Filho, 3000, Bairro L. Nova 59078-970, Natal, Brazil; E-Mail: baseia@cb.ufrn.br

* Author to whom correspondence should be addressed; E-Mail: eddaleite@cb.ufrn.br

External Editor: Isabel C. F. R. Ferreira

Abstract: *Polyporus dermoporus* mushroom, native to Brazil, is produced under natural conditions in the unexplored reserve of Mata da Estrela-Rio Grande do Norte-RN. These mushrooms were delipidated with chloroform:methanol (2:1 v/v), extracted with water at 100 °C, and fractionated with ethanol (one and three volumes) and then centrifuged. The ethanol precipitation showed a high total sugar level of 64.8% and 1% of protein. This precipitate contained a high glucan level, characterized by chemical methods and by NMR of ^{13}C and ^{1}H and spectroscopy. The ^{13}C NMR spectrum of these mushroom extracts showed the presence of β-glucose by a signal at 103.25 ppm. Studies with these glucans were made to elucidate antioxidant and anti-inflammatory activities. This extract of glucans inhibited the lipid peroxidation (42.9%) and superoxide radicals (83.3%) at 67 µg/mL. However, the inhibition of hydroxyl radical by the extract of this mushroom was 96% at 267 µg/mL. The action of this extract on induced pleurisy showed a 92.5% and 68.7% reduction in polymorphonuclears cells and nitric oxide, respectively, at 30 mg/kg. The glucans reduced the croton oil-induced ear edema by 65.6% at 30 mg/kg.

Keywords: antioxidant activities; mushroom; anti-inflammatory effect; glucan-protein; *Polyporus dermoporus*; polysaccharides

---

1. Introduction

Fungal polysaccharides have recently received considerable attention, due to their potential use in a wide variety of industries, including cosmetics, pharmaceuticals and food. [1–3]. Several polysaccharides and polysaccharide-protein complexes have been isolated from fungi (mushrooms) and are being used as a source of therapeutic agents [2,3]. Many, if not all, basidiomycete mushrooms contain biologically active polysaccharides in fruit bodies, culture mycelium and culture broth [1,4].

Oxidation is essential to many living organisms for the production of energy to fuel biological processes. Free radicals are produced in normal and/or pathological cell metabolism [5]. However, the uncontrolled production of oxygen-derived free radicals is involved in the onset of many diseases, such as cancer, rheumatoid arthritis, cirrhosis and arteriosclerosis, as well as in degenerative processes associated with aging [5,6]. It has long been recognized that many naturally occurring substances in plants and fungi have antioxidant activities. Several species of mushroom contain a great variety of molecules, scavenger-free radicals or reactive oxygen species, such as polysaccharides and phenolic compounds [6–9]. This explains why mushrooms have recently become attractive as nutritionally beneficial foods and as a source material for the development of drugs. Several types of inflammatory tissue injury are mediated by reactive oxygen metabolites. The most likely sources of these oxidizing agents are the phagocytic leukocytes (e.g., neutrophils, monocytes, macrophages and eosinophils) that invade the tissue. It is becoming increasingly apparent that in addition to promoting cytotoxicity, reactive oxygen metabolites may also initiate and/or increase inflammation [6,10].

Recent studies have shown that a glucan-protein complex from the fungus *Geastrum saccatum* has strong anti-inflammatory and antioxidant activity [6,7]. Studies with polysaccharides from the mushroom *Polyporus albicans* showed that these compounds have powerful stimulating effects on murine lymphocyte proliferation induced by concanavalin A or lipopolysaccharides and that its branches are extremely important for enhancement expression [8].

In this work, chemical features of a β-glucan-protein complex from *P. dermoporus*, as well as its anti-inflammatory antioxidant and cytotoxic activities were analyzed.

2. Experimental Section

2.1. Mushroom

Fruiting bodies of fungi *Polyporus dermoporus* Pers. (Polyporaceae) were obtained from a private reserve, Mata da Estrela, of the Atlantic Forest in the state of Rio Grande do Norte (06°22′25″ S, 35°01′24″ W) Brazil in July, 2009. The mushroom was identified by Prof. Iuri Goulart Baseia,

Department of Botany, Ecology and Zoology of Universidade Federal do Rio Grande do Norte (UFRN), Natal, Brazil. A voucher specimen was deposited in the herbarium of this department.

2.2. Materials

Thiobarbituric acid, D_2O, sodium salicylate, nitroblue tetrazolium, phenazine methosulfate, EDTA, H_2O_2, indomethacin, croton oil, NADH, Ficoll isopaque, carrageenan, Folin Ciocalteu, gallic acid, ketamine, xylazine and mercaptoethanol were purchased from Sigma Aldrich (St. Louis, MO, USA). RPMI 1640 medium was purchased from Gibco (Grand Island, NY, USA). All other chemicals used were of analytical grade.

2.3. Animals

Male BALBc mice (about 25 g) and male Wistar rats (150–180 g) were housed in temperature-controlled rooms (22–23 °C) until use. Animals were given *ad libitum* access to food and water. Each experimental group contained 6 animals. The mice were used in the croton oil-induced ear edema test and the rat in pleurisy. All of the experiments were performed in accordance with the guidelines and regulations set forth by the Ethics Committee, Centro de Biociencias, UFRN.

2.4. Extract Preparation

Fruiting bodies of *Polyporus dermoporus* were dried at 40 °C and powdered. The resulting powder was delipidated with ether and placed in solution in a proportion of 10 g of powder to 100 mL of distilled water at 100 °C for 3 h. Composition analysis of this extract was made [11]. Then, 1 volume of ethanol (50 mL) was added to the supernatant (50 mL). Subsequently, the extract was centrifuged, and 3 volumes of ethanol (150 mL) were added. The precipitate, rich in polysaccharides, was evaporated in a vacuum to obtain a solid extract.

2.5. Chemical Analysis

The total protein of the extract was determined by Spector's method [12]. Total sugar content was determined by the phenol sulfuric acid colorimetric method for the determination of total sugars (the Dubois method) [13].

2.6. ^{13}C nuclear Magnetic Resonance Spectroscopy (NMR)

^{13}C and 1H NMR spectroscopy experiments were conducted using a 200-MHz Varian Mercury 200 Magneto Oxford spectrometer at 60 °C, and the sample was dissolved in deuterated water (D_2O).

2.7. In Vitro *Antioxidant Tests*

2.7.1. Superoxide Radicals

Superoxide radicals were generated in 3 mL Tris-HCl buffer (16 mM, pH 8.0), which contained 78 μM NADH (reduced form), 50 μM nitroblue tetrazolium, 10 μM phenazine methosulfate [14] and varying concentrations of polysaccharides (67–267 μg/mL). The color reaction of superoxide radical and nitroblue tetrazolium was detected by monitoring the absorbance at 560 nm. In the essential control, NADH was substituted for Tris-HCl buffer [14]. The percentage of scavenging activity on superoxide radicals was determined by this formula: (%) = $[(A_{positive\ control} - A_{sample})/A_{positive\ control} - A_{negative\ control})] \times 100$; where $A_{positive\ control}$ is positive control absorbance (all reagents without sample), A_{sample} is sample absorbance and $A_{negative}$ is negative control absorbance (NADH absence).

2.7.2. Hydroxyl Radical Assay

Hydroxyl radicals were generated by an innovation of the Smirnoff and Cumbes 1989 method [15] in sodium phosphate buffer (150 mM, pH 7.4), which contained 0.15 mM $FeSO_4$-EDTA, 2 mM sodium salicylate, 6 mM H_2O_2 and varying concentrations of polysaccharides (67–267 μg/mL). In the essential control, sodium phosphate buffer replaced H_2O_2. The solutions were incubated at 37 °C for 1 h and detected by monitoring the absorbance at 510 nm. The inhibition rate (%) = $1 - (Abs_1/Abs_2) \times 100\%$, where Abs_1 is the absorbance obtained from the reaction with the samples and Abs_2 is the absorbance of the reaction in which distilled water replaced hydrogen peroxide.

2.7.3. Lipid Peroxides Radical Scavenging Capacity

Liver microsomes were prepared from male Wistar rats according to the method proposed by [16]. The protein content of the microsomes was measured by Spector's method [12]. The product of microsomal lipid peroxidation was malondialdehyde. Microsomes (200–300 μg/mL) were incubated at 37 °C for 1 h with varying amounts of polysaccharides (67–267 μg/mL), 10 μM $FeSO_4$ and 0.1 mM ascorbic acid in 1.0 mL potassium phosphate buffer solution (0.2 M, pH 7.4). The reaction was stopped by 20% trichloroacetic acid (1.0 mL) and 0.67% 2-thiobarbituric acid (1.5 mL) in succession, and the solution was then heated at 100 °C for 15 min. After precipitation, the proteins were removed by centrifugation. The color reaction of the malondialdehyde-thiobarbituric acid complex was detected by monitoring the absorbance at 532 nm [16]. The essential control did not contain $FeSO_4$ or ascorbic acid. The percentage of antioxidant activity of the samples was evaluated according to the following formula: inhibition rate (%) = $(A_0 - A)/(A_0 - A_e) \times 100\%$; where A_0 is the absorbance of the free radical generation system, A is the absorbance of the test sample and A_e is the absorbance of the essential control.

2.8. Assessment Phenolic Compounds

Phenolic compounds extract from *P. dermoporus* were measured by the Folin Ciocalteu method [17] with few modifications using 0.5 mL ethanol, 2.5 mL distilled water, 0.25 mL Folin Ciocalteu reagent, 0.5 mL sodium carbonate and 0.5 mL of polysaccharides (1 mg/mL). Absorbance was determined at 765 nm against the blank, and a gallic acid calibration curve (0–500 mg/mL) was constructed and used to determine the total phenolic content of the samples, which were expressed as gallic acid equivalents (GAE).

2.9. Croton Oil-Induced Ear Edema Test

The croton oil-induced ear edema test was performed as previously described [18,19]. The animals were anesthetized with 80 mg/kg ketamine and 12 mg/kg xylazine (3:1, i.p.). A total of 20 μL (0.4 μg of croton oil in acetone) was applied to the inner surface of the right ear of each mouse (subcutaneous). After 24 h, the extract, rich in polysaccharide, was intravenously administered at doses of 10, 30 and 50 mg/kg by body weight. The left ear remained untreated. Control animals only received the irritant. The animals were killed by cervical dislocation 24 h after the treatment was removed from both the treated and untreated ear. The difference in the thickness between the two plugs before and after the treatment was measured to determine the edematous response.

2.10. Extract Action in Carrageenan-Induced Pleurisy

The animals were anesthetized intraperitoneally with 20 mg/mL solution of xylazine hydrochloride and a 50 mg/mL solution of ketamine hydrochloride at a 1:2 proportion of 100 mg/kg/weight. The animals had been previously treated with 10, 30 and 50 mg/kg i.p. by body weight of the extract, with the exception of the control group animals (carrageenan (control+) and saline (control−)). After 30 min, the animals were submitted to a skin incision at the sixth intercostal space level. The muscles were dissected, and 0.2 mL of saline containing 1% (w/v) carrageenan were injected into the pleural cavity. The skin incision was closed with a suture, and the animals were allowed to recover. The animals were killed 4 h after carrageenan injection. The pleural cavity was carefully opened and washed with 2 mL of saline solution containing heparin (5 U/mL) and diclofenac (10 μg/mL). The exudate was removed by aspiration. The number of leukocytes for the differential leucocyte count (DLC) in the exudates was counted with an optical microscope after Türk's staining.

Measurement of Nitrite-Nitrate Concentration in Pleural Exudates

Total nitrite in exudates, an indicator of NO synthesis, was measured as previously described by Cuzzocrea *et al.* 1998 [19]. Briefly, the nitrate in the samples was first reduced to nitrite by incubation with nitrate reductase (670 mU/mL) and NADPH (160 μM) at room temperature for 3 h. The total nitrite concentration in the samples was then measured using the Griess reaction, by adding 100 μL of Griess reagent ((0.1% (w/v) naphthylethylenediamide dihydrochloride in H_2O and 1% (w/v) sulfanilamide in 5% (v/v) concentrated H_3PO_4(vol 1:1)) to the 100 μL sample [20]. The total nitrite concentration in the samples was then measured using the Griess reaction, by adding 100 μL of Griess reagent. The concentrations were calculated by comparison with the OD_{550} of standard solutions of sodium nitrite prepared in H_2O.

2.11. Histological Examination

For histological examination, ear biopsies were taken 48 h after edema induction by croton oil injection. The tissue slices were fixed in 10% neutral-buffered formaldehyde, embedded in paraffin and sectioned. The sections were stained with hematoxylin and eosin.

2.12. Colorimetric MTT (tetrazolium) Assay

The cytotoxicity of the extract was measured as previously described by Mosmann, 1983 [21]. The leukocytes were obtained from human peripheral blood, previously added with heparin, and centrifuged with Ficoll isopaque (Histopaque-1077). The cells were washed successively with RPMI 1640 medium and supplemented with 50 μM 2-mercaptoethanol and 5%–10% fetal bovine serum, in a 6% CO_2 atmosphere. A total of 100 μL of the solution with 1×10^6 cells/well was added. These cells were incubated for 4 h, and different concentrations (0.5, 1.0 and 1.5 μg/mL) of the extract from *P. dermoporus* and stock MTT solution (10 μL/100 μL medium RPMI) were added to the wells. The plates were incubated at 37 °C for 4 h. Acid-isopropanol (100 μL of 0.04 N HCl) was added to all of the wells and mixed thoroughly to dissolve the dark blue crystals. The plates were read by a MicroElisa reader at 570 nm.

2.13. Statistical Analysis

Values are expressed as the mean ± SEM. Analysis of variance (ANOVA, Tukey–Kramer test) was used to evaluate the data, and $p < 0.05$ was accepted as statistically significant. The statistical analysis involved the use of the software, GraphPad Prism 5 (GraphPad Software Inc., San Diego, CA, USA).

3. Results and Discussion

3.1. Chemical Analysis

In this work, we investigated the chemical composition of the tissue of this native mushroom. The polysaccharides resulting from the extraction of *Polyporus dermoporus* and their cytotoxic, antioxidant, and anti-inflammatory action were studied. The results of the analyses of the composition of the tissue showed levels of carbohydrates, proteins, lipids and ash of 51.8%, 21.4%, 1.66% and 10.8%, respectively; assessment by Association of Official Analytical Chemists (A.O.A.C.), 1984 [11]. When submitted to the fractionation with ethanol, the extracts of this mushroom showed 64.8% carbohydrates and 10.0% proteins. Thus, the polysaccharide extract was composed mainly of glucose polymer (64.6%) (see Table 1), which was confirmed by total sugars by the phenol-H_2SO_4 reaction using D-glucose as the standard [13]. Phenolic compounds have been shown to possess potent antioxidant properties. Several studies have established links between the consumption of foods containing high concentrations of phenolic and antioxidant compounds and a lower incidence of cardiovascular diseases and cancer [5]. However, in extracts of *Polyporus dermoporus*, the antioxidant properties are not attributed to these bioactive compounds, due to the very low content of phenolic compounds and proteins (Table 1).

3.2. ^1H and ^{13}C NMR Spectroscopies of the Extract from Polyporus Dermoporus

The ^1H spectra showed a chemical shift in the anomeric region at 4–6 ppm (Figure 1). In this spectrum, signals 4.1 and 4.6 corresponded to the signals obtained for β glucan [22]. According to Chaveau *et al.*, 1996 [23], the chemical shifts correspond to C2 and C6 [7]. We also observed that the regions between 1.4 and 2.5 are related to the glucan-protein structure. These observations are in accord with other studies [6,7].

Table 1. Chemical composition of natural tissue of mushroom *P. dermoporus* (g/100 g dry weight) and of the polysaccharides and proteins from the extract of this mushroom.

Components	%
Tissue	
Carbohydrates	51.3 ± 1.32
Proteins	21.4 ± 2.11
Moisture	10.01 ± 1.03
Lipids	1.66 ± 0.51
Ashes	10.80 ± 1.81
Extract	
Polysaccharides	64.8 ± 5.23
Proteins	10.2 ± 0.30
Phenolic compounds	0.90 ± 0.02

Figure 1. ^{1}H NMR spectroscopy of *Polyporus dermoporus* glucan protein.

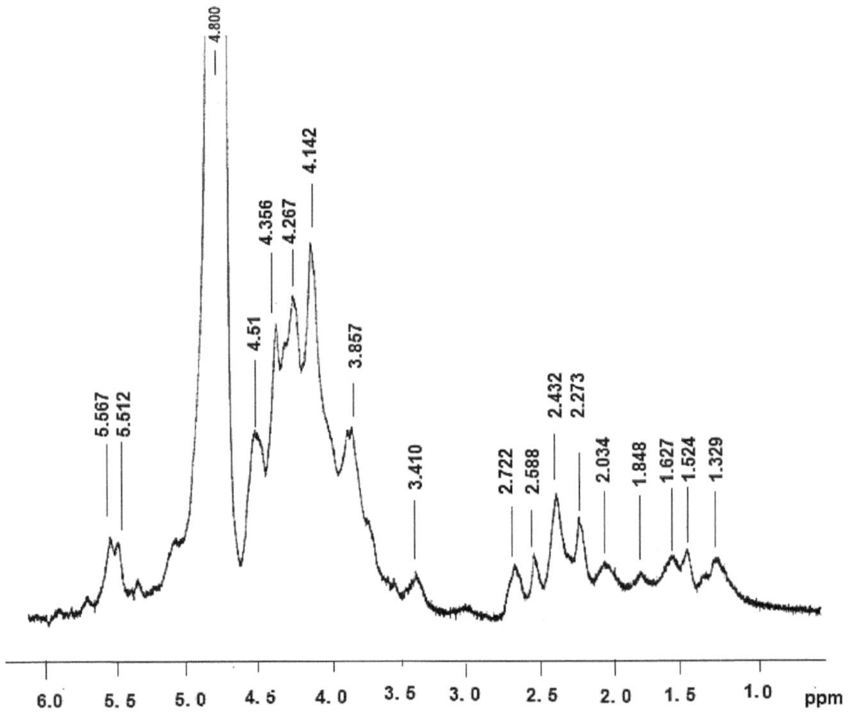

The analysis of ^{13}C NMR spectroscopy showed that the polysaccharides o this mushroom contain a high level of glucose. This fact is due to the type of homopolysaccharides (glucans) present in this organism. Figure 2 shows the NMR spectra of ^{13}C of the extracts from *Polyporus dermoporus.* The presence of glucose can be observed by a signal at 103.2 ppm, characteristic of the β configuration [22–24]. Other important signals in the spectra are those in the 60–80 ppm range, where they are related to C2 (73.9 ppm), C3 (84.0 ppm), C4 (80.0 ppm), C5 (76.24 ppm) and C6 (61.7 ppm) of that carbohydrate [25–28]. A peculiar characteristic of the spectra of β-(1→3) glucans is the presence of at least six signals of similar magnitude (110–60 ppm). The 47.8 ppm signal is related to the $-CH_2N$ group of amino acids.

Figure 2. ^{13}C NMR spectroscopy of *Polyporus dermoporus* glucan-protein.

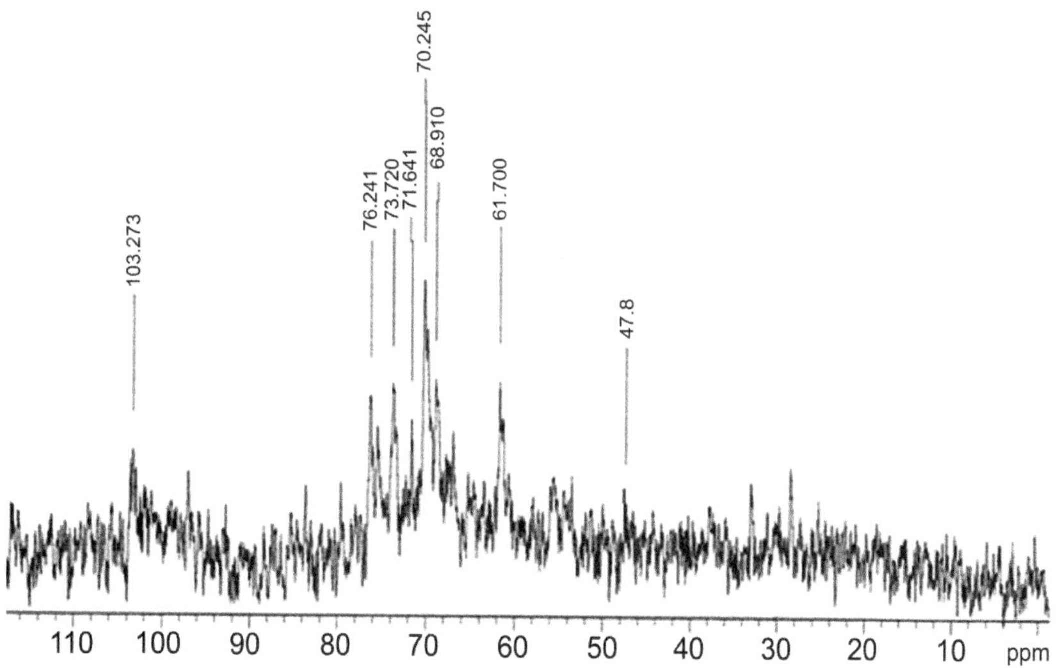

3.3. Superoxide and Hydroxyl Radicals

Glucans are homopolymers of D-glucose, widely distributed in nature, which may have bioactive properties in various organisms [27,28]. As such, they are highlighted for having a direct influence on the stimulation of phagocytic activity, antitumor properties, antioxidant, immunomodulatory and anti-inflammatory activities [10,26]. The antioxidant effect of the carbohydrates did not correlate with the type of intra-chain linkage, molecular weight or degree of polymer branching. The activity did appear to correlate with the monosaccharide composition of the polymer.

The antioxidant effect did appear to have a relation with the monosaccharide composition of the polymer and the composition of the polymer. Korzaski *et al.*, 2011, in studies with mushroom polysaccharides of the species *A. bisporus*, *A. brasiliensis*, *G. lucidum* and *P. linteus*, showed that scavenging ability is a correlation that was found between the EC_{50} values of the chelating and reducing power abilities and the amount of total glucan content in the extracts [25]. The superoxide radicals were generated in a system of NADH-phenazine-methosulfate with different concentrations of the tested polysaccharides; see Zhou and Zheng *et al.*, 1991 [29].

The formation of hydroxyl radicals (•OH) in the biological systems does not occur enzymatically and which need the presence of iron or transition metals [25]. The removing of •OH is important for the protection of living systems. The results obtained (•OH) were expressed as the % inhibition rate; see Table 2. At 267 µg/mL, 96% of the scavenging activity on hydroxyl radicals was exhibited. This shows a high inhibition in the generation of hydroxyl radicals when compared with polysaccharide from *C. montagnei* (38% at 1000 mg) [11], indicating considerable antioxidant activity in this study.

3.4. Determination of Lipid Peroxidation or Malonaldehyde Formation (LPO)

Combined with the generating system of lipid peroxidation using thiobarbituric acid (TBA), a condensation reaction produces malondialdehyde (MDA)-TBA with a pink coloration that can be subsequently quantified [26,30]. The extract of the mushroom used in this work showed efficient inhibition of lipid peroxidation, varying from 24.3% to 42.91% (Table 2). The process of lipid peroxidation results in the formation and propagation of lipid peroxides that act on the cell membrane and microsomes, producing compounds, such as malondialdehyde (MDA) [30,31]. The maximal effect on inhibition microsomal peroxidation occurred at a dose of 67 µg/µL (42.9%). We observed that the inhibition of superoxide radical and peroxidation of this polysaccharide-protein complex is not dose dependent. Polysaccharide of *C. montagnei* at 4 mg/mL produced lipidic peroxidation inhibition of 37.6% ± 0.16%, equivalent to 47% of the activity from α-tocopherol at the same concentration [10].

Table 2. The effect of mushroom polysaccharide extracts from *P. dermoporus* on superoxide and hydroxyl radical generation and microsomal lipid peroxidation.

Concentration (µg/mL)	Inhibition Superoxide Radical (%)	Inhibition Hydroxyl Radical (%)	Inhibition Microsomal Peroxidation (%)
67	83.3	20.0	42.9
113	75.1	37.0	40.3
200	51.5	75.0	31.0
267	48.5	96.0	24.3

Each value represents the mean ± S.E.M. ($n = 6$).

Polysaccharides may be acting similarly to other mechanisms described for chelating, by reducing the redox pro-oxidant potential of metals and stabilizing the oxidized form of the metal, through steric interaction and also by forming soluble and stable complexes with metals, which are eventually excreted in the urine, as related by Ebrahimzadeh *et al.* 2008 [9]. This is strengthened by the description of the effect of other polysaccharides in the chelation of metal ions [10,29]. This may explain the strong inhibition of the peroxyl radical when observed in incubation in the presence of glucans, in addition to its significant effect in the reduction of lipid peroxidation, showing its strong antioxidant effect at low concentrations. This can result in protection against the development of pathologies, such as inflammation, cancer, oxidative stress, and diabetes.

3.5. Action of the Polysaccharides in Carrageenan-Induced Pleurisy

The results obtained showed a high number of leukocytes cells (2854 cell/mm^3) in the pleural infiltrate of the positive control (animals treated with carrageenan). When the animals were submitted to the action of carrageenan and treated previously with the glucans of *P. dermoporus* (10 and 30 mg/kg by body weight), the cells obtained in the pleural infiltrate were 681 and 264 cell/mm^3, respectively (Figure 3). This value corresponds to a reduction of 76% and 92.5% in the number of cells, respectively. At a dose of 50 mg/kg, a 70.6% reduction in infiltrate cells occurs ($p < 0.001$). The nonsteroidal anti-inflammatory drugs (NSAIDs) used in this experimental was indomethacin; $p < 0.001$. ANOVA and the Tukey-Kramer test showed that the difference observed in relation to the control group is highly significant; $p < 0.001$.

Figure 3. Anti-inflammatory effect of *P. dermoporus* extract (PD) on carrageenan-induced pleurisy. The number of pleural exudate leukocytes in carrageenan-induced Wistar rats. The experimental animals were treated with *P. dermoporus* extract at 10 mg/kg (PD 10), 30 mg/kg (PD 30) and 50 mg/kg (PD 50). Data obtained from animal experiments are expressed as the mean ± SD. The differences between treatment and control were tested by ANOVA. A value of (***) $p < 0.001$ was considered statistically significant.

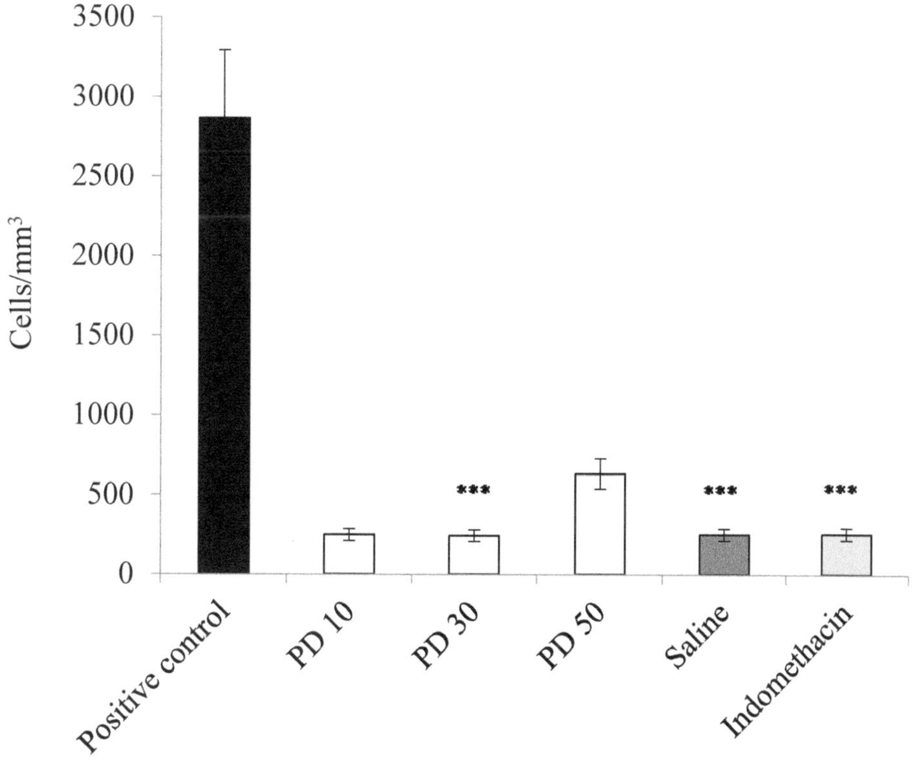

3.6. Effect of Polysaccharide on Nitric Oxide (NO)

Polysaccharides derived from mushrooms have been considered an important class of bioactive compounds, and their anti-inflammatory potential has been widely studied [25,26,32]. Nitric oxide in most body fluids is rapidly metabolized to stable products, such as nitrite and nitrate. Figure 4 shows the value of nitrate/nitrite produced after carrageenan administration and previous treatments with the extracts of *P. dermoporus* at doses of 10 mg/kg (PD 10), 30 mg/kg (PD 30) and 50 mg/kg by body weight (PD 50). The positive control (animals with carrageenan) showed 21.39 ± 1.79 nmol of nitrate/nitrite. In the animals that received *P. dermoporus* (30 mg/kg by body weight), this value was 7.48 ± 0.44 nmol ($p < 0.001$). We observed that at 10 mg/kg and 50 mg/kg, the results were 12.51 ± 0.81 and 7.32 ± 1.36 nmol of nitrate/nitrite ($p < 0.001$). In this study, we found the effect of this extract in the modulation of nitric oxide expression in the inflammation model tested.

Figure 4. Effect of *P. dermoporus* polysaccharides on NO production from the pleural exudate of Wistar rats with carrageenan-induced pleurisy. The animals were treated with 10 mg/kg (PD 10), 30 mg/kg (PD 30) and 50 mg/kg (PD 50) of *P. dermoporus* extract. Control: Wistar rats with carrageenan-induced pleurisy. Data obtained from animal experiments ($n = 7$) are expressed as the mean ± SD. The differences between the treatment and control were tested using ANOVA. A value of (***) $p < 0.001$ was considered statistically significant.

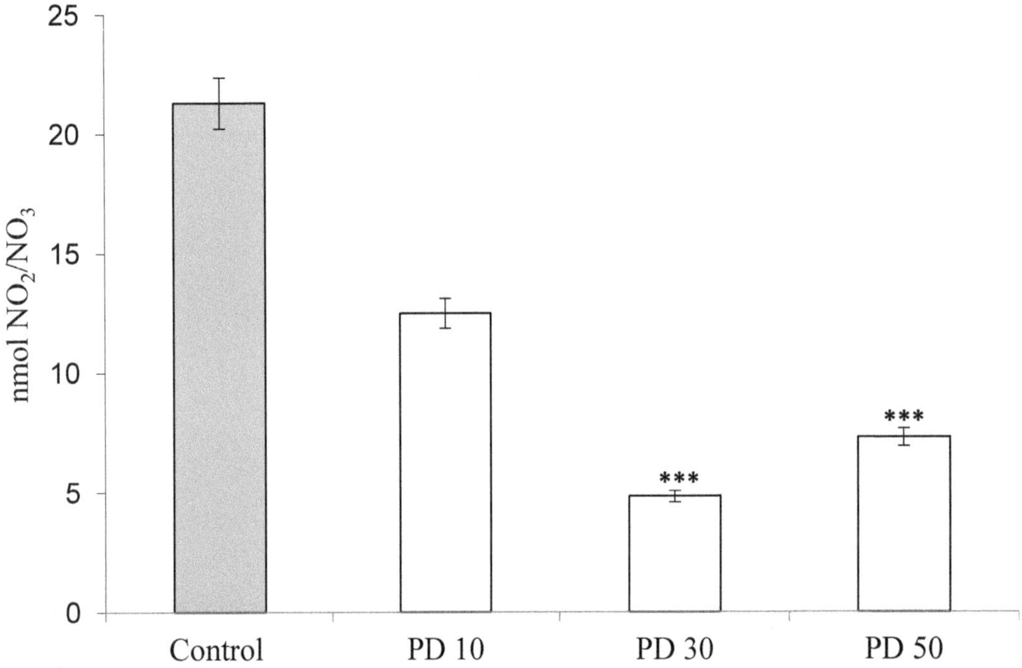

3.7. Croton Oil-Induced Ear Edema Test

The anti-edematogenic activity of the extract from *P. dermoporus* against croton oil-induced ear edema in rats was assessed. For comparison, the effect of nonsteroidal (indomethacin) anti-inflammatory drug was analyzed. The animals received a subcutaneous injection of croton oil (0.1 mL of 1%) solution in acetone. The ear volume was measured in the groups: positive control rats that received croton oil, saline (negative control) and extracts of *P. dermoporus* (10, 30, and 50 mg/kg). The extract of *P. dermoporus* reduced the edema by 65.6% at 30 mg/kg. At 10 mg/kg, the reduction was 58.3%, and at a concentration of 50 mg/kg, the reduction was 55.7% (Figure 5).

3.8. Histological Analysis

To evaluate histological changes, we carried out hematoxylin and eosin (H & E) staining, 200×. Histological examination of ear edema sections (Figure 6A) revealed significant tissue damage to animals treated with croton oil. The slide of animals treated (positive control) presented intense cellular infiltrate, predominantly neutrophilic inflammatory infiltrate (Figure 6A), characteristic of inflammatory reaction. Exposure to saline (Figure 6B) did not lead to significant pathologic alterations. The ear sections of the animals treated with the glucans (Figure 6C–E) were similar to those that received only saline (Figure 6B), showing decreased congestion and infiltration of leukocytes, especially neutrophils into paw tissues.

Figure 5. The effect of *P. dermoporus* polysaccharides on the croton oil-induced ear edema assay in BALBc mice. The animals were treated with 10 mg/kg (PD 10), 30 mg/kg (PD 30) and 50 mg/kg (PD 50) of *P. dermoporus* polysaccharides. Data obtained from animal experiments are expressed as the mean ± SD. The differences between treatment and control were tested by ANOVA. A value of (***) $p < 0.001$ was considered statistically significant.

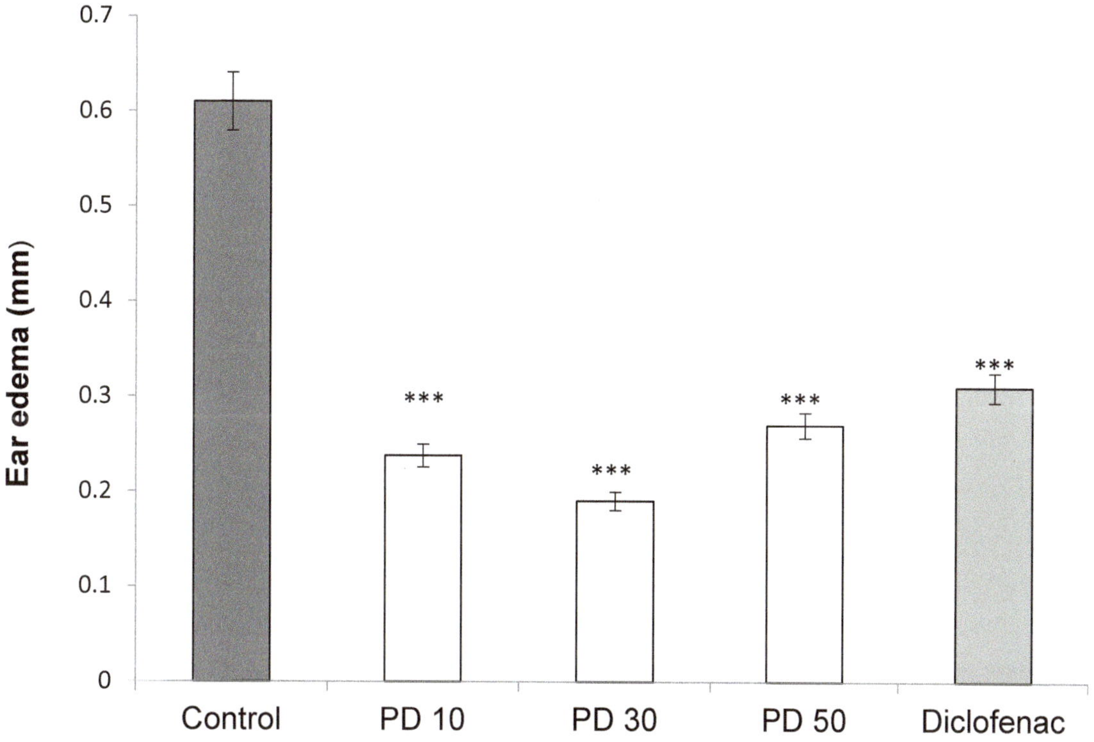

Figure 6. Histological analysis of ear edema with H & E stain 200× from animals submitted to the croton oil-induced ear edema test and treated with *P. dermoporus* polysaccharides: (**A**) positive control (croton oil); (**B**) negative control (saline); (**C**) *P. dermoporus* polysaccharides at 10 mg/kg (PD 10); (**D**) animals treated with 30 mg/kg (PD 30); (**E**) animals treated with 50 mg/kg (PD 50).

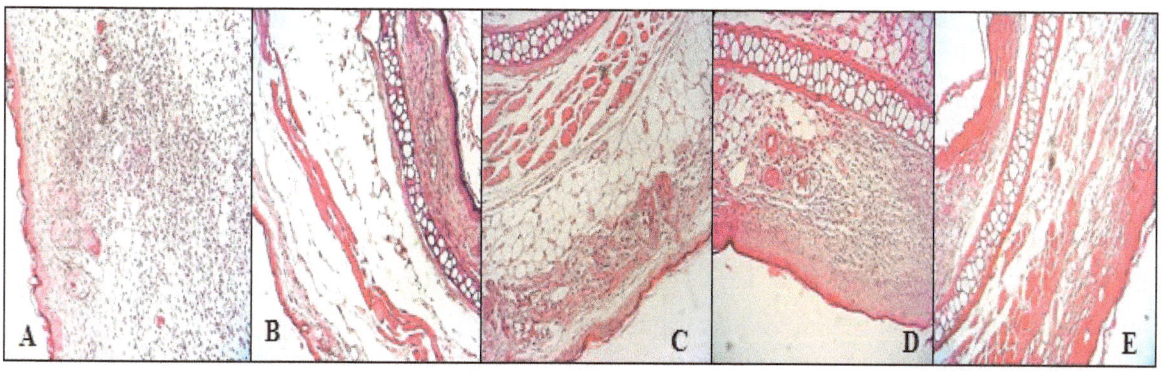

3.9. Cytotoxic Test

The cytotoxicity tests are based on the capacity of the cells to convert to tetrazolium (MTT) in a mixture of blue coloration called formazan [21]. However, only live cells have this capacity. The evaluation of the transformation of MTT in formazan in mononuclear cells of the peripheral blood is accomplished by ELISA, reading at 540 and 620 nm. The cells incubated with the extracts of *P. dermoporus* showed no quantitative difference in relation to the control ($p < 0.001$). Thus, this polysaccharide had no cytotoxic effect.

4. Conclusions

The results obtained in this paper showed that glucans of *Polyporus dermoporus* have inhibitory action in the formation of superoxide and hydroxyl radicals. The effect against this radical is dose dependent and inversely proportional to the increase in concentration. In this study, the intrapleural administration of carrageenan produced an acute inflammatory response. The volume and number of cells in pleural fluid increased in the initial 12 h, followed by a decline. A significant decrease in cell numbers was observed in the pleural fluid compared with the control. The value of nitrate/nitrite produced in the pleural cavity of rats by carrageenan administration and previous treatments of *P. dermoporus* extracts at a concentration of 30 mg/kg (PD 30) showed a 69.5% NO decrease ($p < 0.001$) in relation to the control group (carrageenan). This decrease suggests an anti-inflammatory effect of the *P. dermoporus* extract. Overall, this polysaccharide possessed good antioxidant properties and showed anti-inflammatory activity that is not often studied in mushrooms. We suggest that glucans appear to be related to the inhibition of diapedesis after cell migration to the injury site. Analyses conducted in this study demonstrated that the polysaccharide from *P. dermoporus* has no cytotoxic effect.

Acknowledgments

The authors are grateful for the financial support provided by CAPES and Conselho Nacional de Tecnologia e Pesquisa (CNPq), Process No. 475867/03-3. The authors are grateful to Michael Germain for the English revision.

Author Contributions

Celina Maria P. Guerra Dore participated in the experimental design and completion, as well as interpretation, manuscript design and preparation. Maria da Glória Lima Santos and Monique G. Faustino Alves, Leonardo Augusto R. de Souza participated in the experimental design and data interpretation. Iuri G Baseia collected and classified of mushroom. Finally, Edda Lisboa Leite reviewed all the manuscript and the final version to be submitted participated in manuscript design and preparation. All authors read and approved the final manuscript.

Conflicts of Interest

The authors declare no conflict of interest.

References

1. Vaz, J.A.; Barros, L.; Martins, A.; Santos-Buelga, C.; Vasconcelos, M.H.; Ferreira, I.C.F.R. Chemical composition of wild edible mushrooms and antioxidant properties of their water soluble polysaccharidic and ethanolic fractions. *Food Chem.* **2011**, *126*, 610–616.
2. Chang, S.T. Global impact of edible and medicinal mushrooms on human welfare in the 21st century: Non green evolution. *Int. J. Med. Mushrooms* **1999**, *1*, 1–7.
3. Wasser, S.P.; Weis, A.L. Medicinal mushrooms as a source of antitumor and immunomodulating polysaccharides. *Int. J. Med. Mushrooms* **1999**, *1*, 31–62.
4. Barros, L.; Cruz, T.P.; Baptista, E.L.M.; Ferreira, I.C.F.R. Wild and commercial mushrooms as source of nutrients and nutraceutical. *Food Chem. Tox.* **2008**, *46*, 2742–2747.
5. Shirley, R.; Ord, E.N.J.; Work, L.M. Oxidative stress and the use of antioxidants in stroke. *Antioxidants* **2014**, *3*, 472–501.
6. Dore, C.M.P.G.; Azevedo, T.C.G.; Souza, M.C.R.; Rego, L.A.; Dantas, J.C.M.; Silva, F.R.F.; Rocha, H.A.O.; Baseia, I.G.; Leite, E.L. Antiinflamatory, antioxidant and cytotoxic actions of beta-glucan-rich extract from *Geastrum saccatum* mushroom. *Int. Immunopharm.* **2007**, *7*, 1160–1169.
7. Gonzaga, M.L.C.; Ricardo, N.M.P.S.; Heatley, F.; Soares, S.A. Isolation and characterization of polysaccharides from *Agaricus blazei* Murril. *Carbohydr. Polym.* **2005**, *60*, 43–49.
8. Sun, Y.; Liang, H.; Cai, G.; Guan, S.; Tong, H.; Yang, X.; Liu, J. Sulfated modification of the water-soluble polysaccharides from *Polyporus albicans* mycelia and its potential biological activities. *Int. J. Biol. Macromol.* **2009**, *44*, 14–17.
9. Ebrahimzadeh, M.A.; Pourmorad, F.; Bekhradnia, A.R. Iron chelating activity screening, phenol and flavonoid content of some medicinal plants from Iran. *Afr. J. Biotech.* **2008**, *7*, 3188–3192.
10. Castro, A.J.G.; Castro, L.S.E.P.; Santos, M.S.N.; Faustino, M.G.; Pinheiro, T.S.; Dore, C.M.P.; Baseia, I.G.; Leite, E.L. Anti-inflamatory, anti-angiogenenic and antioxidant activities of polysaccharide-rich extract from fungi *Caripia montagnei. Biomed. Prev. Nutr.* **2014**, *4*, 121–129.
11. Association of Official Analytical Chemists (A.O.A.C.). *Official Methods Analytical Chemists*, 14th ed.; Association of Official Analytical Chemists: Washington, DC, USA, 1984; pp. 988–1141.
12. Spector, J. Refinement of the Coomassie blue method of protein quantification: A simple and linear spectrophotometric assay of 0.5 to 50 mg of protein. *Anal. Biochem.* **1978**, *86*, 142–146.
13. Dubois, M.; Gilles, K.A.; Hamilton, J.K.; Rebers, P.A.; Smith, F. Colorimetric method for determination of sugars, and related substances. *Anal. Chem.* **1956**, *28*, 350–356.
14. Ponti, V.; Dianzani, M.U.; Cheeseman, K.; Slater, T.F. Studies on the reduction of nitroblue tetrazolium chloride mediated through the action of NADH and phenazine methosulphate. *Chem. Biol. Interact.* **1978**, *23*, 281–291.
15. Smirnoff, N.; Cumbes, Q.J. Hydroxyl radical scavenging activity of compatible solutes. *Phytochemistry* **1989**, *28*, 1057–1060.
16. Bueg, J.A.; Aust, S.D. Microsomal lipid peroxidation. In *Methods in Enzymology*; Feischer, S., Packer, L., Eds.; Academic Press: New York, NY, USA, 1978; Volume 52, pp. 302–310.
17. Velioglu, Y.S.; Mazza, G.; Gao, L.; Oomah, B.D. Antioxidant Activity and Total Phenolics in Selected Fruits, Vegetables, and Grain Products. *J. Agric. Food Chem.* **1998**, *46*, 4113–4117.

18. Tubaro, A.; Dri, P.; Delbello, G.; Zilli, C.; Della Loggia, R. The Croton oil ear test revisited. *Agents Actions* **1985**, *17*, 347–349.
19. Cuzzocrea, S.; Costantino, G.; Zingarelli, B.; Caputi, A.P. Beneficial effects of Mn (III) tetrakis (4-benzoic acid) porphyrin (MnTBAP), a superoxide dismutase mimetic, in carrageenan-induced pleurisy. *Free Radic. Biol. Med.* **1998**, *26*, 25–33.
20. Pacher, P.; Beckman, J.S.; Liaudet, L. Nitric oxide and peroxynitrite in health and disease. *Physiol. Rev.* **2007**, *87*, 315–424.
21. Mosmann, T. Rapid colorimetric assay for cellular growth and survival: Application to proliferation and cytotoxicity assays. *J. Immun. Methods* **1983**, *65*, 55–63.
22. Chakraborty, I.; Mondal, S.; Rout, D.; Islam, S.S. A water-insoluble(1→3)-β-D-glucan from the alkaline extract of an edible mushroom *Termitomyces eurhizus*. *Carbohydr. Res.* **2006**, *341*, 2990–2993.
23. Chauveau, C.; Talaga, P.; Wieruszeski, J.M.; Strecker, G.; Chavant, L. A water-soluble beta-D-glucan from *Boletus erythropus*. *Phytochemistry* **1996**, *43*, 413–415.
24. Yalin, W.; Yavanjiang, P.; Cuirong, S. Isolation, purification and structural investigation of a water-soluble polysaccharide from *Solanum lyratum* Thunb. *Int. J. Biol. Macromol.* **2005**, *36*, 241–245.
25. Kozarski, M.; Klaus, A.; Niksic, M.; Jakovljevic, D.; Helsper, .J.P.F.G.; van Griensven, L.J.L.D. Antioxidative and immunomodulating activities of polysaccharide extracts of the medicinal mushrooms *Agaricus bisporus*, *Agaricus brasiliensis*, *Ganoderma lucidum* and *Phellinus linteus*. *Food Chem.* **2011**, *129*, 1667–1675.
26. Santos, M.S.N.; Magalhães, J.E.M.; Castro, L.S.E.P.W.; Pinheiro, T.S.; Sabry, D.A.; Nobre, L.T.D.B.; Lima, J.P.M.S.; Baseia, I.G.; Leite, E.L. Effect of Glucans from *Caripia montagnei* Mushroom on TNBS-Induced Colitis. *Int. J. Mol. Sci.* **2014**, *15*, 2368–2385.
27. Carbonero, E.R.; Gracher, A.H.P.; Smiderle, F.R.; Rosado, F.R.; Sassaki, G.L.; Gorin, P.A.J.; Iacomini, M. A β-glucan from the fruit bodies of an edible mushrooms *Pleurotus eryngii* and *Pleurotus ostreatoroseus*. *Carbohydr. Polym.* **2006**, *66*, 252–257.
28. Synytsya, A.; Novák, M. Structural diversity of fungal glucans. *Carbohydr. Polym.* **2013**, *92*, 792–809.
29. Zhou, Y.C.; Zheng, R.L. Phenolic compounds and analog as superoxide anion scavengers and antioxidants. *Biochem. Pharm.* **1991**, *42*, 1177–1179.
30. Wei, Y.; Li, Z.; Hu, Y.; Xu, Z. Inhibition of mouse liver lipid peroxidation by high molecular weight phlorotannins from *Sargassum kjellmanianum*. *J. App. Phycol.* **2003**, *15*, 507–511.
31. Fulia, F.; Gitto, E.; Cuzzocrea, S.; Reiter, R.J.; Dugo, L.; Gitto, P.; Barberi, S.; Cordaro, S.; Barberi, I. Increased levels of malondialdehyde and nitrite/nitrate in the blood of asphyxiated newborns: Reduction by melatonin. *J. Pin. Res.* **2008**, *1*, 343–349.
32. Zembron-Lacny, A.; Gajewski, M.; Naczk, M.; Siatkowski, I. Effect of shiitake (*Lentinus edodes*) extract on antioxidant and inflammatory response to prolonged eccentric exercise. *J. Physiol. Pharmacol.* **2013**, *64*, 249–254.

7

Evaluation of Antioxidant Activity, Total Flavonoids, Tannins and Phenolic Compounds in *Psychotria* Leaf Extracts

Anelise Samara Nazari Formagio [1,*], Carla Roberta Ferreira Volobuff [2], Matheus Santiago [2], Claudia Andrea Lima Cardoso [3], Maria do Carmo Vieira [1] and Zefa Valdevina Pereira [2]

[1] Faculty of Agricultural Sciences, University Federal of Grande Dourados (UFGD), 79804-970/533 Dourados-MS, Brazil; E-Mail: mariavieira@ufgd.edu.br

[2] Biological and Environmental Sciences, University Federal of Grande Dourados (UFGD), 79804-970/533 Dourados-MS, Brazil; E-Mails: volobuff81@ gmail.com (C.R.F.V.); matheus_ss@hotmail.com (M.S.); zefapereira@ufgd.edu.br (Z.V.P.)

[3] Chemistry Program, State University of Mato Grosso do Sul (UEMS), Mato Grosso do Sul, 79804-970/351 Dourados-MS, Brazil; E-Mail: claudia@uems.br

* Author to whom correspondence should be addressed; E-Mail: aneliseformagio@ufgd.edu.br

External Editor: Luca Sebastiani

Abstract: The antioxidant activity of *Psychotria carthagenensis*, *P. leiocarpa*, *P. capillacea* and *P. deflexa* (Rubiaceae) extracts were investigated, and the concentrations of total phenolics, flavonoids, condensed tannins and flavonols were determined. The chemical compositions of the extracts were investigated using the high performance liquid chromatography (HPLC/PAD) method. We used 1,1-diphenyl-1-picrylhydrazyl free radical (DPPH), β-Carotene bleaching and 2,2-azinobis (3-ethylbenzothiazoline-6-sulfonic acid) (ABTS) radical cations to determine antioxidant activity. The ability to scavenge radical was measured in these experiments by the discoloration of the solution. Concentrations of constituents were measured spectrophotometrically. *P. carthagenensis* and *P. capillacea* exhibited the highest antioxidant activity, in the DPPH test, β-carotene bleaching and ABTS system. The highest phenolic, flavonoid, condensed tannin and flavonol concentration was found in *P. carthagenensis* and *P. capillacea* extracts. HPLC-PDA analysis of *P. carthagenensis* and *P. capillacea* revealed hydroxycinnamic acid (*p*-coumaric acid). This is the first report on the antioxidant properties and constituent analysis of these *Psychotria* extracts.

Keywords: *Psychotria*; antioxidant activity; *p*-coumaric acid

---

1. Introduction

Antioxidants, molecules with a radical-scavenging capacity, are thought to exert a protective effect against free radical damage. These biomolecules may contribute to the prevention of many chronic diseases, such as cancer, cardiovascular disease, atherosclerosis, diabetes, asthma, hepatitis and arthritis [1,2]. The consumption of traditional diets prepared with spices and medicinal and aromatic herbs has attracted increasing interest among consumers and scientists because these spices and herbs exhibit antioxidant properties attributed to a variety of bioactive phytochemicals [3,4].

The species of the genus *Psychotria* are used by the population in the form of infusion and with external application. The internal uses are indicated for diseases of the treat disorders gastrointestinal, bronchial diseases and reproductive disorders. In external use, in applications, skin tumors, ulcers, ocular disorders, such as poultices, and baths for the treatment of fever, sore head and ear [5,6]. Reported phytochemical studies on the *Psychotria* genus showed alkaloids, mainly polypirrolidinoindole [7–13], quinolines [14–19], and monoterpene indole alkaloids [20–26]. Some of these alkaloids display pharmacological effects such as inhibition of human platelet aggregation [27], cytotoxicity [28], and analgesic activity [29], as well as antimalarial and antileishmanial effects [30].

Psychotria leiocarpa Cham. and Schlecht., popularly known as "*grandiúva-de-anta*" or "*cafeeiro-do-mato*", is an understorey woody shrub native to the forests of Southern Brazil [8]. From leaves, was isolated an *N*-glycosylated monoterpenoid indole alkaloid *N*, β-D-glucopyranosyl vincosamide, constitutes up to 2.5% of the dry weight in leaves [31], and iridoid glucosides asperuloside and deacetylasperuloside [32]. Study showed that this alkaloid may act indirectly in *P. leiocarpa* protection against oxidative stress generated upon wounding, UV exposure, and perhaps other environmental stresses [33]. *Psychotria carthagenensis* Jacq., popularly known as "*cafeeiro-do-mato*", "*carne-de-vaca*" or "*erva-de-rato-branca*", occurs in most parts of the southern Brazilian State, Rio Grande do Sul. This plant is one of the components of the hallucinogenic beverage *ayahuasca*, used by the people from the Amazonian Forest [34]. Dimetyl-tryptamine was identified as the major component of a leaf extract [35]. *Psychotria deflexa* DC., popularly known as "*erva-de-rato*" or "*café selvagem*", occurs from Mexico to Argentina. Chemical investigation has previously reported the alkaloid indole chromophore. *Psychotria capillacea* Müll. Arg. Standl., popularly known as "coffee," occurs in the Brazil states Amazonas, Mato Grosso do Sul, and Parana, as well as in Paraguay and Argentina [24]. The pharmacological properties and chemical for this species was not reported on in the consulted literature.

However, the few biological studies for the species indicate the importance of the continuity of phytochemical studies and activities, with these species reported. Thus, this study represents the first antioxidant investigation of *Psychotria carthagenensis*, *P. leiocarpa*, *P. deflexa* and *P. capillacea* with a different analytical methodology. We also determined the concentrations of phenolic compounds, condensed tannins, flavonoids, and flavonol.

2. Experimental Section

2.1. General Information

2,4-Dinitrophenylhydrazine (DNPH), 1,1-Diphenyl-2-picrylhydrazyl (DPPH), butylated hydroxyltoluene (BHT), 2,2-azinobis(3-ethylbenzothiazoline-6-sulfonic acid) (ABTS), 6-hydroxy-2,5,7,8-tetramethylchroman-2-carboxylic acid (trolox), quercetin, catechin, β-carotene and *p*-coumaric acid (98%) were purchased from Sigma Chemical Co., (St. Louis, MO, USA). Potassium persulfate, tween 40, Folin-Ciocalteau, and sodium carbonate were purchased from Dinamina (Diadema, Brazil). Sulfuric acid, methanol, ethanol, hydrochloric acid, ascorbic acid, chloroform, linolenic acid, gallic acid, aluminum chloride, sodium acetate, and vanillin were obtained from Vetec (Duque de Caxias, Brazil). Spectroscopic grade acetonitrile was purchased from Merck (Darmstadt, Germany). Standards (caffeic acid (98%), *p*-coumaric acid (98%), luteolin (98%), quercetin (98%) and apigenin (95%)) were purchased from Sigma Chemical Co. (St. Louis, MO, USA).

2.2. Plant Material

Leaves were collected in Dourados, Mato Grosso do Sul, Brazil. Botanical identification was performed by Dr. Zefa Valdevina Pereira (Faculty of Biological Sciences and Ambiental, Federal University of Grande Dourados (UFGD)). Specimens of *P. carthagenensis* (DDMS 5006), *P. deflexa* (DDMS 5005), *P. leiocarpa* (DDMS 5007) and *P. capillacea* (DDMS 5008) were deposited in the Herbarium of the Faculty of Biological Sciences and Ambiental, Federal University of Grande Dourados (UFGD), Mato Grosso do Sul, Brazil.

2.3. Preparation of Extracts

Dried leaves of *P. carthagenensis* (710 g), *P. deflexa* (620 g), *P. leiocarpa* (560 g) and *P. capillacea* (660 g) were exhaustively extracted by maceration with methanol (8 × 4 L) at room temperature. After filtration, evaporation of the solvent under vacuum furnished the extract. The extract yields (% dry weight) of the *P. carthagenensis*, *P. deflexa*, *P. leiocarpa*, and *P. capillacea* samples were 20.00%, 18.50%, 18.00% and 21.75% (w/w), respectively.

2.4. Measurement of Antioxidant Activity

2.4.1. β-Carotene/Linoleic Acid Method

The β-carotene solution was prepared by dissolving 2 mg β-carotene in 10 mL chloroform; 1 mL of this β-carotene-chloroform solution was mixed with 20 mg linoleic acid and 0.2 g Tween 40. Subsequently, the chloroform was removed by a rotary evaporator at 45 °C. Distilled water (50 mL) was slowly added with vigorous agitation to form an emulsion. Emulsion aliquots (5 mL) were transferred with 0.2 mL of the extracts different concentrations (10–200 μg/mL, sample stock 1.0 mg/mL). Control samples were prepared with 0.2 mL methanol devoid of extract [36–38]. As soon as the emulsion was added to each tube, absorbance was read at 470 nm against blank (zero time). Tubes were placed in a water bath at 50 °C, and oxidation was monitored by absorbance at 15 min intervals until the color of β-carotene in the control sample had disappeared (105 min). BHT was used as reference. The analyses were performed in triplicate. Antioxidant activity (AA) was calculated as percent inhibition relative to the control:

$$\%AA = [1 - (Ai - At)/(A'i - A't)] \times 100 \quad (1)$$

Ai = absorbance of sample at zero time, At = absorbance of sample after incubation (105 min) at 50 °C, $A'i$ = absorbance of control at zero time, and $A't$ = absorbance of control after incubation (105 min) at 50 °C.

2.4.2. Scavenging of 1,1-Diphenyl-2-Picrylhydrazyl (DPPH)

Sample stock solutions (1.0 mg/mL) were diluted to final concentrations of 200, 125, 50, 25, 10 and 5 μg/mL in methanol. Samples were added to 3 mL of methanolic DPPH (0.1 mM), prepared daily. The mixture was shaken and left to stand at room temperature in the dark. After 30 min, absorbance was measured at 517 nm against a blank containing all reagents except the test samples [39]. BHT was used as the positive control. Assays were carried out in triplicate. The percentage of inhibition of DPPH (I%) was calculated using the following equation:

$$I\% = (A_0 - A/A_0) \times 100 \quad (2)$$

A_0 is the absorbance of the blank solution and A is the absorbance of the methanolic extract.

The IC_{50}, the concentration giving 50% inhibition of DPPH, was read off a graph of I% (percentage inhibition) *versus* extract concentration.

2.4.3. $ABTS^+$ Scavenging Activity

Total antioxidant activity was measured using an improved azinobis (ethylbenzothiazoline-6-sulphonic acid) (ABTS) radical scavenging method [40] with minor modifications. Sample stock solutions (1.0 mg/mL) were diluted to final concentrations of 250, 125, 50, 25, 10 and 5 μg/mL in methanol. Briefly, 7.0 mM ABTS and 140 mM potassium persulphate were mixed and kept in the dark for 16 h at ambient temperature. Before usage, the $ABTS^+$ solution was diluted to get an absorbance of 0.700 ± 0.05 at 734 nm with ethanol (P.A.). Then, 3 mL of $ABTS^+$ solution was added to 30 μL of different sample concentrations (5–250 μg/mL). After 30 min, the absorbance was taken at 734 nm using spectrophotometer. The $ABTS^+$ scavenging activity was calculated using the following equation:

$$\text{ABTS radical scavenging activity (\%)} = (A_0 - A/A_0) \times 100 \quad (3)$$

A_0 is the absorbance of the blank solution and A is the absorbance of the methanolics extracts.

2.5. Concentrations of Constituents

2.5.1. Total Phenol Concentration

The total phenol concentration of the samples was determined using folin reagent [41]. Briefly, 100 μL of extract in methanol (1 g/L) were mixed with 1.0 mL of distilled water and 0.5 mL of folin-ciocaleu's reagent (1:10 v/v). After mixing, 1.5 mL of 2% aqueous sodium bicarbonate were added, and the mixture was allowed to stand for 30 min with intermittent shaking. The absorbance was measured at 765 nm using a spectrophotometer. Total phenolic concentration is expressed as gallic acid equivalent in mg per gram of extract. The methanol solution was used as a blank. All assays were carried out in triplicate.

2.5.2. Total Flavonoid Concentration

The amount of total flavonoids in the extracts was measured spectrophotometrically as previously reported [42]. Briefly, 500 μL of each extract was mixed with 1.50 mL of 95% ethanol, 0.10 mL of 10% aluminium chloride ($AlCl_3.6H_2O$), 0.10 mL of sodium acetate ($NaC_2H_3O_2.3H_2O$) (1 M) and 2.80 mL of distilled water. After incubation for 40 min, absorbance was measured at 415 nm using a spectrophotometer. To calculate the concentration of flavonoids, we prepared a calibration curve using quercetin as standard. The flavonoid concentration is expressed as quercetin equivalents in mg per gram of extract. All assays were carried out in triplicate.

2.5.3. Condensed Tannin Concentration (CT)

CT concentrations were determined by a modified version of a method reported previously [41,42]. Samples were mixed with 5 mL vanillin-HCl (8% conc. aq. HCl and 4% vanillin in methanol). Absorbance at 500 nm was read after 20 min. Catechin was used as the standard. The condensed tannin concentration is expressed as catechin equivalents in mg per gram of extract.

2.5.4. Flavonol Concentration

Total flavonols in the plant extracts were estimated using the method reported previously [41,43]. To 2 mL of sample, 2 mL $AlCl_3$ (2%)/ethanol and 3 mL (50 g/L) sodium acetate were added. The mixture was shaken and incubated for 2.5 h at 20 °C. Absorbance was read at 440 nm. Total flavonols are expressed as mg of quercetin equivalents per gram of dry weight (mg QE/g extract) using the calibration curve with quercetin.

2.6. HPLC/PDA Analysis

The *Psychotria* extracts and standards were analyzed using an analytical HPLC (Varian 210, Varian, Sugar Land, TX, USA) system, with a ternary solvent delivery system and an auto-sampler. A photodiode array detector (PAD) was monitored at λ = 200–800 nm. The HPLC column was C-18 (25 cm × 4.6 mm; particle size, 5 μm; Luna, Phenomenex, Torrance, CA, USA), with a small pre-column (2.5 cm × 3 mm) containing the same packing to protect the analytical column. The flow rate and injected volume were 1.0 mL min^{-1} and 10 μL, respectively. All chromatographic analyses were performed at 22 °C. The elution of *P. leicocarpa* and *P. deflexa* extracts and standards was carried out using acetonitrile with formic acid (pH 2.64) (solvent A) and water (solvent B). The solvent gradient program was as follows: 0 min, 70% B; 30 min, 37% B; in 45 min returning to the initial condition. The *P. carthagenensis* and *P. capillacea* extracts and standards were eluted using acetonitrile with formic acid (pH 2.64) (solvent A) and water (solvent B). The solvent gradient program was as follows: 0 min, 84% B; 15 min, 59% B; 20 min, 79% B; 40 min, 0% B, in 50 min returning to the initial condition.

Stock solutions of standards (caffeic acid, *p*-coumaric acid, luteolin, quercetin and apigenin), created by dissolving individual solutions in acetonitrile at 10 μg/mL and were employed in the identification of compounds using a PDA detector (200–800 nm). This did not reveal coeluting substances.

3. Results and Discussion

3.1. Antioxidant Activity

Due to the complexity of some plant extracts, the use of several different methods is recommended for the evaluation of antioxidant activity [43]. Currently used methods include the DPPH (1,1-diphenyl-2-picrylhydrazyl) assay [35], which measures the ability of a substance to scavenge the DPPH radical, reducing it to hydrazine. When a substance that acts as a donor of hydrogen atoms is added to a solution of DPPH, hydrazine is obtained, with a change in color from violet to pale yellow. 2,2-azinobis(3-ethylbenzothiazoline-6-sulfonic acid) ($ABTS^{+}$) is a decolorization technique, in that the radical is generated directly in a stable form prior to reaction with putative antioxidants, which involves the production of the blue/green $ABTS^{+}$ chromophore through the reaction of ABTS with potassium persulfate. The β-carotene bleaching method [44] evaluates the ability of a substance to prevent the oxidation of β-carotene, protecting it from the free radicals generated during the peroxidation of linoleic acid.

The effects of methanolic extracts of *Psychotria* species in the DPPH, linoleic acid peroxidation and ABTS assays are shown in Table 1. *P. carthagenensis* and *P. capillacea* exhibited the highest scavenging activity with an IC_{50} of 16.92 ± 4.58 and 30.05 ± 6.22 μg/mL, respectively (Table 1), which was comparable to that of the standard antioxidant butylated hydroxytoluene (BHT) (IC_{50} = 16.72 ± 1.34 μg/mL) and ascorbic acid (IC_{50} = 22.28 ± 0.53 μg/mL). The comparison of the obtained free-radical scavenging data indicated potent activity for the *P. carthagenensis* and *P. capillacea* at 100 μg/mL, with %FRS values of 93.52% ± 8.41% and 91.78% ± 4.23%, respectively. These results show that the DPPH radical scavenging activity of these extracts was similar to BHT and ascorbic acid (Table 1). Gallic acid (6.56%, standard), *P. carthagenensis* (6.47%) and *P. capillacea* (7.32%) also resulted in a remarkable reduction of the DPPH remaining, compared with *P. leicocarpa* and *P. deflexa* (56.26% and 63.51%), respectively, when recorded after 60 min (Figure 1). It is clear that the more DPPH that remains, the lower the radical-scavenging activity of the tested samples is. All these data clearly indicate that *P. carthagenensis* and *P. capillacea* extracts are effective electrons or hydrogen atoms donor to DPPH.

Higher antioxidant activity was also found for *P. carthagenensis* (79.1% ± 3.70%) with the β-carotene bleaching method compared to quercetin and BHT (Table 1). This higher activity was not observed for ascorbic acid, which has relatively high polarity. We propose that this extract contains lipophilic compounds, which act by inhibiting or retarding oxidation of β-carotene.

The scavenging effect of extracts from *Psychotria* and standard on $ABTS^{+}$ decreased in the order: BHT ~ *P. carthagenensis* > *P. capillacea* ~ ascorbic acid > *P. deflexa* ~ *P. leicocarpa* (96.4% ± 2.44%, 92.5% ± 7.43%, 87.34% ± 8.32%, 80.9% ± 5.56%, 15.58% ± 5.22% and 12.20% ± 4.44%, respectively) at the same concentration (100 μg/mL). *P. carthagenensis* and *P. capillacea* exhibited effective radical cation scavenging activity, when compared to other tested extracts.

Figure 1. Kinetic behavior of methanol extracts (100 μg/mL) against DPPH: *P. carthagenensis*, *P. leiocarpa*, *P. capillacea*, *P. deflexa* and gallic acid standard.

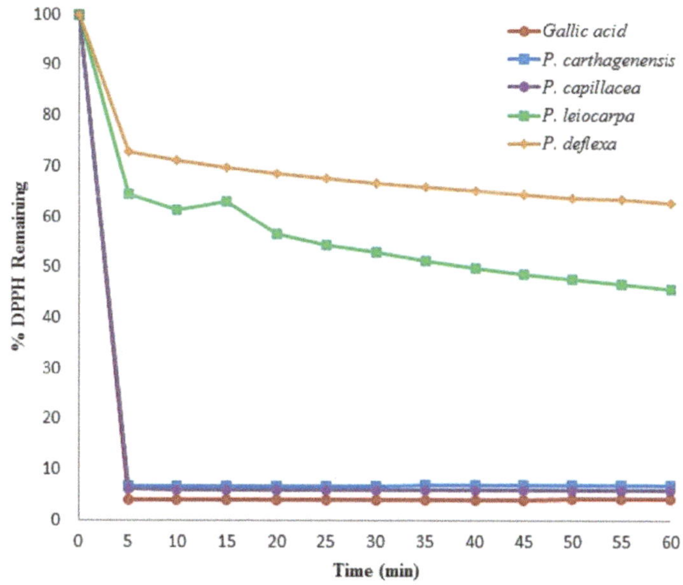

Table 1. Antioxidant activity of *Psychotria* leaf extracts by DPPH, β-carotene/linoleic acid and ABTS assays.

		Test			
		DPPH		β-Carotene/Linoleic Acid	ABTS⁺
		IC_{50} (μg/mL) (Limit Confidence 95%)	%FRS *	(%AA)	(%)
Extracts	*P. carthagenensis*	16.92 ± 4.58 (12.03–22.71)	93.52 ± 8.41	79.12 ± 3.70	92.5 ± 7.43
	P. leiocarpa	127.00 ± 10.55 (123.50–135.94)	54.13 ± 11.10	22.30 ± 7.14	12.20 ± 4.44
	P. capillacea	30.05 ± 6.22 (26.68–37.27)	91.78 ± 4.23	33.40 ± 15.22	87.34 ± 8.32
	P. deflexa	146.40 ± 12.47 (141.81–148.42)	66.37 ± 8.12	26.05 ± 10.60	15.58 ± 5.22
Standards	BHT	16.72 ± 1.34 (14.08–17.22)	92.19 ± 1.29	91.20 ± 4.54	96.4 ± 2.44
	Ascorbic acid	22.28 ± 0.53 (20.20–23.43)	96.40 ± 0.27	4.13 ± 1.42	80.9 ± 5.56
	Quercetin	n.d.	n.d.	80.65 ± 1.25	n.d.

Values are expressed as the mean ± SD ($n = 3$); n.d. = not determined; IC_{50} = concentration resulting in 50% inhibition of DPPH, derived from the graph of I% (inhibition percentage) *versus* concentration in μg/mL; % FRS = free-radical scavenging percentage (* antioxidant activity evaluated by DPPH free-radical scavenging at a final concentration equivalent to 100 μg/mL of extract); %AA = antioxidant activity, evaluated by the β-carotene/linoleic acid method. (%) = ABTS radical scavenging activity.

3.2. Levels of Constituents

The total concentrations of phenolic compounds in the extracts are shown in Figure 2A. *P. capillacea* had the highest total concentration (148.42 ± 4.69 mg gallic acid equivalents (GAE)/g extract), followed by *P. deflexa* and *P. leiocarpa* (111.42 ± 8.12 and 78.45 ± 5.20 mg GAE /g extract, respectively). *P. carthagenensis*, with 182.07 ± 2.78 mg quercetin equivalents (QE)/g extract, exhibited the highest flavonoid concentration (Figure 2B). The flavonoid concentrations of *P. capillacea*, *P. leiocarpa*, and *P. deflexa* were 91.58 ± 3.74, 59.80 ± 6.45, and 37.64 ± 10.14 mg QE/g extract, respectively (Figure 2B). The highest flavonol concentrations were found in extracts *P. deflexa* and *P. carthagenensis* yielded 275.07 ± 8.40 and 241.19 ± 9.48. *P. leiocarpa* and *P. capillacea* also yielded 189.20 ± 6.44, and 185.54 ± 5.33 mg QE/g extract, respectively (Figure 2C). High levels of condensed tannins were also found in *P. carthagenensis* (632.39 ± 5.63 mg catechin equivalents (CE)/g extract) and *P. capillacea* (571.95 ± 7.22 mg CE/g extract), while the lowest concentrations were observed in *P. deflexa* and *P. leiocarpa* (194.67 ± 9.02 and 60.97 ± 10.45 mg CE/g extract, respectively (Figure 2D).

Figure 2. Total phenols (**A**), Total flavonoids (**B**), Total flavonols (**C**) and condensed tannins (**D**) in *P. carthagenis*, *P. leiocarpa*, *P. capillacea* and *P. deflexa* extracts. The data represent the mean ± SD.

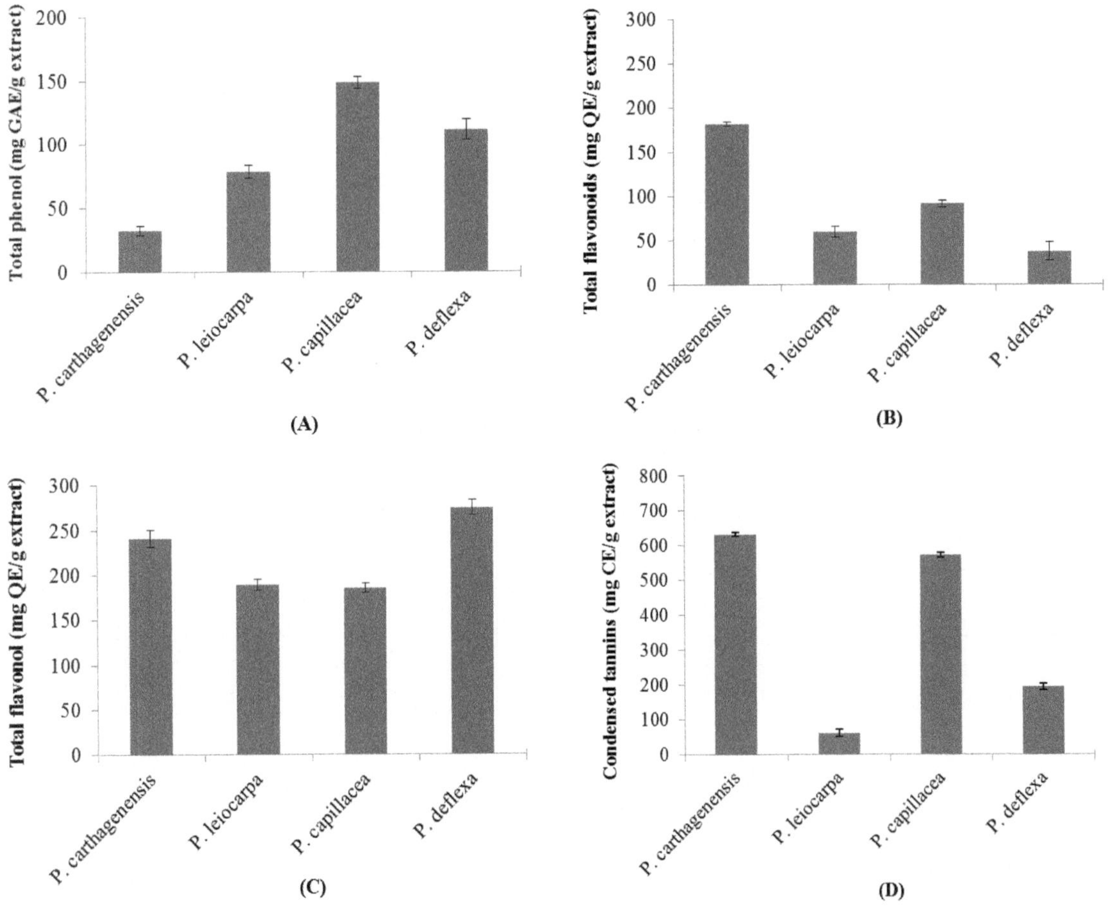

In plants, flavonoids occasionally occur as aglycones, although the most common forms are glycoside derivatives. These compounds account for 60% of total dietary phenolic compounds [45,46]. Flavonols are the most common flavonoids in the plant kingdom, and glycosides of quercetin are the most common naturally occurring flavonols [45].

Phenolic compounds are known as high-level antioxidants because of their ability to scavenge free radicals and active oxygen species, such as singlet oxygen, superoxide free radicals and hydroxyl radicals [47]. The radical-scavenging activity is attributed to replacement of hydroxyl groups in the aromatic ring systems of the phenolic compounds as a result of their hydrogen donating ability [44].

3.3. HPLC/PAD Analysis from Psychotria Extracts

After determination of the antioxidant potential and levels of constituents, we analyzed the methanolic extracts obtained from *Psychotria* in an analytical LC (Figure 3). The standards were easily identified based on their UV absorption spectra and retention times. The substances found in the extracts were unambiguously identified by performing co-injection experiments in which aliquots of extract and standard were mixed and diluted to a known volume, and analyzed by HPLC. Only *p*-coumaric acid (*tr* = 24.91 min) was found in *P. carthagenensis* and *P. capillacea* (Figure 3A,B).

Figure 3. Chromatogram of *P. carthagenensis* (**A**), *P. capillacea* (**B**), *P. leiocarpa* (**C**) and *P. deflexa* (**D**) extracts by HPLC/PDA analysis.

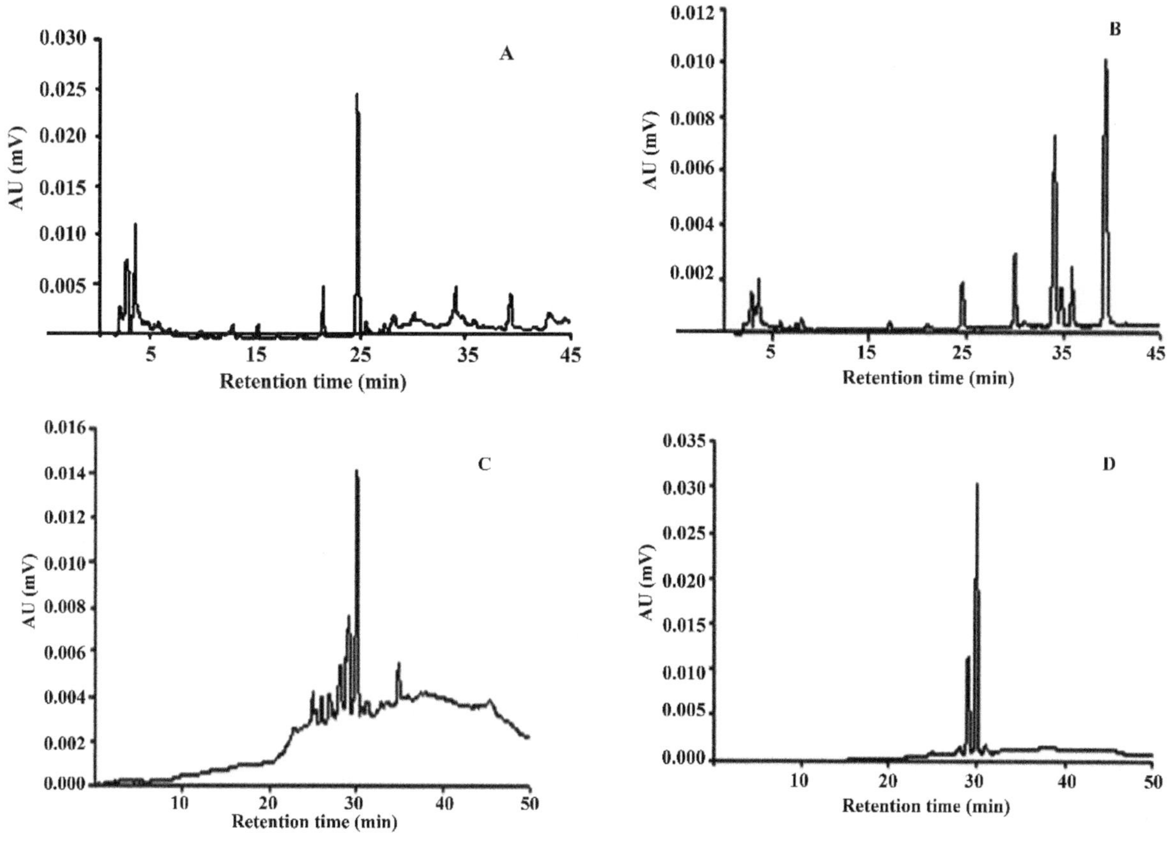

The antioxidant activity of the pure *p*-coumaric acid at 40 μg/mL concentration was determined. DPPH radical scavenging activity of *p*-coumaric acid was found to be 55.6% ± 3.4%; the β-carotene bleaching with 78.6% ± 2%; and 87.14% ± 5.7% in ABTS radical scavenging activity. Comparison of the obtained of *P. carthagenensis* and *P. capillacea* data indicated the potent activity for the pure *p*-coumaric acid.

p-Coumaric acid (4-hydroxycinnamic acid), a phenolic acid, hydroxyl derivative of cinnamic acid, is widely used in the chemical, food, health, cosmetic, and pharmaceutical industries. Studies reported that antioxidants, such as *p*-coumaric acid and others hydroxycinnamic acids, function as chemoprotective agents by quenching carcinogenic nitrosating agents in several biological compartments, including salivary and gastric fluids [48]. However, minor attention has also been directed to the activity of simple phenolic acids, such as benzoic or cinnamic acids, and their derivatives.

4. Conclusions

According to data obtained from the present study, *P. carthagenensis* and *P. capillacea* were found to be effectives antioxidants in different *in vitro* assay including DPPH radical, ABTS radical and β-carotene bleaching activities when they are compared to standard antioxidant compounds, such as BHT, gallic acid, quercetin and ascorbic acid. The results indicate that the antioxidant activity to these plants can be attributed the presence the *p*-coumaric acid. Further, more detailed, studies on the chemical composition of those extracts, as well as studies with other models, such *in vivo* assays, are essential to characterize them as biological antioxidants.

Acknowledgments

We are grateful to Fundect, Capes and CNPq for fellowships.

Author Contributions

Anelise Samara Nazari Formagio, Carla Roberta Ferreira Volobuff and Matheus Santiago designed the study, performed extractions, determined the levels of total phenols, flavonoids, tannins and flavonols, assessed antioxidant activity and helped in writing and editing. Claudia Andrea Lima Cardoso participated in the HPLC-DAD analysis. Maria do Carmo Vieira and Zefa Valdevina Pereira participated in the collection and identification of plant material. All authors read and approved the final manuscript.

Conflicts of Interest

The authors declare that they have no competing interests.

References

1. Lindley, M. The impact of food processing on antioxidants in vegetable oils, fruits and vegetables. *Trends Food Sci. Technol.* **1998**, *9*, 336–340.
2. Middleton, E.; Kandaswamy, C.; Theoharides, T.C. The effects of plant flavonoids on mammalian cells: Implications for inflammation, heart disease, and cancer. *Pharmacol. Rev.* **2000**, *52*, 673–751.

3. Salah, N.; Miller, N.J.; Paganga, G.; Tijburg, L.; Rice-Evans, C.A. Polyphenolic flavonols as scavengers of aqueous phase radicals and as chain-breaking antioxidants. *Arch. Biochem. Biophys.* **1995**, *322*, 339–346.
4. Whitehead, T.P.; Robinson, D.; Allaway, S.; Syms, J.; Hale, A. Effect of red wine ingestion on the antioxidant capacity of serum. *Clin. Chem.* **1995**, *41*, 32–35.
5. Perry, L.M.; Metzger, J. *Medicinal Plants of East and Southeast*; MIT Press: Cambridge, UK, 1980.
6. Lajis, N.H.; Mahmud, Z.; Toia, R.F. The alkaloids of *Psychotria rostrata*. *Planta Med.* **1993**, *59*, 383–384.
7. Hart, N.K.; Johns, S.R.; Lamberton, J.A.; Summons, R.E. Psychotridine, a $C_{55}H_{62}N_{10}$ alkaloid from *Psychotria beccarioides* (Rubiaceae). *Austrian J. Chem.* **1974**, *27*, 639–646.
8. Roth, A.; Kuballa, B.; Bounthanh, C.; Cabalion, P.; Sévenet, T.; Beck, J.P.; Anton, R. Cytotoxic activity of polyindoline alkaloids of *Psychotria fosteriana* (Rubiaceae). *Planta Med.* **1986**, *5*, 450–453.
9. Libot, F.; Miet, C.; Kunesch, N.; Poisson, J.E.; Pusset, J.; Sévenet, T. Rubiacées d'océanie: Alcaloïdes de *Psychotria oleoides* de Nouvelle-Calédonie et de *Calycodendron milnei* du Vanuatu (Nouvelle-Hébrides). *J. Nat. Prod.* **1987**, *50*, 468–473.
10. Adjibade, Y.; Weniger, B.; Quirion, J.C.; Kuballa, B.; Cabalion, P.; Anton, R. Dimeric alkaloids from *Psychotria fosteriana*. *Phytochemistry* **1992**, *31*, 317–319.
11. Verotta, L.; Pilati, T.; Tato, M.; Elisabetsky, E.; Amador, T.A.; Nunes, D.S. Pyrrolidinoindoline alkaloids from *Psychotria colorata*. *J. Nat. Prod.* **1998**, *61*, 392–396.
12. Jannic, V.; Gueritte, F.; Laprevote, O.; Serani, L.; Martin, M.T.; Sevenet, T.; Potier, P. Pyrrolidinoindoline alkaloids from *Psychotria oleoides* and *Psychotria lyciiflora*. *J. Nat. Prod.* **1999**, *62*, 838–843.
13. Takayama, M.; Hamano, H.; Ochi, Y.; Seaegusa, H.; Komatsu, K.; Murati, T.; Arakura, N.; Imai, Y.; Hasebe, O.; Kawa, S.; *et al.* Recurrent attacks of autoimune pancreatites results in pancreatic stone formation. *Am. J. Gastrienterol.* **2004**, *99*, 932–937.
14. Itoh, A.; Tanahashi, T.; Nagakura, N. Six tetrahydroisoquinoline-monoterpene glucosides from *Cephaelis ipecacuanha*. *Phytochemistry* **1991**, *30*, 3117–3123.
15. Itoh, A.; Tanahashi, T.; Nagakura, N.; Nayeshiro, H. Tetrahydroisoquinoline-monoterpene glucosides from *Alangium lamarckii* and *Cephaelis ipecacuanha*. *Phytochemistry* **1994**, *36*, 383–387.
16. Itoh, A.; Ikuta, Y.; Baba, Y.; Tanahashi, T.; Nagakura, N. Ipecac alkaloids from *Cephaelis acuminata*. *Phytochemistry* **1999**, *52*, 1169–1176.
17. Nagakura, N.; Itoh, A.; Tanahashi, T. Four tetrahydroisoquinoline-monoterpene glucosides from *Cephaelis ipecacuanha*. *Phytochemistry* **1993**, *32*, 761–765.
18. Solis, P.; Ravelo, A.G.; Palenzuela, J.A.; Gupta, M.P.; Gonzalez, A.; Phillipson, J.D. Quinoline alkaloids from *Psychotria glomerata*. *Phytochemistry* **1997**, *44*, 963–969.
19. Muhammad, I.; Dunbar, C.D.; Shabana, I.K.; Tekwani, B.L.; Bedir, E.; Takamatsu, S.; Ferreira, D.; Walker, L.A. Antiparasitic alkaloids from *Psychotria klugii*. *J. Nat. Prod.* **2003**, *66*, 962–967.
20. Achenbach, H.; Lottes, M.; Waibel, R.; Karias, G.A.; Correa, M.D.; Gupta, M.P. Alkaloids and other compounds from *Psychotria correae*. *Phytochemistry* **1995**, *38*, 1537–1545.

21. Kerber, V.A.; Gregianni, T.S.; Paranhos, J.S.; Schwanbach, J.; Farias, F.; Fett, J.P.; Fett-Neto, A.G.; Zuanazzi, J.A.; Quirion, J.C.; Elisabetsky, E.; *et al*. Brachycerine, a novel monoterpene indole alkaloid from *Psychotria brachyceras*. *J. Nat. Prod.* **2001**, *64*, 677–679.
22. Santos, L.V.; Fett-Neto, A.G.; Kerber, V.A.; Elisabetsky, E.; Quirion, J.C.; Henriques, A.T. Indole monoterpene alkaloids from leaves of *Psychotria suterella* Müll. Arg. (Rubiaceae). *Biochem. Syst. Ecol.* **2001**, *29*, 1185–1187.
23. Paul, J.H.A.; Maxwell, A.R.; Reynolds, W.F. Novel bis(monoterpenoid) indole alkaloids from *Psychotria bahiensis*. *J. Nat. Prod.* **2003**, *66*, 752–754.
24. Henriques, A.T.; Lopes, S.O.; Paranhos, J.T.; Gregianini, T.S.; von Poser, G.L.; Fett-Neto, A.G.; Schripsema, J. *N*,β-D-Glucopyranosyl vincosamide, a light regulated indole alkaloid from the shoots of *Psychotria leiocarpa*. *Phytochemistry* **2004**, *65*, 449–454.
25. Simoes-Pires, C.A.; Farias, F.N.; Marston, A.; Queiroz, E.F.; Chaves, C.G.; Henriques, A.T.; Hostettman, K. Indole monoterpenes with antichemotactic activity from *Psychotria myriantha*: Chemotaxonomic significance. *Nat. Prod. Commun.* **2006**, *1*, 1101–1106.
26. Nomura, T.; Quesada, A.L.; Kutchan, T.M. The New β-D-glucosidade in terpenoide isoquinoline alkaloid biosynthesis in *Psychotria ipecacuanha*. *J. Biol. Chem.* **2008**, *283*, 34650–34659.
27. Beretz, A.; Roth-Georger, A.; Corre, G.; Kuballa, B.; Anton. R.; Cazenale, J.P. Polyindolinic alkaloids from *Psychotria fosteriana*. Potent inhibitors of the aggregation of human platelets. *Planta Med.* **1985**, *4*, 300–303.
28. Leal, M.B.; Elisabetsky, E. Opioid-like activity of *Psychotria brachypoda*. *Int. J. Pharmacogn.* **1996**, *34*, 267–272.
29. Lopes, S.O. Análise Química e Cultivo *in Vitro* de *Psychotria leiocarpa* Cham. et Schlecht. e *Psychotria carthagenensis* Jacq. (Rubiaceae). Master's Thesis, Universidade Federal do Rio Grande do Sul, Porto Alegre, Brazil, **1998**.
30. Lopes, S.; Poser, G.L.; Kerber, V.A.; Farias, F.M.; Konrath, E.L.; Moreno, P.; Sobral, M.E.; Zuanazzi, J.A.S.; Henriques, A.T. Taxonomic significance of alkaloids and iridoid glucosides in the tribe *Psychotrieae* (Rubiaceae). *Biochem. Syst. Ecol.* **2004**, *32*, 1187–1195.
31. Dillenburg, C.R.; Porto, M.L. Rubiaceae-tribo *Psychotriae*. *Boletim Instituto de Biociências—UFRGS* **1985**, *39*, 1–76.
32. Rivier, L.; Lindgren, J.E. "Ayahuasca", the South American hallucinogenic drink: An ethnobotanical and chemical investigation. *Econ. Bot.* **1972**, *26*, 101–129.
33. McKenna, D.J.; Towers, G.N.T.; Abbot, F. Monoamine oxidase inhibitors in South American hallucinogenic plants: Trypatamine and β-carboline constituents of Ayahusca. *J. Ethnopharmacol.* **1984**, *10*, 195–223.
34. Adedapo, A.A.; Jimoh, F.O.; Koduru, S.; Afolayan, A.J.; Masika, P.J. Antibacterial and antioxidant properties of the methanol extracts of the leaves and stems of *Calpurnia aurea*. *BMC Complement. Altern. Med.* **2008**, *53*, 1–8.
35. Blois, M.S. Antioxidant determinations by the use of a stable free radical. *Nature* **1958**, *181*, 1199–1200.
36. Kaur, C.; Kapoor, H.C. Antioxidant activity and total phenolic content of some Asian vegetables. *Int. J. Food Sci. Technol.* **2002**, *37*, 153–161.

37. Shahidi, F.; Chavan, U.D.; Naczk, M.; Amarowicz, R. Nutrient distribution and phenolic antioxidants in air-classified fractions of beach pea (*Lathyrus maritimus* L.). *J. Agric. Food Chem.* **2001**, *49*, 926–933.
38. Rufino, M.S.M.; Alves, R.E.; Brito, E.S.; Morais, S.M.; Sampaio, C.G.; Pérez-Jiménez, J.; Saura-Calixto, F.D. Determinação da atividade antioxidante total em frutas pela captura do radical livre ABTS. *Embrapa Agroindústria Tropical/Comunicado Técnico* **2007**, *128*, 1–4.
39. Harborne, J.B.; Williams, C.A. Advances in flavonoid research since 1992. *Phytochemistry* **2000**, *55*, 481–504.
40. Djeridane, A.; Yousfi, M.; Nadjemi, B.; Boutassouna, D.; Stocker, P.; Vidal, N. Antioxidant activity of some algerian medicinal plants extracts containing phenolic compounds. *Food Chem.* **2006**, *97*, 654–660.
41. Maxson, E.D.; Rooney, L.W. Evaluation of methods for tannin analysis in sorghum grain. *Cereal Chem.* **1972**, *49*, 719–729.
42. Lin, J.Y.; Tang, C.Y. Determination of total phenolic and flavonoid contents in selected fruits and vegetables, as well as their stimulatory effects on mouse splenocyte proliferation. *Food Chem.* **2007**, *101*, 140–147.
43. Brand-Williams, W.; Cuvelier, M.E.; Benset, C. Use of free radical method to evaluate antioxidant activity. *Lebensm. Wiss. Technol.* **1995**, *28*, 25–30.
44. Naczk, M.; Shahidi, F. Extraction and analysis of phenolics in food. *J. Chromatogr.* **2004**, *1054*, 95–111.
45. Hall, C.A.; Cuppett, S.L. Structure-activities of natural antioxidants. In *Antioxidant Methodology: in Vivo and in Vitro Concepts*; Aruoma, O.I., Cuppett, S.L., Eds.; AOCS Press: Champaign, IL, USA, 1997; pp. 141–172.
46. Robbins, R.J. Phenolic acids in foods: An overview of analytical methodology. *J. Agric. Food Chem.* **2003**, *51*, 2866–2887.
47. Torres, J.L.; Rosazza, J.P.N. Microbial transformations of *p*-coumaric acid by *Bacillus megaterium* and *Curvularia lunata*. *J. Nat. Prod.* **2001**, *64*, 1408–1414.
48. Jayaprakasha, G.K.; Singh, R.P.; Sakariah, K.K. Antioxidant activity of grape seed (*Vitis vinifera*) extracts on peroxidation models *in vitro*. *Food Chem.* **2001**, *73*, 285–290.

8

Antioxidant Content, Antioxidant Activity, and Antibacterial Activity of Five Plants from the Commelinaceae Family

Joash Ban Lee Tan *, Wei Jin Yap, Shen Yeng Tan, Yau Yan Lim, and Sui Mae Lee

School of Science, Monash University Sunway Campus, Bandar Sunway 46150, Malaysia; E-Mails: wjyap4@student.monash.edu (W.J.Y.); sytan89@student.monash.edu (S.Y.T.); lim.yau.yan@monash.edu (Y.Y.L.); lee.sui.mae@monash.edu (S.M.L.)

* Author to whom correspondence should be addressed; E-Mail: tan.ban.lee@monash.edu

External Editors: Maria G. Miguel and João Rocha

Abstract: Commelinaceae is a family of herbaceous flowering plants with many species used in ethnobotany, particularly in South America. However, thus far reports of their bioactivity are few and far between. The primary aim of this study was to quantify the antioxidant and antibacterial activity of five Commelinaceae methanolic leaf extracts. The antioxidant content was evaluated by the total phenolic content (TPC), total tannin content (TTC), and total flavonoid content (TFC) assays. The antioxidant activities measured were DPPH free radical scavenging (FRS), ferric reducing power (FRP), and ferrous ion chelating (FIC); of the five plants, the methanolic leaf extract of *Tradescantia zebrina* showed the highest antioxidant content and activity, and exhibited antibacterial activity against six species of Gram-positive and two species of Gram-negative bacteria in a range of 5–10 mg/mL based on the broth microdilution method.

Keywords: Commelinaceae; antioxidant; antibacterial; methanolic extracts

---

1. Introduction

The Commelinaceae family comprises 37 genera and over 600 species [1] of monocotyledonous herbaceous flowering plants [2,3], notably the spiderworts (*Tradescantia* sp.). They are believed to have originated from the old world tropics, but now they are widely distributed throughout the sub-tropics and tropics of both hemispheres, with some species surviving even in more temperate climates [4]. These plants are often grown for ornamental purposes due to their bluish or purplish leaves and/or flowers, but are also known to be used ethnobotanically to treat many diseases, including mycosal infections [5], venereal diseases [6], wounds [7], gastrointestinal disorders, and cancer [8], which may be linked to their antibacterial and antioxidant properties. However, reports on the antibacterial and antioxidant properties of these plants have remained rare so far.

Plants of this family are a good source of renewable bioactive compounds, as members of the Commelinaceae are generally evergreen, perennial, hardy, and prolific. Like other silicon-accumulating plant families such as the Gramineae (rice/wheat) and Cucurbitales (pumpkin/squash), Commelinaceae are less prone to growth, development, and reproduction abnormalities than plants with less efficient silicon accumulation [9]. In fact, many species of the Commelinaceae are considered weeds and pests due to their rapid growth, hardiness, and ability to root at the nodes. They are also resilient to most herbicides, and able to rapidly regenerate if left untreated [4]. However, these same traits also make them an ideal renewable and abundant resource for bioactive compounds.

The antioxidant and antibacterial activity of *Rhoeo spathacea* (Swartz) Stearn leaves had been previously reported by our group [10]. In the present paper, two other cultivars of *R. spathacea* are reported; this represents the first time the bioactivity of these cultivars has been studied. The first is *Rhoeo bermudensis*, a dwarf cultivar of *R. spathacea* that is also known as *Tradescantia spathacea* "Hawaiian Dwarf", or *Rhoeo spathacea nana.* It is physically similar to *R. spathacea* (Swartz) Stearn, albeit much smaller—its leaves are a mere 10–20 cm long, compared to 15–30 cm for *R. spathacea* (Swartz) Stearn. The overall plant grows to 30–45 cm tall, approximately 60% of the height of *R. spathacea* (Swartz) Stearn. Unlike *R. spathacea,* it lacks flowers, but is able to root readily at the nodes and spread across the ground rapidly, like many other species of the Commelinaceae family [4]. It is colloquially known as the Hawaiian dwarf, or dwarf oyster plant [11]. The second cultivar, *Rhoeo spathacea* var. variegata, is also known as *Rhoeo spathacea* vittata and bears resemblance to *R. spathacea* (Swartz) Stearn, but with yellow striations on the upper side of the leaves instead of the stark green upper side of the original cultivar [12].

Tradescantia pallida (Rose) D. R. Hunt., also known as purple heart or wandering jew, is a low-growing tetraploid plant that is shade-tolerant, able to thrive on various soil conditions, and possesses strong resistance to insects and parasites. These traits allow it to rapidly colonize various environments, acting as an invasive weed [13,14]. Therefore, it is well adapted and widely distributed in tropical and subtropical regions [15,16] and is commonly being grown as an ornamental, ground-covering, or hanging plant. Although also renowned for its ability to effectively remove volatile organic pollutants from the air [17], *T. pallida* has also been traditionally used as an anti-inflammatory and anti-toxic supplement, and to improve blood circulation [18].

Similarly, *Tradescantia zebrina* Bosse var. *zebrina*, a close relative to *T. pallida*, is similarly categorized as an invasive plant species, and commonly distributed in tropical and subtropical regions [19]. It has oval, pointed leaves, variegated green and silver stripes on the upper surface, with a purple underside [20,21]. *T. zebrina* is traditionally used to treat gastrointestinal disorders [22].Other studies have also reported that *T. zebrina* exhibits insecticidal properties [23] and can inhibit the 15-lipoxygenase pathway involved in asthmatic attacks [24].

Callisia fragrans Wood. (basket plant) is an all-green plant of the Commelinaceae family. Its leaves and runners contain phenolic compounds such as flavonoids and phytosteroids. *C. fragrans* is traditionally used to treat burns, arthritis, skin and oncological diseases, tuberculosis, and asthma [25]. Amongst the phenolic compounds reported to be present in *C. fragrans* are gallic acid, caffeic acid, quercetin, scopoletin, and chicoric acid [26].

Thus far, antioxidant and antibacterial activity studies on plants from the Commelinaceae family have been relatively limited. The antioxidant and antibacterial activity of *R. spathacea* variegata, *R. bermudensis*, *T. pallida*, and *T. zebrina* leaves have never been reported. The antioxidant activity of the juice pressed from *C. fragrans* has been reported by Misin and Sazhina [27] and Olennikov *et al.* [28]. However, the antioxidant and antibacterial activity of *C. fragrans* methanolic leaf extracts has never been reported. This is therefore the first report of the antioxidant and antibacterial activity of the methanolic leaf extracts for all five Commelinaceae plants.

2. Materials and Methods

2.1. Samples

Fresh leaf samples of all five species were obtained from Selangor, Malaysia, within a 10 km radius. All plants were grown under similar conditions: a monthly mean minimum temperature from 20.8 °C to 25.0 °C, with a monthly mean maximum temperature from 29.6 °C to 32.8 °C; grown on soil; exposed to sunlight (mean daily solar radiation of 19.70 MJ/m^2) and rain (300–400 mm/month) [29]; and with no fertilizer. For the antioxidant tests, for each species, leaves of similar size were collected from three different individual plants ($n = 3$).

Bacterial isolates were obtained from American Type Culture Collection (ATCC), with the exception of *Proteus vulgaris,* which was obtained from the Institute of Medical Research (IMR), Malaysia. A total of 12 strains of bacteria were used: six Gram-positive bacteria (*Bacillus cereus* (ATCC 14579), *Bacillus subtilis* (ATCC 8188), *Enterococcus faecalis* (ATCC 29212), *Micrococcus luteus* (ATCC 4698), methicillin-resistant *Staphylococcus aureus* (ATCC 33591), and *Staphylococcus epidermidis* (ATCC 12228)) and six Gram-negative bacteria (*Aeromonas hydrophila* (ATCC 49140), *Enterobacter aerogenes* (ATCC 13048), *Pseudomonas aeruginosa* (ATCC 10145), *Proteus mirabilis* (ATCC 12453), *Proteus vulgaris* (clinical), and *Salmonella enterica* Typhimurium (ATCC 14028)). All bacteria were grown on nutrient agar at 37 °C.

2.2. Chemicals and Reagents

The various reagents used throughout this project were purchased from suppliers as follows. TPC analysis: Folin-Ciocalteu's phenol reagent (2N, R and M Chemicals, Essex, UK), gallic acid (98%, Fluka, Steinheim, France), anhydrous sodium carbonate (99%, J. Kollin, UK), diphenyl-2-picrylhydrazyl (DPPH) assay: 2,2-diphenyl-1-picrylhydrazyl (90%, Sigma, St. Louis, MO, USA), ferric reducing power (FRP) assay: ferric chloride hexa-hydrate (100%, Fisher Scientific, Loughborough, UK), potassium ferricyanide (99%, Unilab, Auburn, Australia), trichloroacetic acid (99.8%, HmbG Chemicals, Barcelona, Spain), potassium dihydrogen orthophosphate (99.5%, Fisher Scientific), dipotassium hydrogen phosphate (99%, Merck, Darmstadt, Germany), iron chloride (99%, RandM Chemicals), ferrous ion chelating (FIC) assay: ferrozine (98%, Acros Organics, Morris Plains, NJ, USA), ferrous sulphate hepta-hydrate (HmbG Chemicals), ethylenediaminetetraacetic (99.5%, Bendosen Laboratory Chemicals, Bendosen, Norway), potassium acetate (99%, R and M chemicals), rutin (98%, Sigma), phytochemical screening: sulfuric acid (95%–97%, HmBG Chemicals), hydrochloric acid (37%, Merck, Darmstadt, Germany), Dragendorff reagent (Fluka), α-naphthol (99%, Sigma), antimicrobial activity: nutrient broth (Oxoid, Hampshire, England), nutrient agar (Oxoid, Hampshire, England), and vancomycin (Sigma).

2.3. Extraction of Samples

The fresh leaves were gently washed and dabbed dry before being processed. One gram of leaf sample was subjected to liquid nitrogen-aided crushing with a mortar and pestle, followed by extraction with 50 mL of solvent for an hour at room temperature. Different compositions of methanol (50%, 70%, or 100%) were chosen for the extraction from the different species based on the best extraction efficiency (as determined by TPC). *R. bermudensis* and *C. fragrans* were extracted with 50% methanol (84.2% and 88.6% extraction efficiency, respectively), *R. spathacea* variegata with 70% methanol (89.9% extraction efficiency), and *T. zebrina* and *T. pallida* with 100% methanol (92.3% and 83.1% extraction efficiency, respectively). Extracts were filtered and stored at 8 °C when not in use. All analyses were conducted in triplicates.

2.4. Determination of Total Phenolic Content (TPC)

The TPC assay was modified from [30] utilizing the Folin–Ciocalteu reagent. Samples (300 μL, in triplicate) were mixed with 1.5 mL of the 10% Folin–Ciocalteu reagent, followed by an addition of 1.2 mL of 7.5% (w/v) sodium carbonate (Na_2CO_3) solution. The test tubes were then left to stand for 30 min in the dark at room temperature before the absorbance values were measured at 765 nm with a Hitachi U-1800 spectrophotometer (Shimadzu Corporation, Kyoto, Japan). The total phenolic content was expressed as mg gallic acid equivalent per 100 g of fresh sample (mg GAE/100 g).

2.5. DPPH Free Radical Scavenging (FRS) Assay

The DPPH assay was based on the procedures described in Leong and Shui [31] and Miliauskas *et al.* [32], where the reduction of the DPPH (2,2-diphenyl-1-picrylhydrazyl) radical was measured spectrophotometrically to determine the radical scavenging activity of the extract. Two mL of DPPH solution (5.9 mg in 100 mL methanol) was added to 1 mL of three different concentrations of the sample extract. The absorbance of the solution was measured at 517 nm after a 30 min incubation time. The free radical scavenging (FRS) activity was expressed as ascorbic acid (AA) equivalent antioxidant capacity (mg AA/100 g) using the equation: $FRS = IC_{50(AA)}/IC_{50(sample)} \times 10^5$. $IC_{50(AA)}$ was 0.00387 mg/mL [33].

2.6. Ferric Reducing Power (FRP) Assay

The FRP was determined with potassium ferricyanide, as in the procedure described by Tan and Chan [34]. This assay assessed the ability of any antioxidants present in the extracts to reduce ferric ions (Fe^{3+}) to ferrous ions (Fe^{2+}). One mL of sample extract of different dilutions was added with 2.5 mL of

0.2 M phosphate buffer (pH 6.6) and 2.5 mL of 1% (w/v) potassium ferricyanide. The mixture was incubated in a 50 °C water bath for 20 min. Subsequently, 2.5 mL of 10% (w/v) trichloroacetic acid was added to stop the reaction. Next, the mixture in each test tube was separated into aliquots of 2.5 mL, added to 2.5 mL of miliQ water and 0.5 mL of 0.1% (w/v) $FeCl_3$. The mixtures were incubated at room temperature for 30 min before the absorbance was measured at 700 nm. FRP was expressed as mg gallic acid equivalent per gram of sample, mg GAE/g using a gallic acid standard curve.

2.7. Ferrous Ion Chelating (FIC) Assay

The ferrous ion chelating activity of the extract was based on the procedures described in Mau *et al.* [35], and Singh and Rajini [36]. One mL of 0.1 mM $FeSO_4$ was added to 1 mL of sample of different dilutions (0.2, 0.5, and 1 mL of extract, corresponding to 4 mg, 10 mg, and 20 mg of leaf material, respectively), followed by 1 mL of 0.25 mM ferrozine. The mixtures were incubated at room temperature for 10 min before the absorbance was measured at 562 nm. It was expressed as the percentage of iron chelating activity. EDTA (0.017–0.067 mg/mL) was used as a positive control.

2.8. Determination of Total Flavonoid Content (TFC)

The flavonoid content was determined with the aluminum chloride colorimetric method as described in Chew *et al.* [37]. Equal volumes of 10% aluminum chloride and 1.0 M potassium acetate (0.1 mL each) were added to 0.5 mL of extract, followed by 2.8 mL of distilled water. The solutions were mixed well and incubated at room temperature for 30 min before the absorbance was taken at 435 nm. The flavonoid concentration was expressed as mg rutin equivalent per 100 g sample, mg RE/100 g.

2.9. Total Tannin Content (TTC)

Total tannins were determined based on the method described by Makkar *et al.* [38], which compares the TPC extracts treated with polyvinylpolypyrrolidone (PVPP) with those untreated. Tannins, being protein-binding phenolic compounds, would bind to the PVPP.

The TPC was determined via the Folin–Ciocalteau method as detailed previously. For PVPP treatment, 1 mL of distilled water was added to 100 mg PVPP before adding 1 mL of extract. The mixture was vortexed, incubated at 4 °C for 15 min, vortexed again, and centrifuged at 3000 g for 10 min. The supernatant (consisting of simple phenolics other than tannins) was collected and the phenolic content was determined using the TPC assay. Tannic acid was used to plot the standard curve. Results were expressed as mg equivalent per 100 g sample (mg TAE/100 g).

2.10. Determination of Antimicrobial Activity

The minimum inhibitory concentration (MIC) of the samples was determined using the broth microdilution technique in 96-well flat bottom microtiter plates as described by the Clinical and Laboratory Standards Institute [39], with a few modifications. Nutrient broth (180 µL) was loaded into all the wells of the first column of the 96-well plate, followed by 100 µL of nutrient broth in all the other wells. Twenty µL of each sample type (200 mg/mL stock concentration of leaf extract) was loaded into the first column in triplicate, enabling two samples to be run concurrently on a single plate.

Serial doubling dilution was then performed nine times, keeping the volume of each well at 100 μL. One hundred μL of nutrient broth inoculated with bacteria the day before was standardized with the McFarland standard and then loaded into these wells for a final working concentration of sample ranging from 10 mg/mL to 0.02 mg/mL. The plates were then incubated overnight. The lowest concentration, where complete inhibition was observed with the unaided eye, was noted as the MIC. Vancomycin (10–0.02 mg/mL) was used as a positive control.

2.11. Statistical Analysis

One-way analysis of variance (ANOVA) and a post-hoc Tukey range test were performed to determine significance. A probability value of $p < 0.05$ was considered significant. Analysis was done using SPSS 16 (SPSS Inc., Chicago, IL, USA).

3. Results and Discussion

According to Table 1, *T. zebrina* exhibited the highest TPC, TTC, TFC, FRS, and FRP of all the plants tested in this study. *T. zebrina* methanolic leaf extracts also contained higher TPC, FRS, FRP, and TFC than *R. spathacea* (Swartz) Stearn methanolic leaf extracts [10], showing the highest antioxidant content amongst the Commelinaceae reported to date. Moreover, *T. zebrina* has a TPC comparable to lemon myrtle tea, which was previously reported as the herbal tea with the highest TPC in a study comparing 18 tropical and temperate herbal teas [33]; our extract was prepared from 1 g fresh leaf while the lemon myrtle decoction was prepared from 1 g powder. This finding is relevant, as *T. zebrina* is traditionally used to treat gastrointestinal cancer and disorders [8], and consumed as a tea called "Matali" in Mexico. It is therefore worthy of further investigation and promotion as an herbal tea. In addition, approximately 82% of its phenolic content consisted of tannins, a class of phenolic compounds with potent antioxidant and antibacterial activities [40].

Table 1. Total phenolic content (TPC), total tannin content (TTC), total flavonoid content (TFC), free radical scavenging (FRS), and ferric reducing power (FRP) of the various Commelinaceae leaf extracts.

Species	TPC (mg GAE/100 g) *	TTC (mg TAE/100 g) *	TFC (mg RE/100 g)	FRS (mg AA/100 g)	FRP (mg GAE/g)
R. spathacea variegata	203.9 ± 16.3 ^{a,d} (24.8 ± 2.2) ^{a}	20.6 ± 2.3 ^{a}	10.8 ± 2.9 ^{a}	177.3 ± 15.3 ^{a}	1.4 ± 0.1 ^{a}
R. bermudensis	296.6 ± 21.7 ^{b} (33.5 ± 2.4) ^{b}	11.5 ± 1.7 ^{b}	18.4 ± 2.4 ^{b}	462.2 ± 97.3 ^{b}	2.7 ± 0.3 ^{b}
T. zebrina	620.9 ± 39.7 ^{c} (70.1 ± 4.5) ^{c}	57.6 ± 3.5 ^{c}	17.1 ± 2.8 ^{b}	906.5 ± 88.2 ^{c}	4.8 ± 0.3 ^{c}
T. pallida	153.1 ± 21.8 ^{d} (17.8 ± 2.3) ^{d}	13.6 ± 2.1 ^{b}	10.6 ± 4.0 ^{a}	103.0 ± 36.9 ^{a}	0.9 ± 0.1 ^{a}
C. fragrans	269.3 ± 33.8 ^{a,b} (29.3 ± 2.9) ^{a,b}	19.7 ± 1.5 ^{a}	2.5 ± 1.4 ^{a}	262.5 ± 67.2 ^{a}	1.5 ± 0.3 ^{a}

Results are expressed as mean ± S.D. ($n = 3$); For each column, values followed by the same letter are not significantly different at $p < 0.05$ as measured by the Tukey HSD test; All gram measurements are expressed in terms of fresh weight; * Values in parentheses are in terms of mg TAE/100 g.

C. fragrans and *R. bermudensis* had the second-highest TPC, which was half that of *T. zebrina*. The antioxidant content of the juice pressed from *C. fragrans* leaves has been previously reported as being 73.2 mg/L of juice sample [27] with an IC_{50} of 1.07 mg/mL, based on the DPPH radical scavenging assay [28]. In comparison, the methanolic leaf extract of *C. fragrans* in our study had an IC_{50} of 1.5 ± 0.4; a comparable result with the previously reported value. On the other hand, comparing the antioxidant content was less straightforward because the juice yield was not reported by Misin and Sazhina [27], and it was unclear if the values reported had factored in the 100-fold dilution of their sample. However, using the juice yield reported by Olennikov *et al.* [28] (0.8 L juice per kilogram of leaves), and assuming that the value reported by Misin and Sazhina [27] was based on the undiluted juice, the antioxidant content of the juice would have been 5.85 mg/100 g. Although this value took into account all antioxidants, including non-phenolic antioxidants, it was over 45-fold lower than our methanolic leaf extract. The TPC of the juice has also been reported as 7.12 μg GAE/mL [26], which is equivalent to 0.6 mg GAE/100 g—almost 450 times lower than the TPC of the methanolic leaf extract. Methanolic extraction of *C. fragrans* is therefore more efficient than squeezing tissues for juice, and the leaves of *C. fragrans* contain a higher phenolic content than the runners reported by Olennikov *et al.* [26].

The methanolic extract of *C. fragrans* also exhibited by far the highest FIC activity (Figure 1), with a chelating EC_{50} of 17.3 mg leaf material. *R. spathacea* variegata exhibited the second highest FIC, but did not achieve EC_{50} at the concentrations tested, thus indicating reasonably weak FIC activity. The other species exhibited practically no FIC activity. This indicates that the antioxidants present in some of the Commelinaceae leaves lack the ability to chelate ferrous ions when compared to ferrozine and thus have low or no secondary antioxidant activity, which implies the low ability of the extracts to prevent the Fenton reaction. This shows that the primary antioxidant activity of a plant (as measured by the FRS and FRP assays) does not necessarily reflect its secondary antioxidant activity due to the differences in mechanisms [41]. The primary antioxidant activities of the five species did, however, show a strong positive correlation with the TPC (FRS: $R = 0.98$, $p < 0.001$; FRP: $R = 0.97$, $p < 0.001$) and a good positive correlation with the TTC (FRS: $R = 0.85$, $p < 0.001$; FRP: $R = 0.85$, $p < 0.001$), but not with the TFC (FRS: $R = 0.61$, $p = 0.020$; FRP: $R = 0.65$, $p = 0.012$). This shows that tannins play a stronger role than flavonoids in single electron transfer-based primary antioxidant activity.

Figure 1. Ferrous ion chelating (FIC) activity of the various Commelinaceae leaf extracts.

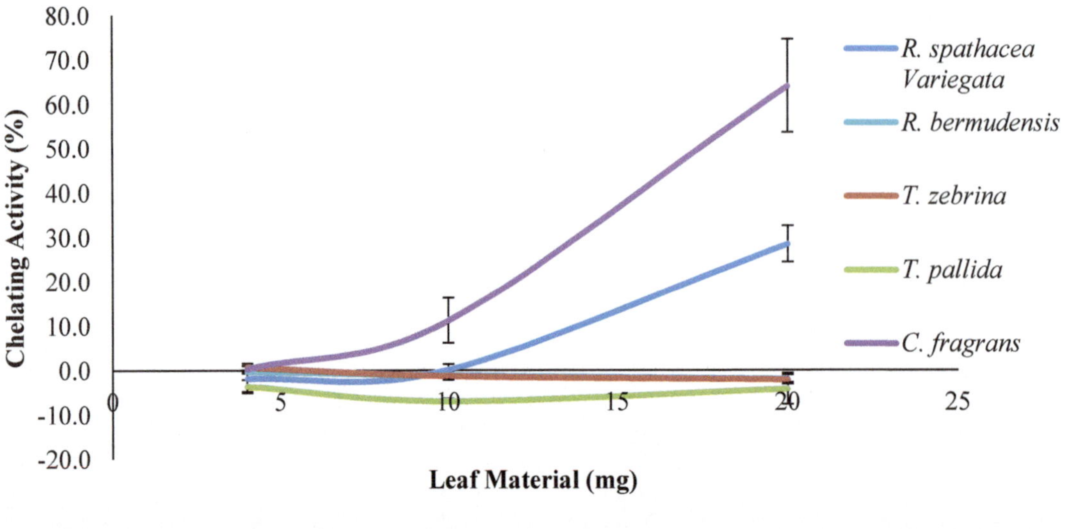

R. bermudensis and *R. spathacea* variegata showed: (1) lower TPC, FRS, and FRP and (2) higher TFC than *R. spathacea* (Swartz) Stearn [10], despite being closely related. Almost 80% of *R. spathacea* variegata's phenolic compounds are in the form of tannins, while only a third of the phenolic compounds in *R. bermudensis* were tannins. Although the TPC of *R. bermudensis* was similar to *C. fragrans* and not much higher than *R. spathacea* variegata, *R. bermudensis* exhibited exceptionally high FRS and FRP activity. It is therefore likely that the phenolic compounds present in *R. bermudensis* are more effective antioxidants and free radical scavengers than those present in the other species.

All five Commelinaceae species demonstrated antibacterial activity against most of the strains, although the six Gram-negative strains proved more resistant to treatment, with only two showing susceptibility at an extract concentration of 10 mg/mL or below (Table 2). The lower susceptibility of the Gram-negative strains can be attributed to their higher outer layer impermeability [42], an observation consistent with the antibacterial activity of *R. spathacea* (Swartz) Stearn [10]. Of the five Commelinaceae species, *T. zebrina* exhibited the best antibacterial activity against Gram-positive bacteria, with a MIC of 5 mg/mL against three strains, while the other four Commelinaceae leaf extracts had a MIC of 5 mg/mL on fewer than three strains. *T. zebrina* also showed antibacterial activity against two of the six Gram-negative strains, with a MIC of 5 mg/mL. The superior antibacterial activity of *T. zebrina* compared to the other tested species may be attributed to its exceptionally high phenolic and tannin content. The superior antioxidant and antibacterial activity of *T. zebrina* compared to the other Commelinaceae may explain several of its ethnopharmalogical applications, particularly its anti-inflammatory and anti-infective activity [43,44]. Surprisingly, *T. pallida* was similarly active against Gram-negative bacteria (MIC of 5 mg/mL against the same two Gram-negative strains) despite having the lowest antioxidant content and activity. It is likely that the phenolic compounds present may be more potent anti-Gram-negative agents in *T. zebrina* than in *T. pallida*, and/or the anti-Gram-negative compounds present may not necessarily be phenolics (for example, terpenoids or alkaloids). The leaves of *T. zebrina* and *T. pallida* warrant future isolation and identification work given their activity against Gram-negative bacteria, as the isolated compound(s) are likely to have a lower MIC than the crude extracts used in this experiment.

Table 2. MIC of Commelinaceae leaf extracts against 12 species of bacteria.

Bacterium	*R. spathacea* Variegata	*R. bermudensis*	*T. zebrina*	*T. pallida*	*C. fragrans*
Gram-Positive					
Bacillus cereus (ATCC 14579)	>10	10	5	5	5
Bacillus subtilis (ATCC 8188)	>10	10	10	10	5
Micrococcus luteus (ATCC 4698)	>10	10	5	10	10
Methicillin-Resistant *Staphylococcus aureus* (ATCC 33591)	>10	10	5	10	10
Staphylococcus epidermidis (ATCC 12228)	5	10	>10	>10	>10

Table 2. *Cont.*

Enterococcus faecalis (ATCC 29212)	10	>10	>10	5	>10
Gram-Negative					
Aeromonas hydrophila (ATCC 49140)	5	>10	5	5	5
Proteus vulgaris (Clinical)	>10	>10	5	5	>10

MIC expressed as mg/mL; Vancomycin MIC < 0.02 mg/mL; *P. aeruginosa* (ATCC 10145), *P. mirabilis* (ATCC 12453), *S. typhimurium* (ATCC 14028), and *E. aerogenes* (ATCC 13048) have MIC exceeding 10 mg/mL for all the crude fractions.

4. Conclusions

T. zebrina exhibited the highest antioxidant content and antioxidant activity of the five Commelinaceae plants studied, higher than the previously-reported antioxidant content and activity of *R. spathacea* (Swartz) Stearn leaves. All five leaf extracts exhibited varying degrees of antibacterial activity against many of the bacteria tested, with *T. zebrina* being the most active against Gram-positive and Gram-negative bacteria.

Acknowledgments

The authors wish to thank Monash University Malaysia for the financial support.

Author Contributions

This project was conceived and designed by Tan Joash Ban Lee and Lim Yau Yan. Tan Joash Ban Lee, Tan Shen Yeng, and Yap Wei Jin performed the experiments. Tan Joash Ban Lee and Lim Yau Yan analyzed the data. Tan Joash Ban Lee, Lim Yau Yan and Lee Sui Mae jointly supervised Tan Shen Yeng and Yap Wei Jin, and provided the resources required for the experiments (reagents/facilities/materials). Tan Joash Ban Lee wrote the paper, and all parties contributed to the editing process.

Conflicts of Interest

The authors declare no conflict of interest.

References

1. Edeoga, H.O.; Ogbebor, N.O. Distribution of calcium oxalate crystals in some Nigerian species of *Aneilema* R. Br. (Commelinaceae). *Plant Biosyst.* **1999**, *133*, 193–198.
2. Satterfield, S.K.; Mertens, T.R. *Rhoeo spathacea*: A tool for teaching meiosis and mitosis. *J. Hered.* **1972**, *63*, 375–378.
3. Faden, R.B. Commelinaceae. In *Flowering Plants—Monocotyledons*; Kubitzki, K., Ed.; Springer: Berlin-Heidelberg, Germany, 1998; Volume 4, pp. 109–128.
4. Wilson, A.K. Commelinaceae—A review of the distribution, biology and control of the important weeds belonging to this family. *Trop. Pest Manag.* **1981**, *27*, 405–418.

5. González-Avila, M.; Arriaga-Alba, M.; de la Garza, M.; del Carmen HernándezPretelín, M.; Domínguez-Ortíz, M.A.; Fattel-Fazenda, S.; Villa-Treviño, S. Antigenotoxic, antimutagenic and ROS scavenging activities of a *Rhoeo discolor* ethanolic crude extract. *Toxicol. In Vitro* **2003**, *17*, 77–83.
6. Arriaga-Alba, M.; Blasco, J.L.; Ruíz-Pérez, N.J.; Sánchez-Navarrete, J.; Rivera-Sánchez, R.; González-Avila, M. Antimutagenicity mechanisms of the *Rhoeo discolor* ethanolic extract. *Exp. Toxicol. Pathol.* **2011**, *63*, 243–248.
7. Mensah, A.Y.; Houghton, P.J.; Dickson, R.A.; Fleischer, T.C.; Heinrich, M.; Bremner, P. *In vitro* evaluation of effects of two Ghanaian plants relevant to wound healing. *Phytother. Res.* **2006**, *20*, 941–944.
8. Alonso-Castro, A.J.; Villarreal, M.L.; Salazar-Olivo, L.A.; Gomez-Sanchez, M.; Dominguez, F.; Garcia-Carranca, A. Mexican medicinal plants used for cancer treatment: Pharmacological, phytochemical and ethnobotanical studies. *J. Ethnopharmacol.* **2011**, *133*, 945–972.
9. Perry, C.C. An overview of silica in biology: Its chemistry and recent technological advances. In *Biosilica in Evolution, Morphogenesis, and Nanobiotechnology*; Müller, W.E.G., Grachev, M.A., Eds.; Springer: Berlin-Heidelberg, Germany, 2009; Volume 47, pp. 295–313.
10. Tan, J.B.L.; Lim, Y.Y.; Lee, S.M. Antioxidant and antibacterial activity of *Rhoeo spathacea* (Swartz) Stearn leaves. *J. Food Sci. Technol.* **2013**, doi:10.1007/s13197-013-1236-z.
11. Xeriscape Plants. Available online: http://www.ctahr.hawaii.edu/oc/freepubs/pdf/OF-42.pdf (accessed on 10 May 2014).
12. Golczyk, H. Cytogenetics of the permanent translocation heterozygote *Rhoeo spathacea* var. *variegata*. Implications for complex chromosome rearrangements in *Rhoeo*. *Caryologia* **2011**, *64*, 325–334.
13. Paiva, E.A.S.; Isaias, R.M.S.; Vale, F.H.A.; Queiroz, C.G.S. The influence of light intensity on anatomical structure and pigment contents of *Tradescantia pallida* (Rose) Hunt. cv. *purpurea* Boom (Commelinaceae) leaves. *Braz. Arch. Biol. Technol.* **2003**, *46*, 617–624.
14. Rainho, C.R.D.; Kaezer, A.; Aiub, C.A.F.; Felzenszwalb, I. Ability of *Allium cepa* L. root tips and *Tradescantia pallida* var. *purpurea* in *N*-nitrosodiethylamine genotoxicity and mutagenicity evaluation. *An. Acad. Bras. Ciênc.* **2010**, *82*, 925–932.
15. Misik, M.; Ma, T.H.; Nersesyan, A.; Monarca, S.; Kim, J.K.; Knasmueller, S. Micronucleus assays with *Tradescantia* pollen tetrads: An update. *Mutagenesis* **2011**, *26*, 215–221.
16. Thewes, M.R.; Junior, D.E.; Droste, A. Genotoxicity biomonitoring of sewage in two municipal wastewater treatment plants using the *Tradescantia pallida* var. *purpurea* bioassay. *Genet. Mol. Biol.* **2011**, *34*, 689–693.
17. Yang, D.S.; Pennisi, S.V.; Son, K.C.; Kays, S.J. Screening indoor plants for volatile organic pollutant removal efficiency. *HortScience* **2009**, *44*, 1377–1381.
18. Li, T.S.C. Taiwanese Native Medicinal Plant: Phytopharmacology and Therapeutic Values; CRC Press: Boca Raton, FL, USA, 2006; pp. 1–328.
19. Faden, R.B. The author and typification of *Tradescantia zebrina* (Commelinaceae). *Kew Bull.* **2008**, *63*, 679–680.
20. Yanzhi, M. Pigment content and anatomical structure of leaves of several species of red-leafed plants. *J. N. E. Forest. Univ.* **2009**, *37*, 51-53.

21. Glimn-Lacy, J.; Kaufman, P.B. Spiderwort family (Commelinaceae). In *Botany Illustrated*; Springer US: New York, NY, USA, 2006; pp.121–121.
22. Amaral, F.M.M.; Ribeiro, M.N.S.; Barbosa-Filho, J.M.; Reis, A.S.; Nascimento, F.R.F.; Macedo, R.O. Plants and chemical constituents with giardicidal activity. *Braz. J. Pharmacog.* **2006**, *16*, 696–720.
23. González-Coloma, A.; Reina, M.; Sáenz, C.; Lacret, R.; Ruiz-Mesia, L.; Arán, V.J.; Sanz, J.; Martinez-Diaz, R.A. Antileishmanial, antitrypanosomal, and cytotoxic screening of ethnopharmacologically selected Peruvian plants. *Parasitol. Res.* **2012**, *110*, 1381–1392.
24. Alaba, C.S.M.; Chichioco-Hernandez, C.L. 15-Lipoxygenase inhibition of Commelina benghalensis, Tradescantia fluminensis, Tradescantia zebrina. *Asian. Pac. J. Trop. Biomed.* **2014**, 4, 184–188.
25. Chernenko, T.V.; Ul'chenko, N.T.; Glushenkova, A.I.; Redzhepov, D. Chemical investigation of *Callisia fragrans*. *Chem. Nat. Compd.* **2007**, *43*, 253–255.
26. Olennikov, D.N.; Ibragimov, T.A.; Zilfikarov, I.N.; Chelombit'ko, V.A. Chemical composition of *Callisia fragrans* juice 1. Phenolic compounds. *Chem. Nat. Compd.* **2008**, *44*, 776–777.
27. Misin, V.M.; Sazhina, N.N. Content and activity of low-molecular antioxidants in juices of medicinal plants. *Russ. J. Phys. Chem. B* **2010**, *4*, 797–800.
28. Olennikov, D.N.; Zilfikarov, I.N.; Toropova, A.A.; Ibragimov, T.A. Chemical composition of *Callisia fragrans* wood. juice and its antioxidative activity (*in vitro*). *Chem. Plant Raw Mater.* **2008**, *4*, 95–100.
29. Malaysian Meteorological Department. Available online: http://www.met.gov.my/index.php?option=com_content&task=view&id=34&Itemid=1586 (accessed on 22 October 2014).
30. Kähkönen, M.P.; Hopia, A.I.; Vuorela, H.J.; Rauha, J.P.; Pihlaja, K.; Kujala, T.S.; Heinonen, M. Antioxidant activity of plant extracts containing phenolic compounds. *J. Agric. Food Chem.* **1999**, *47*, 3954–3962.
31. Leong, L.P.; Shui, G. An investigation of antioxidant capacity of fruits in Singapore markets. *Food Chem.* **2002**, *76*, 69–75.
32. Miliauskas, G.; Venskutonis, P.R.; van Beek, T.A. Screening of radical scavenging activity of some medicinal and aromatic plant extracts. *Food Chem.* **2004**, *85*, 231–237.
33. Chan, E.W.C.; Lim, Y.Y.; Chong, K.L.; Tan, J.B.L.; Wong, S.K. Antioxidant properties of tropical and temperate herbal teas. *J. Food Comp. Anal.* **2010**, *23*, 185–189.
34. Tan, Y.P.; Chan, E.W.C. Antioxidant, antityrosinase and antibacterial properties of fresh and processed leaves of *Anacardium occidentale* and *Piper betle*. *Food Biosci.* **2014**, *6*, 17–23.
35. Mau, J.L.; Lai, E.Y.C.; Wang, N.P.; Chen, C.C.; Chang, C.H.; Chyau, C.C. Composition and antioxidant activity of the essential oil from *Curcuma zedoaria*. *Food Chem.* **2003**, *82*, 583–591.
36. Singh, N.; Rajini, P.S. Free radical scavenging activity of an aqueous extract of potato peel. *Food Chem.* **2004**, *85*, 611–616.
37. Chew, Y.L.; Goh, J.K.; Lim, Y.Y. Assessment of *in vitro* antioxidant capacity and polyphenolic composition of selected medicinal herbs from Leguminosae family in peninsular Malaysia. *Food Chem.* **2009**, *116*, 13–18.
38. Makkar, H.P.S.; Siddhuraju, P.; Becker, K. *Plant Secondary Metabolites*; Humana Press: Totowa, NJ, USA, 2007; pp. 67–82.

39. Clinical and Laboratory Standards Institute. Methods for dilution antimicrobial susceptibility tests for bacteria that grow aerobically; approved standard—Eight edition. *CLSI Doc.* **2009**, *29*, 15–18.
40. Michel, T.; Destandau, E.; Le Floch, G.; Lucchesi, M.E.; Elfakir, C. Antimicrobial, antioxidant and phytochemical investigations of sea buckthorn (*Hippophaë rhamnoides* L.) leaf, stem, root and seed. *Food Chem.* **2012**, *131*, 754–760.
41. Lai, H.Y.; Lim, Y.Y. Evaluation of antioxidant activities of the methanolic extracts of selected ferns in Malaysia. *Int. J. Environ. Sci. Dev.* **2011**, *2*, 442–447.
42. Russell, A.D. Bacterial resistance to disinfectants: Present knowledge and future problems. *J. Hosp. Infect.* **1999**, *43*, S57–68.
43. Frei, B.; Heinrich, M.; Bork, P.M.; Herrmann, D.; Jaki, B.; Kato, T.; Kuhnt, M.; Schmitt, J.; Schühly, W.; Volken, C.; *et al.* Multiple screening of medicinal plants from Oaxaca, Mexico: Ethnobotany and bioassays as a basis for phytochemical investigation. *Phytomedicine* **1998**, *5*, 177–186.
44. Tene, V.; Malagón, O.; Finzi, P.V.; Vidari, G.; Armijos, C.; Zaragoza, T. An ethnobotanical survey of medicinal plants used in Loja and Zamora-Chinchipe, Ecuador. *J. Ethnopharmacol.* **2007**, *111*, 63–81.

9

An Optimised Aqueous Extract of Phenolic Compounds from Bitter Melon with High Antioxidant Capacity

Sing Pei Tan [1,*], Costas Stathopoulos [2], Sophie Parks [1,3] and Paul Roach [1]

[1] School of Environmental and Life Sciences, University of Newcastle, Ourimbah, NSW 2258, Australia; E-Mails: sophie.parks@dpi.nsw.gov.au (S.P.); paul.roach@newcastle.edu.au (P.R.)

[2] Faculty of Bioscience Engineering, Ghent University Global Campus, Incheon 406-840, Korea; E-Mail: costas.stathopoulos@ghent.ac.kr

[3] Central Coast Primary Industries Centre, NSW Department of Primary Industries, Ourimbah, NSW 2258, Australia

* Author to whom correspondence should be addressed; E-Mail: sing.p.tan@uon.edu.au

External Editors: Antonio Segura-Carretero and David Arráez-Román

---

Abstract: Bitter melon (*Momordica charantia* L.) is a tropical fruit claimed to have medicinal properties associated with its content of phenolic compounds (TPC). The aim of the study was to compare water with several organic solvents (acetone, butanol, methanol and 80% ethanol) for its efficiency at extracting the TPC from freeze-dried bitter melon powder. The TPC of the extracts was measured using the Folin-Ciocalteu reagent and their antioxidant capacity (AC) was evaluated using three assays. Before optimisation, the TPC and AC of the aqueous extract were 63% and 20% lower, respectively, than for the best organic solvent, 80% ethanol. However, after optimising for temperature (80 °C), time (5 min), water-to-powder ratio (40:1 mL/g), particle size (1 mm) and the number of extractions of the same sample (1×), the TPC and the AC of the aqueous extract were equal or higher than for 80% ethanol. Furthermore, less solvent (40 mL water/g) and less time (5 min) were needed than was used for the 80% ethanol extract (100 mL/g for 1 h). Therefore, this study provides evidence to recommend the use of water as the solvent of choice for the extraction of the phenolic compounds and their associated antioxidant activities from bitter melon.

Keywords: bitter melon; phenolic compounds; antioxidant capacity; aqueous extraction; organic solvents

1. Introduction

Bitter melon (*Momordica charantia* L.) is a popular medicinal fruit particularly in Asia and Africa, where many varieties are grown. For example, bitter melon has been associated with anti-cancer [1], anti-microbial [2], anti-inflammatory [3] and anti-diabetic properties [4]. The potential medicinal values of the fruit have been linked to its content of phenolics and their antioxidant properties [4–7].

The phenolic structure is comprised of an aromatic group with one or more hydroxyl groups [8]. Owing to the hydroxyl groups, most phenolics in bitter melon are hydrophilic compounds, such as gallic acid, gentisic acid, catehcin, chlorogenic acid, epicatechin, vanillin acid, protocatechuic acid, *p*-coumaric acid, *o*-coumaric acid, and *t*-cinnamic acid [9,10].

Studies have reported that phenolics have potent antioxidant and free radical-scavenging activities [11,12]. Whole bitter melon (flesh, aril and seeds) has been shown to be a good source of phenolic compounds [9] and one study demonstrated that the flesh, aril and seeds all had very high antioxidant activity [9]. Another study suggested that the antioxidant and free radical scavenging properties in the extracts of bitter melon could be attributed to flavonoids and other phenolic compounds [11]. The results also revealed that, on a weight per weight basis, the bitter melon had better 2,2-diphenyl-1-picrylhydrazyl (DPPH) radical scavenging and ferric reducing antioxidant power (FRAP) than vitamin E [11].

The phenolic compounds extracted from plant matrices and their associated antioxidant activities are dependent on the parameters of the method used, including the solvent type and polarity, the temperature, the length of the extraction, the number of times a sample is extracted, and on the condition of the raw material, including its particle size [13,14]. To date, several solvents, including methanol [15], water [12] and the combination of ethanol and water [10], have been used for extracting phenolic compounds from bitter melon.

Although studies have reported that organic solvents can be more effective than water at extracting phenolic compounds from plant materials [16,17], water is a more desirable extractant for the food industry because it is non-toxic, environmentally friendly and inexpensive compared to organic solvents [18].

To our knowledge, no published study has comprehensively investigated and optimised the aqueous extraction of phenolic compounds and their associated antioxidant activities from bitter melon and compared the optimised aqueous extraction of these compounds with that achieved with organic solvents. Therefore, this study aimed to maximise the aqueous extraction of the phenolic compounds from freeze-dried and powdered bitter melon and to compare the optimised aqueous extract with that obtained with the best organic solvent from methanol, 80% ethanol, butanol and acetone.

The conditions investigated for the aqueous extraction of the phenolic compounds from the powdered bitter melon were temperature, time, water-to-powder ratio, powder particle size and the number of extractions of the same sample. The optimised aqueous extract was then compared to the best of the organic solvent extract in terms of total phenolic content (TPC) and antioxidant capacity (AC). The correlation between the TPC and the AC of the extracts was also determined.

2. Materials and Methods

2.1. Plant Material

Bitter melon (Moonlight variety) fruits (20 kg), grown in the Northern Territory (Darwin, NT, Australia), were purchased from the Sydney markets (Sydney, NSW, Australia) and frozen at −20 °C until used. The frozen bitter melons were cut into slices (~1 to 2 mm) and placed in liquid nitrogen before drying in a FD3 freeze dryer (Rietschle Thomas, Seven Hills, NSW, Australia) at −40 °C for 72 h at 2×10^{-1} mbar. After freeze-drying, samples were ground into a powder using a commercial blender (John Morris Scientific, Chatswood, NSW, Australia). The powders were mixed thoroughly and sorted into six particle sizes (<0.25, 0.5–0.25, 1–0.5, 1–2, 2–2.8 and >2.8 mm) by sieving through a series of EFL 2000 stainless steel sieves (Endecotts Limited, London, England). The powders were then kept in sealed containers at −20 °C until used.

2.2. Chemicals

Methanol and acetone were obtained from Merck (Kilsyth, VIC, Australia). Ethanol was purchased from Fronine (Taren Point, NSW, Australia). Hydrochloric acid (36%) was obtained from Ajax Finechem (North Ryde, NSW, Australia) and *n*-butanol was purchased from Swift Australia Chemical Supplier (Mulgrave, VIC, Australia). Folin-Ciocalteau (FC) reagent, 2,2′-azinobis-(3-ethylbenzothiozoline-6-sulfonic acid (ABTS), 2,2′-diphenyl-1-picrylhydrazyl (DPPH), sodium carbonate, sodium nitrite, aluminium chloride, sodium hydroxide, vanillin, sodium acetate trihydrate, acetic acid, 2,4,6-tripyridyl-*s*-triazine (TPTZ), ferric (III) chloride hexahydrate, sulphuric acid and standards (trolox, gallic acid, rutin and aecsin) were purchased from Sigma-Aldrich (Castle Hill, NSW, Australia).

2.3. Initial Solvent Extractions

The overall experimental design for the study is illustrated in Figure 1. Initially, extractions were done with five solvents: acetone, *n*-butanol, 80% ethanol (80 ethanol/20 water, v/v), methanol and deionised water. Bitter melon extracts were prepared by adding 1 g of freeze-dried bitter melon powder to 100 mL of solvent and extracting for 1 h using a shaking water bath (Ratek Instruments, Boronia, VIC, Australia). The extraction temperatures were set below the boiling point of each solvent: 50 °C for acetone, 60 °C for methanol and 80 °C for *n*-butanol, 80% ethanol and deionised water. After extraction, the samples were allowed to cool and settle on ice for 10 min. The extracts were then centrifuged at 4350× *g* for 10 min at 10 °C (Beckman Instruments Inc., Palo Alto, CA, USA) and the supernatant from each sample was filtered through a 0.45 μm syringe filter (Phenomenex, Pennants Hills, NSW, Australia) prior to analysis. All extractions were conducted in triplicate.

Figure 1. Diagram for the experimental design. Initially, the total phenolic content (TPC) and antioxidant capacity (AC) of the extracts of freeze-dried bitter melon powder obtained with five different solvents, including water, were compared. The aqueous extraction conditions were then optimised in terms of TPC and AC and compared to the 80% ethanol extract, which had the highest values in the initial solvent screening experiment.

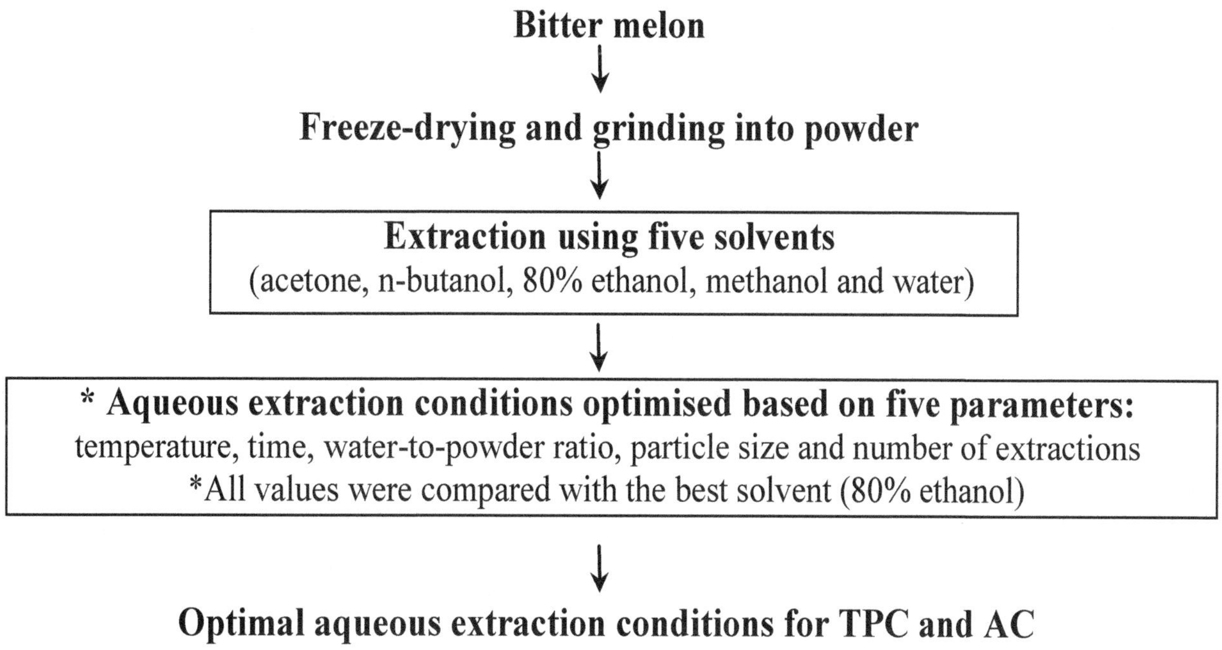

2.4. Optimising the Aqueous Extraction

The aqueous extraction was then optimised by testing, in sequential order, a range of temperatures, times, water-to-powder ratios, particle sizes and the number of times the same sample was extracted.

To determine the effect of the extraction temperature on the aqueous extraction of phenolic compounds, 1 g of freeze-dried bitter melon powder was extracted with 100 mL of deionised water at 5 °C and at intervals of 10° from 10 to 90 °C for 1 h using the shaking water bath.

The optimal temperatures for extraction (40 and 80 °C) were then used to determine the effect of the extraction time; 1 g of freeze-dried bitter melon powder was extracted with 100 mL of deionised water at 40 and 80 °C for 5, 10, 15, 20, 25, 30, 40, 50 and 60 min.

The optimal combination of time (5 min) and temperature (80 °C) was then used to determine the effect of the water-to-powder ratio; 1 g of freeze-dried powder was extracted in 10, 20, 25, 30, 40, 50 and 100 mL of deionised water.

To determine the effect of the freeze-dried particle size, the optimum conditions for temperature (80 °C), time (5 min) and water-to-powder ratio (40:1 mL/g) were used to extract 2.5 g of ground freeze-dried bitter melon having various sizes (<0.25, 0.25–0.50, 0.50–1.0, 1.0–2.0, 2.0–2.8 and >2.8 mm in diameter) with 100 mL of deionised water.

Finally, to evaluate the effect of extracting the same sample several times, 2.5 g of ground freeze-dried powder (1 mm) was extracted one, two or three times with 100 mL of deionised water at 80 °C for 5 min.

After each of the aqueous extraction experiments, the samples were allowed to cool on ice for 10 min before they were centrifuged and their supernatant was filtered through a 0.45 μm syringe filter (Phenomenex, Pennants Hills, NSW, Australia) prior to analysis.. All experiments were conducted in triplicate.

2.5. Extraction Efficiency

From the initial extractions done with the five solvents (acetone, *n*-butanol, 80% ethanol, methanol and deionised water), it was found that the 80% ethanol extract had the highest TPC and AC. Therefore, the 80% ethanol extract was chosen as the control extract for comparing the extraction efficiency for the various aqueous extractions—in terms of TPC and antioxidant capacity.

The extraction efficiency for the aqueous extractions, with respect to their TPC and AC, was calculated and expressed as a percentage of the 80% ethanol (EtOH) extract as follows:

Extraction efficiency % = (TPC or AC for aqueous extract/TPC or AC for EtOH extract) × 100.

2.5.1. Total Phenolic Content

The TPC of samples was determined according to [19] with some modifications. Briefly, 300 μL of appropriately diluted bitter melon extract or standard solution (appropriate for each of the five extracting solvents, including a blank) was mixed and incubated with 300 μL of FC solution for 2 min before 2400 μL of a 50 g/L sodium carbonate solution was added. The solutions were mixed well and placed in the dark at room temperature for 2 h before the absorption was measured at 765 nm using a spectrophotometer (Carry 50 Bio, Varian Pty. Ltd., Mulgrave, VIC, Australia). Gallic acid was used as the standard and the TPC was expressed as mg gallic acid equivalents (GAE) per g of dry powder weight (mg GAE/g).

2.5.2. ABTS Assay

The ABTS assay was conducted according Re *et al.* [20] with some modifications. Stock solutions of 7.4 mmol/L ABTS and 2.6 mmol/L potassium persulfate in deionised water were prepared and kept at 4 °C until used. Fresh working solution was prepared for each assay by mixing the two stock solutions in equal quantities and incubating for 12–16 h in the dark at room temperature. Then, 1 mL of the working solution was mixed with 60 mL methanol to obtain an absorbance of 1.1 ± 0.02 units at 734 nm. Each appropriately diluted bitter melon extract and standard solution (150 μL) was mixed and incubated with 2850 μL of the working solution for 2 h in the dark at room temperature before the absorption was measured at 734 nm. Trolox was used as a standard and the AC was expressed as μmol of trolox equivalents (TE) per g of dry powder weight (μmol TE/g).

2.5.3. DPPH Assay

The DPPH assay was conducted according to Brand-Williams *et al.* [21] with some modifications. In brief, a stock solution of 0.6 mol/L DPPH in methanol was prepared and kept at −20 °C until used. Fresh working solution was prepared for each assay by mixing 10 mL of stock solution with 45 mL of methanol to obtain an absorbance of 1.1 ± 0.02 units at 515 nm. Each appropriately diluted bitter melon extract and standard solution (150 μL) was mixed and incubated with 2850 μL of working solution for 30 min at room temperature before the absorption was measured at 515 nm. Trolox was used as a standard and the AC was expressed as μmol trolox equivalents (TE) per g of dry powder weight (μmol TE/g).

2.5.4. FRAP Assay

The FRAP assay was conducted according to Benzie and Strain [22] with some modifications. The stock solutions of 300 mmol/L acetate buffer pH 3.6, 10 mmol/L TPTZ in 40 mmol/L HCl and 20 mmol/L ferric (III) chloride hexahydrate were prepared and kept at 4 °C until used. Fresh working solution was prepared for each assay by mixing 100 mL of acetate buffer, 10 mL of TPTZ and 10 mL of ferric (III) chloride hexahydrate in a ratio of 10:1:1 and incubated at 37 °C before it was used. Each appropriately diluted bitter melon extract and standard solution (150 μL) was mixed and incubated with 2850 μL of the working solution for 30 min in the dark at room temperature before the absorption was measured at 593 nm. Trolox was used as a standard and AC was expressed as μmol trolox equivalents (TE) per g of dry powder weight (μmol TE/g).

2.6. Statistical Analysis

The Statistical Package for Social Science (SPSS) Version 19 (IBM Australia Limited, St Leonards, NSW, Australia) was used for data analyses. All experiments were conducted in triplicate and the one-way ANOVA and the Bonferroni *post-hoc* test were used to determine any significant differences between the mean values for the different treatments within an experiment. The Pearson correlation test was performed to establish the significance of correlations between the TPC and AC.

3. Results

3.1. Initial Solvent Extractions

Initially, the freeze-dried bitter melon powder was extracted with five solvents: acetone, *n*-butanol, 80% ethanol, methanol and deionised water. The 80% ethanol extract had the highest TPC (Figure 2) and the highest AC in all three antioxidant assays (Figure 3). The methanol extract also exhibited the highest TPC (Figure 2) and the highest AC in the DPPH and FRAP antioxidant assays but had a significantly lower AC than the 80% ethanol extract in the ABTS assay (Figure 3). The extraction of phenolic compounds by water was significantly less effective than for 80% ethanol and methanol but was about three times more effective than for butanol and acetone.

The AC for the aqueous extract was also much higher than for the butanol and acetone extracts in all three antioxidant assays (Figure 3). There was no significant difference between the ABTS values for the aqueous and the methanol extracts but the AC for the aqueous extract was lower than for the 80% ethanol extract (Figure 3). In the DPPH and FRAP assays, the aqueous extract had a significantly lower AC than both the methanol and 80% ethanol extracts (Figure 3). Overall, strong correlations were observed between the TPC of the various solvent extracts (Figure 2) and their AC (Figure 3) as measured with the three assays: ABTS ($R = 0.94$), DPPH ($R = 0.95$) and FRAP ($R = 0.99$).

Figure 2. Total phenolic content of bitter melon extracts obtained with five solvents. Values are means ± standard deviations (n = 3) and those not sharing a superscript letter are significantly different ($p < 0.05$) from each other.

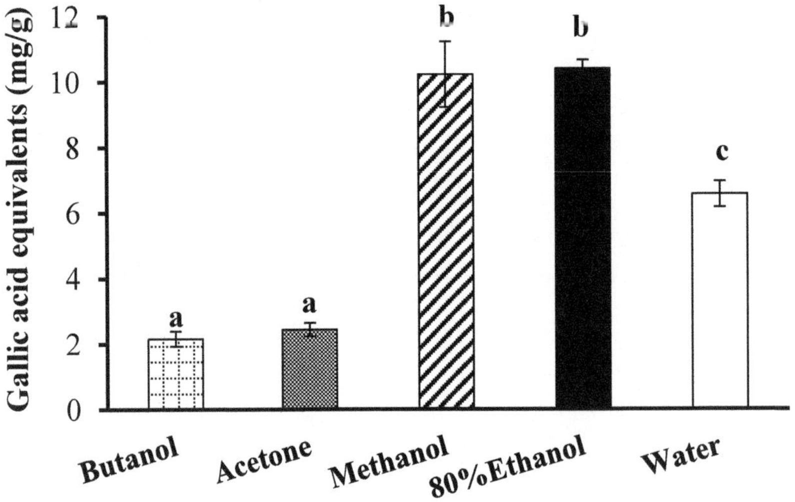

Figure 3. The 2,2′-azinobis-(3-ethylbenzothiozoline-6-sulfonic acid (ABTS), 2,2′-diphenyl-1-picrylhydrazyl (DPPH) and ferric reducing antioxidant power (FRAP) assays were used to determine the antioxidant capacity. Values are means ± standard deviations (n = 3) and those not sharing a superscript letter are significantly different ($p < 0.05$) from each other.

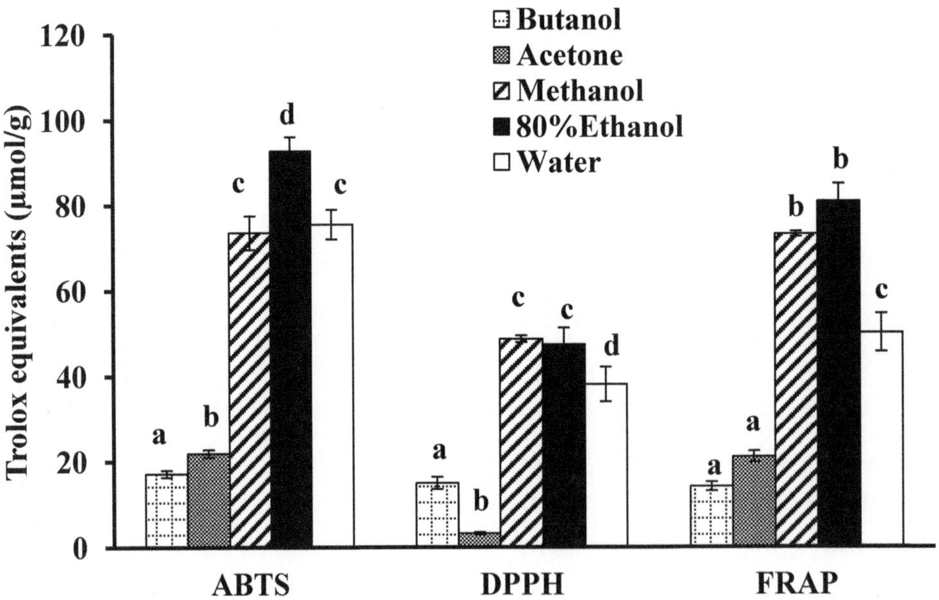

3.2. Optimising the Aqueous Extraction

The optimal conditions for the aqueous extraction of the bitter melon phenolic compounds and their antioxidant activities were then determined by testing, in sequential order, a range of temperatures, times, water-to-powder ratios, particle sizes and the number of times the same sample was extracted. The results, expressed as extraction efficiency relative to the 80% ethanol extract, for TPC are presented in Figure 4 and those for AC in Figure 5.

Figure 4. The extraction efficiency for phenolic compounds from bitter melon using water. The aqueous extraction was optimised using one-variable-at-a-time method and each variable was tested in sequential order: (**A**) temperature; (**B**) time; (**C**) water-to-powder ratio; (**D**) powder particle size and (**E**) number of times the same sample is extracted. The powder particle size categories were: (1) <0.25, (2) 0.25–0.05, (3) 0.5–1.0, (4) 1.0–2.0, (5) 2.0–2.8 and (6) >2.8 mm. The extraction efficiency was relative to the 80% ethanol extract and data are means ± standard deviations ($n = 3$) and those not sharing a letter are significantly different ($p < 0.05$) from each other. * This value is also significantly different ($p < 0.05$) from the values obtained at 40 °C.

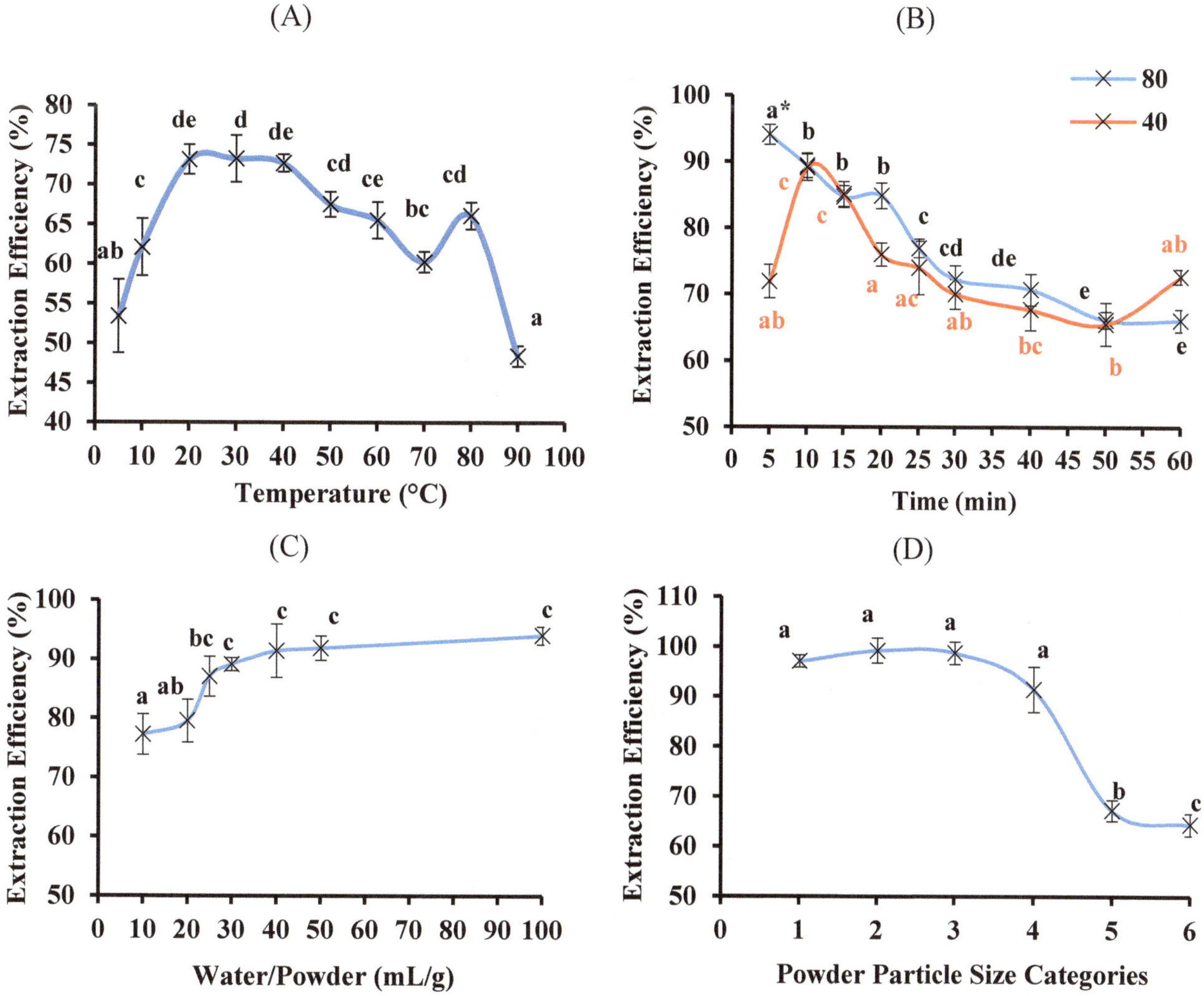

Figure 4. *Cont.*

(E)

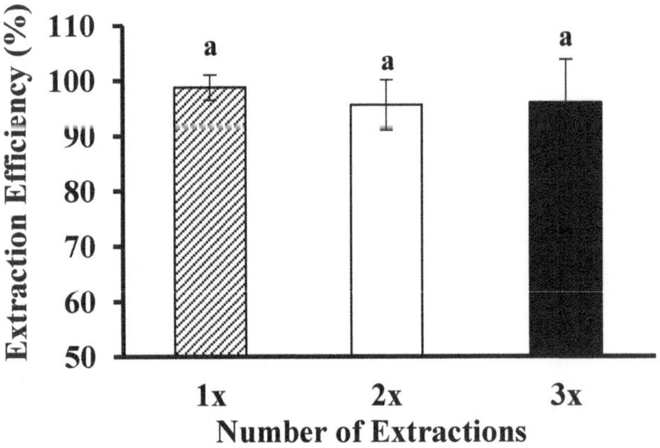

The extraction efficiency for TPC was highest at temperatures between 20 and 50 °C and at 80 °C (Figure 4A). Therefore, the extraction time experiment was done at two temperatures, 40 and 80 °C, which showed that 80 °C for 5 min was the optimal combination of temperature and time (Figure 4B). Notably, the extraction efficiency for TPC obtained for 5 min at 80 °C was higher than the highest efficiency at 40 °C, which was obtained at 10 min (Figure 4B). Importantly, only 5 min was needed for the extraction, which is a much shorter time than the 1 h usually used for organic solvent extractions.

Using 80 °C for 5 min, it was then found that values between 25 and 100 mL/g for the water-to-powder ratio gave the highest extraction efficiency for TPC (Figure 4C). From these results, a water-to-powder ratio of 40 mL/g was chosen as being in the optimal range. Therefore, less water was needed than 100 mL/g, which is usually used for organic solvent extractions.

The extraction conditions of 80 °C for 5 min and a water-to-powder ratio of 40 mL/g were then used for testing the effect of powder particle size. As seen in Figure 4D, the extraction efficiency for TPC was higher at powder particle sizes ≤2.0 mm in diameter compared to sizes ≥2.0 mm. Therefore, a powder particle size of ≤1.0 mm was chosen as being in the optimal range.

Finally, the number of times the same powder same is extracted was tested using a powder particle diameter size of ≤1.0 mm with a water-to-powder ratio of 40 mL/g at 80 °C for 5 min. There was no significant difference in the extraction efficiency for TPC whether a sample for extracted once, twice or thrice (Figure 4E) and thus, a single step was considered optimal.

Similarly, for the AC of the aqueous extracts, the highest extraction efficiency was observed at 80 °C for the DPPH and FRAP assays (Figure 5A) while the extraction temperature had less of an effect on the AC measured using the ABTS assay. Therefore, these results supported the choice of 80 °C as the optimal temperature. Furthermore, doing the aqueous extraction at 80 °C for 5 min also gave optimal AC values (Figure 5B). Subsequently, a water-to-powder ratio of 40 mL/g (Figure 5C), a powder particle diameter size of ≤1.0 mm (Figure 5D) and one extraction (Figure 5D) proved to be optimal for the aqueous extraction done at 80 °C for 5 min, in terms of AC measured by the three antioxidant assays, the same optimal conditions as for TPC (Figure 4).

Figure 5. The extraction efficiency for antioxidant capacity from bitter melon using water. The aqueous extraction was optimised using one-variable-at-a-time method and each variable was tested in sequential order: (**A**) temperature; (**B**) time; (**C**) water-to-powder ratio; (**D**) powder particle size and (**E**) number of times the same sample is extracted. The powder particle size categories were: (1) <0.25, (2) 0.25–0.05, (3) 0.5–1.0, (4) 1.0–2.0, (5) 2.0–2.8 and (6) >2.8 mm. The extraction efficiency was relative to the 80% ethanol extract and data are means ± standard deviations (n = 3) and those not sharing a letter are significantly different ($p < 0.05$) from each other.

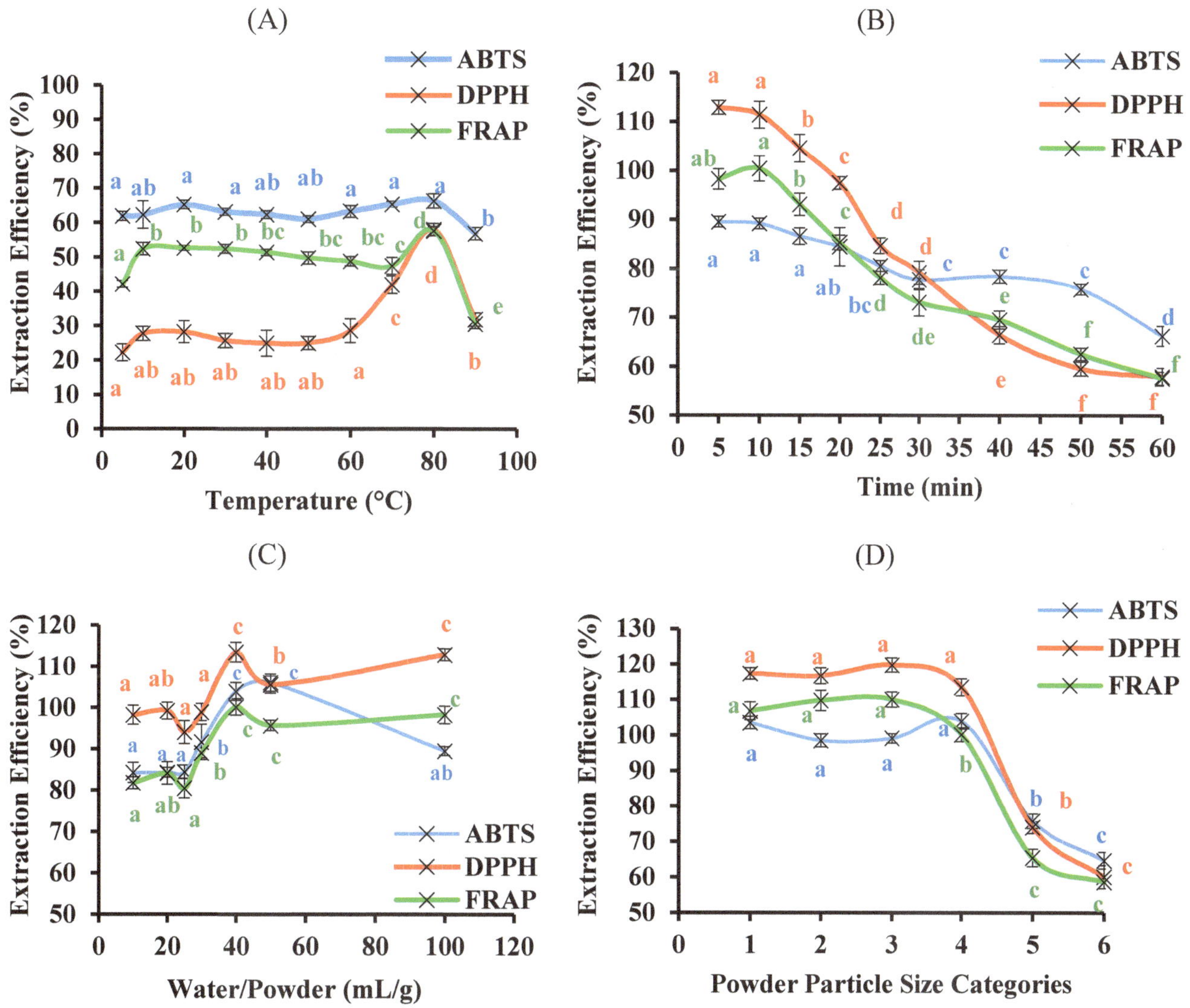

Figure 5. *Cont.*

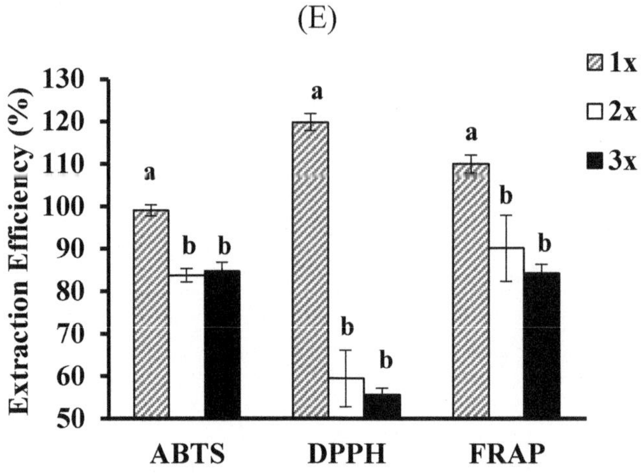

As seen in Table 1, the TPC and the AC of the aqueous extract were substantially improved under the optimised conditions compared to the values shown in Figures 2 and 3, respectively. Importantly, this meant that the TPC and the AC (measured with the ABTS assay) of the optimised aqueous extract were also the same as for the 80% ethanol extract (Table 1). Furthermore, the AC of the aqueous extract was 20% and 10% higher for the DPPH and FRAP assays, respectively, compared to the 80% ethanol extract (Table 1).

Table 1. The phenolic compound content (TPC) and the antioxidant capacity of the optimised aqueous extract of bitter melon compared with the 80% ethanol extract.

Chemical Properties	Aqueous Extract	80% Ethanol Extract
TPC (mg GAE/g dry basis)	10.6 ± 0.2^{a}	10.7 ± 0.3^{a}
ABTS (μmol TE/g dry basis)	94.8 ± 1.3^{a}	95.7 ± 3.5^{a}
DPPH (μmol TE/g dry basis)	58.6 ± 1.0^{a}	49.0 ± 4.0^{b}
FRAP (μmol TE/g dry basis)	91.9 ± 1.8^{a}	83.5 ± 4.3^{b}

Values are means ± standard deviations ($n = 3$) and those in a row not sharing a superscript letter are significantly different from each other ($p < 0.05$).

Finally, the TPC of the various aqueous extracts was shown to have a high association with AC measured with the three antioxidant assays (Table 2). However, the association with TPC was 10% and 15% lower for the AC measured with the ABTS and the DPPH assays, respectively, compared to with the FRAP assay. This was despite the AC being highly correlated between the three assays (≥0.90).

Table 2. Pearson correlation coefficients (R) for the total phenolic content (TPC) of aqueous extracts and their total antioxidant capacity measured with the three assays.

Correlation	ABTS	DPPH	FRAP
TPC	0.830 **	0.784 **	0.922 **
FRAP	0.919 **	0.937 **	
DPPH	0.901 **		

** All correlations were significant at $p < 0.01$ ($n = 132$).

4. Discussion

This study demonstrated that the aqueous extraction of phenolic compounds from bitter melon was as efficient as organic solvent extractions, when the extraction conditions were optimised. In the present study, before optimisation, 80% ethanol was shown to be the most suitable solvent for extracting phenolics and their associated AC from bitter melon (Figures 2 and 3). This finding is consistent with the study on phenolics and antioxidant activity of bitter melon by Horax *et al.* (2010), which demonstrated that a mixture of 80 ethanol/20 water (v/v) was the optimal ethanol: water solvent [10]. The TPC of the aqueous extract was 63% lower than for the best organic solvent, 80% ethanol (Figure 2), and the AC was at least 20% lower (Figure 3). However, after optimisation, the TPC and the AC of the aqueous extract were equal or higher than for 80% ethanol (Table 1). Furthermore, less solvent (40 mL water/g) and less time (5 min) were needed than was used for the 80% ethanol extract (100 mL/g for 1 h).

In general, low temperatures (5 and 10 °C) reduced the extraction of phenolic compounds from bitter melon whereas higher temperatures (20–80 °C) improved the extraction (Figures 4A and 5A). Increasing the temperature of water is known to decrease its viscosity, which can increase the diffusion coefficients of solutes, including water soluble phenolic compounds, and thereby increase the extraction efficiency of the phenolic compounds from the bitter melon by the aqueous solvent [23]. Furthermore, heating allows the cell wall of plant materials to be more permeable, thereby allowing the solvent and the solutes, from inside the plant cell walls, to diffuse in and out of the plant cells more easily [18,24]. However, temperatures above 80 °C lowered the extraction efficiency; this may have reflected degradation of the phenolic compounds, which can occur at these high temperatures [25].

The length of the extraction at 80 °C also had a significant impact (Figures 4B and 5B); interestingly, only 5 min was needed to reach the maximum for the extraction. This result indicates that the phenolic compounds in bitter melon may be highly soluble in water at 80 °C [26]. They are fairly polar compounds because, the polar organic solvents, 80% ethanol and methanol, were also more efficacious at extracting the phenolic compounds from bitter melon compared to the less polar solvents, acetone and butanol (Figure 2). Phenolic compounds have also been extracted from bitter melon in previous studies using polar solvents, including water [10,12].

Doing more than one extraction of the same sample also did not improve the extraction of the phenolic compounds or the antioxidant activity of the aqueous extract (Figures 4E and 5E). This supports the suggestion that the predominant phenolic compounds in bitter melon are very hydrophilic; they were very efficiently extracted within 5 min during the first extraction. Other studies have also reported that the phenolic compounds were successfully extracted from bitter melon using water [11,12].

Although less likely, exposure of the phenolic compounds to temperatures higher than 80 °C (Figures 4A and 5A) and for longer than 5 min at 80 °C (Figures 4B and 5B) may have led to their degradation [25]. In the present study, great care was taken to reduce the length of time the extracts were exposed to the extraction temperature being tested; all samples were placed on ice immediately after the intended extraction time and allowed to cool down for 10 min. This is important because it has previously been noted that phenolic compounds in hot water extracts can substantially degrade when the extracts are left to cool down at room temperature [27].

The present finding on the length of the aqueous extraction are in agreement with a study on peanut skins [28] and on papaya leaves [29]. In contrast, the extraction efficiency for TPC and AC of olive seeds [30] and pomegranate leaves [31] extracts increased with longer extraction times. This suggests that the type of phenolic compounds in bitter melon are likely to be more like those in peanut skins and papaya leaves than in olive seeds and pomegranate leaves.

Increasing the water-to-powder ratio resulted in a higher extraction of the phenolic compounds (Figures 4C and 5C), which was consistent with the extraction of water soluble components from plant material in general [32]. This is due to the high osmotic pressure, which results from the steep concentration gradient of the components in the plant material *vs.* the solvent that is generated at the high water-to-powder ratios [32]. The current finding was also in agreement with the results of studies on the extraction of other solutes from plant materials [33,34].

The powder particle sizes ≤2 mm, also resulted in higher extraction efficiencies than the powder sizes ≥2 mm (Figures 4D and 5D). This was consistent with previous studies, which have shown that a smaller powder particle size results in a higher extraction efficiency of bioactive compounds from powdered plan material because the surface area of the powder in contact with the solvent is increased [18]. Furthermore, the diffusion distance within the powder particle is shorter and the concentration gradient is steeper between the plant material and the solvent, resulting in a higher extraction efficiency [35].

Finally, the TPC of the various aqueous extracts was highly correlated with the AC of the extracts as measured with the FRAP, DPPH and ABTS assays (Table 2). Similarly high correlations ($R \geq 0.94$) were also observed between the TPC and the AC of the extracts in the original experiment with the five different solvents (Figures 2 and 3). These findings are in agreement with a previous study on greenhouse-grown bitter melons of different varieties, which showed strong positive correlations between TPC and the AC measured with the three assays [12]. Therefore, these high correlations strongly suggest that the extracted phenolic compounds, including those in the aqueous extracts of bitter melon, possess potent antioxidant and free radical scavenging activities.

This study supports the use of water as the solvent of choice for the extraction of the phenolic compounds from bitter melon. However, whether the efficiency of extraction for phenolic compounds and their associated AC from bitter melon can be further improved using water with other emerging modern assisted-extraction technologies, such as ultrasonics [36], microwave [37], pulsed electric field [38], supercritical fluid [39] and pressurized liquid [40], needs to be investigated in future studies.

5. Conclusions

This study demonstrated that water could be used to effectively extract phenolic compounds and their associated antioxidant activities from bitter melon. It was shown that the aqueous method could be optimised to give an extraction equivalent to that obtained with the best organic solvent tested (80% ethanol) and less solvent and less time was required with water than with 80% ethanol. The optimal conditions for the aqueous extraction were a single extraction at 80 °C for 5 min at a water-to-powder ratio of 40:1 mL/g and a powder particle size of 1 mm. Therefore, this study provides evidence to recommend the use of water as the solvent of choice for the extraction of the phenolic compounds from bitter melon.

Acknowledgments

The authors gratefully acknowledge the University of Newcastle for granting Sing P. Tan a University of Newcastle Postgraduate Research Scholarship (UNPRS).

Author Contributions

Sing Pei Tan, Costas Stathopoulos, Sophie Parks and Paul Roach contributed in experimental design and data analysis. Sing Pei Tan, Sophie Parks and Paul Roach have done manuscript preparation and data interpretation. All authors read and approved the final manuscript.

Conflicts of Interest

The authors declare no conflict of interest.

References

1. Fang, E.F.; Zhang, C.Z.; Zhang, L.; Fong, W.P.; Ng, T.B. *In vitro* and *in vivo* anticarcinogenic effects of RNase MC2, a ribonuclease isolated from dietary bitter gourd, toward human liver cancer cells. *Int. J. Biochem. Cell Biol.* **2012**, *44*, 1351–1360.
2. Patel, S.; Patel, T.; Parmar, K.; Bhatt, Y.; Patel, Y.; Patel, N.M. Isolation, characterization and antimicrobial activity of charantin from *Momordica charantia* linn. fruit. *Int. J. Drug Dev. Res.* **2010**, *2*, 629–634.
3. Hsu, C.; Tsai, T.-H.; Li, Y.-Y.; Wu, W.-H.; Huang, C.-J.; Tsai, P.-J. Wild bitter melon (*Momordica charantia* linn. var. *abbreviata* Ser.) extract and its bioactive components suppress *Propionibacterium acnes*-induced inflammation. *Food Chem.* **2012**, *135*, 976–984.
4. Grover, J.K.; Yadav, S.P. Pharmacological actions and potential uses of *Momordica charantia*: A review. *J. Ethnopharmacol.* **2004**, *93*, 123–132.
5. Sathishsekar, D.; Subramanian, S. Antioxidant properties of *Momordica charantia* (bitter gourd) seeds on streptozotocin induced diabetic rats. *Asia Pac. J. Clin. Nutr.* **2005**, *14*, 153–158.
6. Semiz, A.; Sen, A. Antioxidant and chemoprotective properties of *Momordica charantia* L. (bitter melon) fruit extract. *Afr. J. Biotechnol.* **2007**, *6*, 273–277.
7. Lii, C.K.; Chen, H.W.; Yun, W.T.; Liu, K.L. Suppressive effects of wild bitter gourd (*Momordica charantia* linn. var. *abbreviata* Ser.) fruit extracts on inflammatory responses in raw 264.7 macrophages. *J. Ethnopharmacol.* **2009**, *122*, 227–233.
8. Haminiuk, C.W.I.; Maciel, G.M.; Plata-Oviedo, M.S.V.; Peralta, R.M. Phenolic compounds in fruits—An overview. *Int. J. Food Sci. Tech.* **2012**, *47*, 2023–2044.
9. Horax, R.; Hettiarachchy, N.; Islam, S. Total phenolic contents and phenolic acid constituents in 4 varieties of bitter melons (*Momordica charantia*) and antioxidant activities of their extracts. *J. Food Sci.* **2005**, *70*, C275–C280.
10. Horax, R.; Hettiarachchy, N.; Chen, P. Extraction, quantification, and antioxidant activities of phenolics from pericarp and seeds of bitter melons (*Momordica charantia*) harvested at three maturity stages (immature, mature, and ripe). *J. Agric. Food Chem.* **2010**, *58*, 4428–4433.

11. Wu, S.J.; Ng, L.T. Antioxidant and free radical scavenging activities of wild bitter melon (*Momordica charantia* linn. var. *abbreviata* Ser.) in Taiwan. *LWT-Food Sci. Technol.* **2008**, *41*, 323–330.
12. Tan, S.P.; Parks, S.E.; Stathopoulos, C.E.; Roach, P.D. Greenhouse-grown bitter melon: Production and quality characteristics. *J. Sci. Food Agric.* **2014**, *94*, 1896–1903.
13. Cerda, A.; Martínez, M.E.; Soto, C.; Poirrier, P.; Perez-Correa, J.R.; Vergara-Salinas, J.R.; Zúñiga, M.E. The enhancement of antioxidant compounds extracted from *Thymus vulgaris* using enzymes and the effect of extracting solvent. *Food Chem.* **2013**, *139*, 138–143.
14. Negrão Murakami, A.N.; de Mello Castanho Amboni, R.D.; Prudêncio, E.S.; Amante, E.R.; de Moraes Zanotta, L.; Maraschin, M.; Cunha Petrus, J.C.; Teófilo, R.F. Concentration of phenolic compounds in aqueous mate (*Ilex paraguariensis* A. St. Hil) extract through nanofiltration. *LWT-Food Sci. Technol.* **2011**, *44*, 2211–2216.
15. Kubola, J.; Siriamornpun, S. Phenolic contents and antioxidant activities of bitter gourd (*Momordica charantia* L.) leaf, stem and fruit fraction extracts *in vitro*. *Food Chem.* **2008**, *110*, 881–890.
16. Meneses, N.G.T.; Martins, S.; Teixeira, J.A.; Mussatto, S.I. Influence of extraction solvents on the recovery of antioxidant phenolic compounds from brewer's spent grains. *Sep. Purif. Technol.* **2013**, *108*, 152–158.
17. Alothman, M.; Bhat, R.; Karim, A.A. Antioxidant capacity and phenolic content of selected tropical fruits from malaysia, extracted with different solvents. *Food Chem.* **2009**, *115*, 785–788.
18. Vuong, Q.V.; Golding, J.B.; Stathopoulos, C.E.; Nguyen, M.H.; Roach, P.D. Optimizing conditions for the extraction of catechins from green tea using hot water. *J. Sep. Sci.* **2011**, *34*, 3099–3106.
19. Cicco, N.; Lanorte, M.T.; Paraggio, M.; Viggiano, M.; Lattanzio, V. A reproducible, rapid and inexpensive folin-ciocalteu micro-method in determining phenolics of plant methanol extracts. *Microchem. J.* **2009**, *91*, 107–110.
20. Re, R.; Pellegrini, N.; Proteggente, A.; Pannala, A.; Yang, M.; Rice-Evans, C. Antioxidant activity applying an improved ABTS radical cation decolorization assay. *Free Radic. Biol. Med.* **1999**, *26*, 1231–1237.
21. Brand-Williams, W.; Cuvelier, M.E.; Berset, C. Use of a free radical method to evaluate antioxidant activity. *LWT-Food Sci. Technol.* **1995**, *28*, 25–30.
22. Benzie, I.F.F.; Strain, J.J. The ferric reducing ability of plasma (FRAP) as a measure of "antioxidant power": The FRAP assay. *Anal. Biochem.* **1996**, *239*, 70–76.
23. Spigno, G.; de Faveri, D.M. Antioxidants from grape stalks and marc: Influence of extraction procedure on yield, purity and antioxidant power of the extracts. *J. Food Eng.* **2007**, *78*, 793–801.
24. Wissam, Z.; Ghada, B.; Wassim, A.; Warid, K. Effective extraction of polyphenols and proanthocyanidins from pomegranate's peel. *Int. J. Pharm. Pharm. Sci.* **2012**, *4*, 675–682.
25. Sólyom, K.; Solá, R.; Cocero, M.J.; Mato, R.B. Thermal degradation of grape marc polyphenols. *Food Chem.* **2014**, *159*, 361–366.
26. Rispail, N.; Morris, P.; Webb, K. Phenolic compounds: Extraction and analysis. In *Lotus Japonicus Handbook*; Márquez, A., Ed.; Springer: Dordrecht, The Netherlands, 2005; pp. 349–354.
27. Hirun, S.; Roach, P.D. An improved solvent extraction method for the analysis of catechins and caffeine in green tea. *J. Food Nutr. Res.* **2011**, *50*, 160–166.

28. Nepote, V.; Grosso, N.R.; Guzmán, C.A. Optimization of extraction of phenolic antioxidants from peanut skins. *J. Sci. Food Agric.* **2005**, *85*, 33–38.
29. Vuong, Q.V.; Hirun, S.; Roach, P.D.; Bowyer, M.C.; Phillips, P.A.; Scarlett, C.J. Effect of extraction conditions on total phenolic compounds and antioxidant activities of *Carica papaya* leaf aqueous extracts. *J. Herbal Med.* **2013**, *3*, 104–111.
30. Alu'datt, M.H.; Alli, I.; Ereifej, K.; Alhamad, M.N.; Alsaad, A.; Rababeh, T. Optimisation and characterisation of various extraction conditions of phenolic compounds and antioxidant activity in olive seeds. *Nat. Prod. Res.* **2011**, *25*, 876–889.
31. Wang, C.; Shi, L.; Fan, L.; Ding, Y.; Zhao, S.; Liu, Y.; Ma, C. Optimization of extraction and enrichment of phenolics from pomegranate (*Punica granatum* L.) leaves. *Ind. Crop. Prod.* **2013**, *42*, 587–594.
32. Gertenbach, D.D. Solid-liquid extraction technologies for manufacturing nutraceuticals. In *Functional Foods: Bioachemical and Processing Aspects*; Shi, J., Mazza, G., Maguer, M.L., Eds.; CRC Press: Boca Raton, FL, USA, 2002; pp. 332–365.
33. You, Q.; Yin, X.; Zhang, S.; Jiang, Z. Extraction, purification, and antioxidant activities of polysaccharides from *Tricholoma mongolicum* Imai. *Carbohydr. Polym.* **2014**, *99*, 1–10.
34. Tan, S.P.; Parks, S.E.; Stathopoulos, C.E.; Roach, P.D. Extraction of flavonoids from bitter melon. *Food Nutr. Sci.* **2014**, *5*, 458–465.
35. Wang, Z.; Pan, Z.; Ma, H.; Atungulu, G.G. Extraction of phenolics from pomegranate peels. *Open Food Sci. J.* **2011**, *5*, 17–25.
36. Zu, G.; Zhang, R.; Yang, L.; Ma, C.; Zu, Y.; Wang, W.; Zhao, C. Ultrasound-assisted extraction of carnosic acid and rosmarinic acid using ionic liquid solution from *Rosmarinus officinalis*. *Int. J. Mol. Sci.* **2012**, *13*, 11027–11043.
37. Chiremba, C.; Rooney, L.W.; Beta, T. Microwave-assisted extraction of bound phenolic acids in bran and flour fractions from sorghum and maize cultivars varying in hardness. *J. Agric. Food Chem.* **2012**, *60*, 4735–4742.
38. Delsart, C.; Ghidossi, R.; Poupot, C.; Cholet, C.; Grimi, N.; Vorobiev, E.; Milisic, V.; Peuchot, M.M. Enhanced extraction of phenolic compounds from merlot grapes by pulsed electric field treatment. *Am. J. Enol. Vitic.* **2012**, *63*, 205–211.
39. Giannuzzo, A.N.; Boggetti, H.J.; Nazareno, M.A.; Mishima, H.T. Supercritical fluid extraction of naringin from the peel of *Citrus paradisi*. *Phytochem. Anal.* **2003**, *14*, 221–223.
40. Erdogan, S.; Ates, B.; Durmaz, G.; Yilmaz, I.; Seckin, T. Pressurized liquid extraction of phenolic compounds from *Anatolia propolis* and their radical scavenging capacities. *Food Chem. Toxicol.* **2011**, *49*, 1592–1597.

10

Rosmarinic Acid, a New Polyphenol from *Baccaurea ramiflora* Lour. Leaf: A Probable Compound for Its Anti-Inflammatory Activity

Talambedu Usha [1], Sushil Kumar Middha [1], Malay Bhattacharya [2], Prakash Lokesh [3] and Arvind Kumar Goyal [4,*]

[1] Department of Biotechnology, Maharani Lakshmi Ammanni College for Women, Bangalore-560012, India; E-Mails: ushatalambedu@gmail.com (T.U.); sushil.middha@gmail.com (S.K.M.)

[2] Department of Botany, Kalimpong College, Kalimpong-734301, India; E-Mail: malaykpgc@gmail.com

[3] Padmashree Institute of Management and Science, Bangalore-560060, India; E-Mail: connectloki@gmail.com

[4] Bamboo Technology, Department of Biotechnology, Bodoland University, Kokrajhar-783370, India

* Author to whom correspondence should be addressed; E-Mail: arvindgoyal210883@gmail.com

External Editors: Antonio Segura-Carretero and David Arráez-Román

---

Abstract: Despite several pharmacological applications of *Baccaurea ramiflora* Lour., studies on the influence of its polyphenol content on pharmacological activity such as anti-inflammatory properties have been scarce. Here we evaluated *in vitro* antioxidant activity, poyphenolics by HPLC and the anti-inflammatory potential of the methanolic leaf extract of *Baccaurea ramiflora* (BME) and its protective effects in carrageenan-induced paw edema model of inflammation in rats. The BME extract contained 79.06 ± 0.03 mg gallic acid equivalent (GAE)/g total polyphenols, 28.80 ± 0.01 mg quercetin equivalent (QE)/g flavonoid and 29.42 ± 0.01 µg cathechin equivalent/g proanthocyanidin respectively and rosmarinic acid (8 mg/kg) as a main component was identified by HPLC. Results demonstrate that administration of BME at the dose of 200 mg/kg can reduce paw edema by over 63%, and it exhibits a dose-response effect. Depending on concentration, the extract exerted scavenging activity on DPPH radical (IC_{50} 36.4 μg/mL), significantly inhibited IL-1β (4.4 pg/mg protein) and TNF-α (0.21 ng/μg protein). Therefore, we conclude BME causes a substantial reduction of inflammation in *in vivo* models. We propose that rosmarinic acid and similar phenolic compounds may be useful in the therapy of inflammation-related injuries.

Keywords: *Baccaurea ramiflora*; rosmarinic acid; anti-inflammatory activity; cytokines

---

1. Introduction

Baccaurea ramiflora Lour. syn. *B. sapida* (Roxb.) Muell. Arg. popularly known as Burmese grapes belongs to the family Euphorbiaceae and is native to Southeast Asia. It is mainly encountered in the sub-Himalayan tract, from Nepal to Sikkim, the Darjeeling hills, Arunachal Pradesh, Tripura, Assam, Bhutan, Burma, Penninsular Malaysia, Tibet and the Andaman islands [1]. It is a slow growing semi-evergreen tree reaching a height of about 25 m [2]. Fruit is 2–3 cm in diameter and is yellowish to red in color with leathery pericarp, three seeded arills embedded in pinkish white pulp [1]. It is variously named in different languages like Mafia in Thai, Latkan in Hindi, Leteku in Assamese, Bhupi in Bengali, *etc.* [2]. The whole plant is traditionally used in Chinese Dai medicine [3]. In Bangladesh, young leaves are used as a vegetable, flavoring agent with curries and minced meat [4]. In India, juice is used orally for constipation, proven to have antioxidant properties [1]. The antiviral and antioxidant properties of the fruit and diuretic activity of stem bark have been reported [1,4,5]. So far fourteen compounds have been isolated from the leaves by various workers [6,7]. A review of the literature showed that though works on hypoglycemic and hypolipidemic activity of *B. ramiflora* leaf have been reported [8], its anti-inflammatory activity is yet to be explored. Keeping this in mind the present study was designed to evaluate *in vivo* anti-inflammatory, *in vitro* antioxidant activity and HPLC analysis of the methanolic extract of the leaf of *B. ramiflora* (BME).

2. Experimental Section

2.1. Chemicals and Reagents

Rosmarinic acid (HPLC grade), carrageenan were obtained from Sigma (St. Louis, MO, USA), 2,2-diphenyl-1-picryl-hydrazyl (DPPH), quercetin, sodium nitrite ($NaNO_2$), trichloroacetic acid (TCA), ascorbic acid, ferric chloride ($FeCl_3$), gallic acid were obtained from HiMedia Laboratories Pvt. Ltd., Mumbai, India. Potassium di-hydrogen phosphate (KH_2PO_4), di-potassium hydrogen phosphate (K_2HPO_4), sodium hydroxide (NaOH), ammonium molybdate, sulfuric acid, potassium ferricyanide ($K_2Fe(CN)_6$), sodium carbonate (Na_2CO_3), hydrogen peroxide (H_2O_2) and methanol were procured from Merck, Mumbai, India. Diclofenac sodium from Cipla, Mumbai, India and Folin-Ciocalteu (FC) reagent from Sisco Research Laboratory, Mumbai, India. Aluminum chloride ($AlCl_3$), Greiss reagent was obtained from Sd fine Chemicals Ltd., Mumbai, India. All chemicals and solvents were of analytical grade.

2.2. Plant Material Collection and Extraction

Fresh leaves of *B. ramiflora* were collected from Jalpaiguri, West Bengal, India. The plant material was authenticated by a plant taxonomist and a voucher specimen (Voucher No. KPGC/MB/74) was deposited at Kalimpong Government College, Kalimpong, West Bengal, India.

The leaves were air dried and powdered using a mechanical grinder. 10 g of the powdered leaf was extracted in a Soxhlet apparatus using 80% aqueous methanol (the ratio of plant material to solvent was 1:15 w/v) [9]. The extraction was carried out at boiling temperature for 6 h. The extract obtained was evaporated under pressure at 50 °C to a constant weight and stored at 4 °C until required. The extract was dissolved in double-distilled water (DDW) and dimethyl sulphoxide (DMSO) in the desired concentrations according to the protocols just before use.

2.3. In vitro Assay

Determination of Biochemical Constituents

The total soluble phenolics (TPC) was determined by the Singleton and Rossi [10] method with a slight modification [11]. Briefly the leaf extract (0.5 mL) was mixed with 0.5 mL of FC reagent (previously diluted to 1:1 ratio with double distilled water) and incubated for 5 min at room temperature (RT), then 1 mL of 2% Na_2CO_3 solution was added. After incubation at RT for 10 min, the absorbance of the blue color that developed was read at 730 nm using a Themo UV1-Vis spectrophotometer (Thermo Electron Corporation, England, UK). Gallic acid was used as a standard. The concentration of total phenolic compounds was determined in µg of gallic acid equivalent using an equation obtained from the standard gallic acid graph. The total flavonoid content (TFC) was determined according to Zhishen *et al.* [12] using quercetin as a standard. The plant extract (0.25 mL) was added to 1.25 mL DDW followed by 75 µL of 5% $NaNO_2$.and was incubated at RT for 5 min, $AlCl_3$ (0.15 mL, 10%) was then added. After a further incubation for 6 min at RT, the reaction mixture was treated with 0.5 mL of 1 mM NaOH. Finally, the reaction mixture was diluted with 275 µL of DDW. The reaction mixture was incubated at RT for 20 min and the absorbance maxima was measured at 510 nm. The total proanthocyanidin content (TPrC) was determined as per the protocol previously reported by Vuong *et al.* [13]. To 0.5 mL of plant extract 3 mL vanillin (4%) was added followed by 1.5 mL of conc. HCl and incubated for 15 min at RT. The absorbance was measured at 500 nm. Total proanthocyanidin content was expressed as cathechin equivalent.

2.4. In vitro Antioxidant Properties of the Extracts

2.4.1. Free Radical Scavenging Activity (DPPH Method)

The antioxidant activity of the *B. ramiflora* methanolic leaf extract (BME) and standard were assessed on the basis of the radical scavenging effect of the stable DPPH free radical as per the modified protocol by Goyal *et al.* [11]. The DPPH solution (0.006% w/v) was prepared in 95% methanol. Freshly prepared DPPH solution was taken in a test tube and extract was added in different alliquots (100–1000 µg) and the final volume was made up to 2 mL. The discoloration of the reaction mixture was measured at 517 nm for 30 min of incubation in the dark. The experiment was performed at least in triplicate.

Ascorbic acid was used as a reference standard and dissolved in DDW to make the stock solution with the same concentration (1 mg/mL). Methanol (95%) was used as blank. Percentage scavenging of the DPPH free radical was measured using the equation:

$$\text{Percentage scavenging} = A_0 - A_1/A_1 \times 100 \tag{1}$$

where A_0 = Absorbance of the control and A_1 = Absorbance in the presence of the sample. The actual decrease in absorption induced by the test compounds was compared with the positive controls. The IC50 value was calculated using the dose inhibition curve.

2.4.2. Nitric Oxide Scavenging Activity

NO generated from sodium nitroprusside was measured by the Greiss reaction [14] as per Goyal *et al.* [15]. The absorbance of the chromophore formed due to diazotization of sulphanilamide with nitrite and subsequent coupling with napthylethylene diamine was read at 540 nm in a UV-visible spectrophotometer. Percentage scavenging was measured as per Equation (1).

2.4.3. Determination of Total Antioxidant Capacity by Phosphomolybdenum Assay

The total antioxidant capacity of BME was assessed by the phosphomolybdenum method [16]. To 0.3 mL of extract (1 mg/mL), 3 mL reagent solution (28 mM sodium phosphate, 6 M sulfuric acid and 4 mM ammonium molybdate) was added and incubated at 95 °C for 90 min. Then the absorbance of the solution was measured at 695 nm against a blank. The antioxidant capacity of extract was evaluated as equivalents of ascorbic acid.

2.5. Quantitative Analysis of Antioxidant Compounds Using High-Performance Liquid Chromatography (HPLC)

2.5.1. Preparation of Standard and Sample Solutions for HPLC

Stock solution of rosmarinic acid was prepared at a concentration of 25 µg/mL just prior to use and used as reference standard. BME was dissolved in HPLC grade methanol to get the desired concentration and used as sample. Prior to injection both the sample and standard were filtered through a 0.22 µm millipore filter.

2.5.2. Chromatographic Conditions for HPLC

The HPLC system (Waters, Singapore) consisted of photodiode array detector (W2998), dual pump system (515-waters), temperature control module II (TC2-waters), pump control module (PC2-waters), system controller (EMOAA01712) and a reverse phase HPLC analytical column-waters Spherisorb C8, 4.6 × 100 mm, 5 µm particle size. The extract and standard were passed through a 0.22 µm filter (Merck Millipore, Darmstadt, Germany) before injection into a reverse phase NOVA-PAK C18 (Waters, Singapore) column (4.6 × 100 mm, particle size 4 µm) at ambient temperature (20 °C). A Waters 515 system controller coupled with a photodiode array detector (Waters 2998 series) was used. The mobile phase was Acetonitrile (A) and water containing 0.1% Phosphoric acid (B). The gradient was as follows: 0 min, 5% A; 10 min, 15% A; 30 min, 25% A; 35 min, 30% A; 50 min, 55% A; 55 min, 90% A; 57 min,

100% A and then held for 10 min before returning to the initial conditions. The flow rate was adjusted to 1.0 mL/min, sample run time was 30 min and the detector was set at 320 nm at 1.2 nm resolution with the mobile phase methanol: water (50:50 v/v, isocratic) [17]. Active constituent in the sample was identified by comparison of the retention time with a standard. Data was analyzed using Empower software (Waters, Singapore).

2.6. In vivo Assay

2.6.1. Animals

Male *Wistar albino* rats (140–160 g) were housed under standard laboratory conditions of light and dark cycles of 7:00 am to 7:00 pm, temperature of 25 °C ± 2 °C and 68% ± 1% relative humidity. The animals were provided standard rat pellet (Lipton India Ltd., Bangalore, India) and tap water *ad libitum*. The study protocol was approved by Maharani Lakshmi Ammanni College Ethical Committee, clearance from ethical committee (1368/ac/10/CPCSEA), Bangalore.

2.6.2. Acute Toxicity Test

Swiss albino mice (25–30 g) of both sexes were divided into six groups of 10 each. Animals were fasted overnight and were free to access water preceding the experiment. BME was administered to each group at different dose levels of (0.5, 1.0, 1.5, 2.0, 2.5, 3.0 g/kg BW/mL). The mice were critically observed for 24 h, mortality was recorded and LD_{50} (Median lethal dose) was determined as per author's previous studies [18].

2.6.3. Experimental Groups

Rats were randomly divided into five groups of six rats each. NL: Normal (1 mL/kg *p.o* dimethyl sulphoxide, DMSO), IC: carrageenan injected control; IC + low dosage of BME (LBME): control rats treated with 100 mg/mL of *B. ramiflora* methanolic leaf extract; IC + high dosage of BME (HBME): control rats treated with 200 mg/mL of BME; IC + diclofenac: control rats treated with diclofenac (10 mg/kg) (Sigma Chemical Co., St. Louis, MO, USA) dissolved in DMSO.

2.7. Anti-Inflammatory Activity

The carrageenan-induced rat paw oedema test [19] has been used as an experimental model for testing the anti-inflammatory activity as per previous reports [20]. One hour after the oral administration of LBME and HBME or diclofenac, carrageenan-saline solution (0.5%, w/v) and saline were injected in a volume of 0.1 mL into the plantar surface of the right and left hind paw, respectively. The left paw served as the control (non-inflamed) paw. The experimental animals were observed for 0 h, 1 h, 2 h and 4 h; the paw volume was measured using a plethysmometer (PanLab, Barcelona, Spain). The anti-inflammatory effect was calculated using Equation (2):

$$\text{Anti-inflammatory effect (\%)} = k - e/k \times 100 \qquad (2)$$

where *k* is the difference in the paw weight in the control group, and *e* is the difference in the paw weight in the treatment group. Blood samples were collected by retro-orbital puncture from all rats and serum was separated for further biochemical estimations.

2.8. Measurement of IL-1β, TNF-α Level

The serum concentration of interleukin 1 beta (IL-1β) and tumor necrosis factor alpha (TNF-α) (Endogen, Woburn, MA, USA) were measured using a commercial enzyme-linked immunosorbent assay (ELISA) kit method (Biosource, San Diego, CA, USA) [21].

2.9. Statistical Analysis

All the experiments were repeated six times and expressed as mean ± standard error of means (SEM) and statistically analyzed by two-way analysis of variance followed by Tukey's multiple range using Graph pad prism. $p < 0.05$ and $p < 0.01$ was considered to be statistically significant.

3. Results and Discussion

3.1. Yield of the Extract

Total yield of the extract was found to be 8.4%.

3.2. Determination of Biochemical Constituents

The total phenol, flavonoid and proanthocyanidin content of the BME were found to be 79.06 ± 0.03 mg GAE/g, 28.80 ± 0.01 mg QE/g and 29.42 ± 0.01 µg cathechin/g respectively. The methanolic fraction of *B. ramiflora* was found to be rich in phenols, flavonoid and proanthocyanidin as compared to other polar and non-polar fractions. The higher polyphenolic content of BME might be due to the potential of methanol to release the bound polyphenols present in the cell wall of the plants. Thus, methanol proves to be a suitable solvent for extraction of the plant [22,23].

3.3. In Vitro Antioxidant Properties of the Extracts

3.3.1. Free Radical Scavenging Activity (DPPH Method)

BME showed radical scavenging activity as measured by DPPH, across concentrations ranging from 0.1 to 1.0 mg/mL (Figure 1). The value for maximum % inhibition was noted at a concentration of 1.0 mg/mL with 36.4% inhibition, whereas that of standard (ascorbic acid) was found to be 64.76%. The radical scavenging activity, using DPPH exhibited the highest radical scavenging activity reveals that BME contains powerful inhibitor compounds, which act as potential antioxidants and thus scavenge the DPPH radicals to form stable reduced DPPH molecules [24].

The high DPPH scavenging activities of the methanolic extract have already been reported in comparison to the aqueous and acetone extract by other researchers [22,23].

3.3.2. Nitric Oxide Scavenging Activity

A dose dependent increase in NO scavenging activity was noted in BME (Figure 2). At the concentration of 1.0 mg/mL both BME and the standard ascorbic acid showed comparable NO scavenging ability of 61.33% and 60.32% respectively. Nitric oxide is a powerful intermediary of a number of physiological processes [15]. It is a diffusible free radical that can act as an effective molecule in several biological systems like vasodilation, neuronal messenger, antimicrobial and antitumor activities [25]. The strong NO scavenging activity in a dose dependent manner depicts BME as a potential source of natural antioxidants.

Figure 1. 2,2-Diphenyl-1-picryl-hydrazyl (DPPH) scavenging activity of methanolic extract of *Baccaurea ramiflora* leaf.

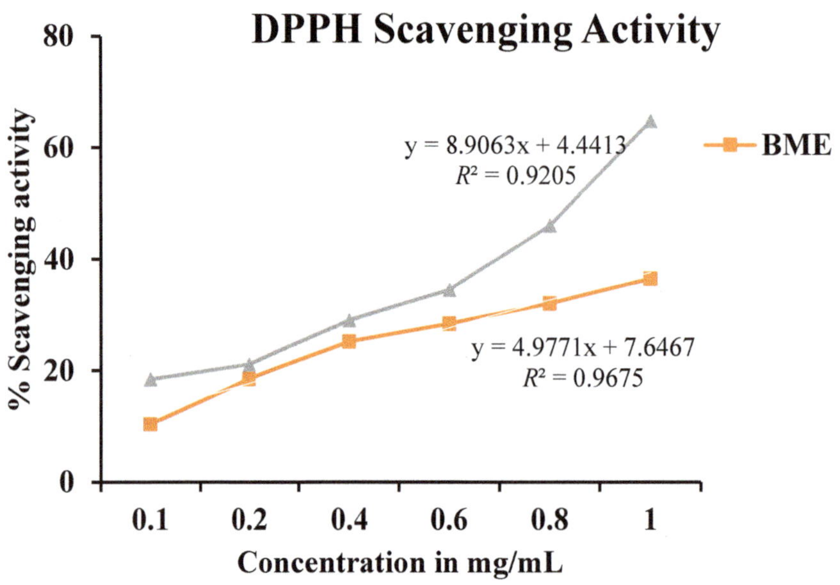

Figure 2. Nitric oxide scavenging activity of methanolic extract of *Baccaurea ramiflora* leaf.

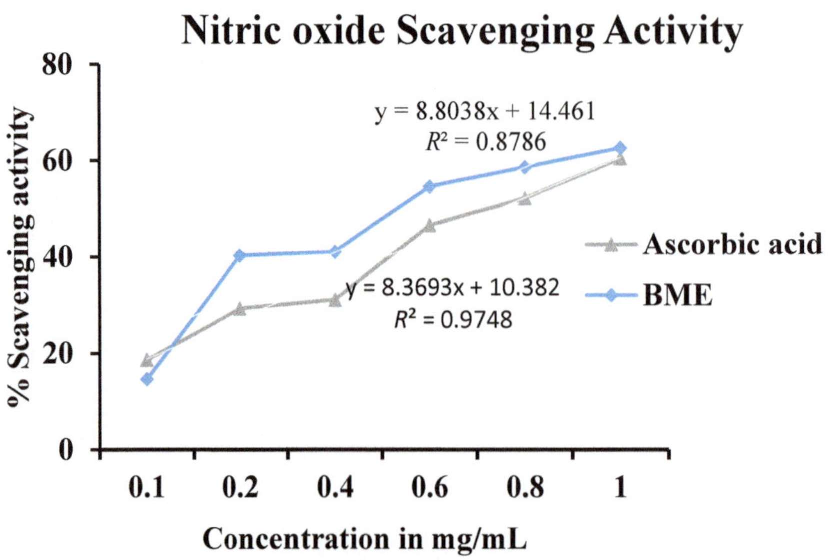

3.3.3. Phosphomolybdenum Assay

This assay deals with the reduction of Mo (VI) to Mo (V) either in the presence of the extract or standard ascorbic acid. BME was more effective in reducing Mo (VI) to Mo (V) (IC_{50} 58.47 ± 1.17 µg/mL) than ascorbic acid (IC_{50} 81.3 ± 4.12 µg/mL), suggesting the presence of effective antioxidants in BME, which was also dose dependent, which increased in a concentration-dependent manner.

3.4. HPLC Analysis

Due to high polyphenolic content and strong antioxidative properties BME was subjected to HPLC. HPLC being simple and easy to use has been found to be an effective technique in identification and quantification of major phenolic compounds present in plants. The chromatogram tested at 320 nm is depicted in Figure 3. The chromatogram shows the presence of rosmarinic acid with retention time of 5.087 when compared to standard rosmarinic acid with RT of 5.084. The content of rosmarinic acid was found to be 0.0008% w/w or 8 mg/kg. This was the first ever study to detect the presence of rosmarinic acid in BME. Rosmarinic acid is an ester of caffeic acid and 3,4-dihydroxyphenyllactic acid which is commonly encountered in species of Boraginaceae and the subfamily Nepetoideae of Lamiaceae along with other higher plants [26].

Figure 3. HPLC Chromatogram of methanolic extract of *Baccaurea ramiflora* leaf (BME) showing the presence of rosmarinic acid.

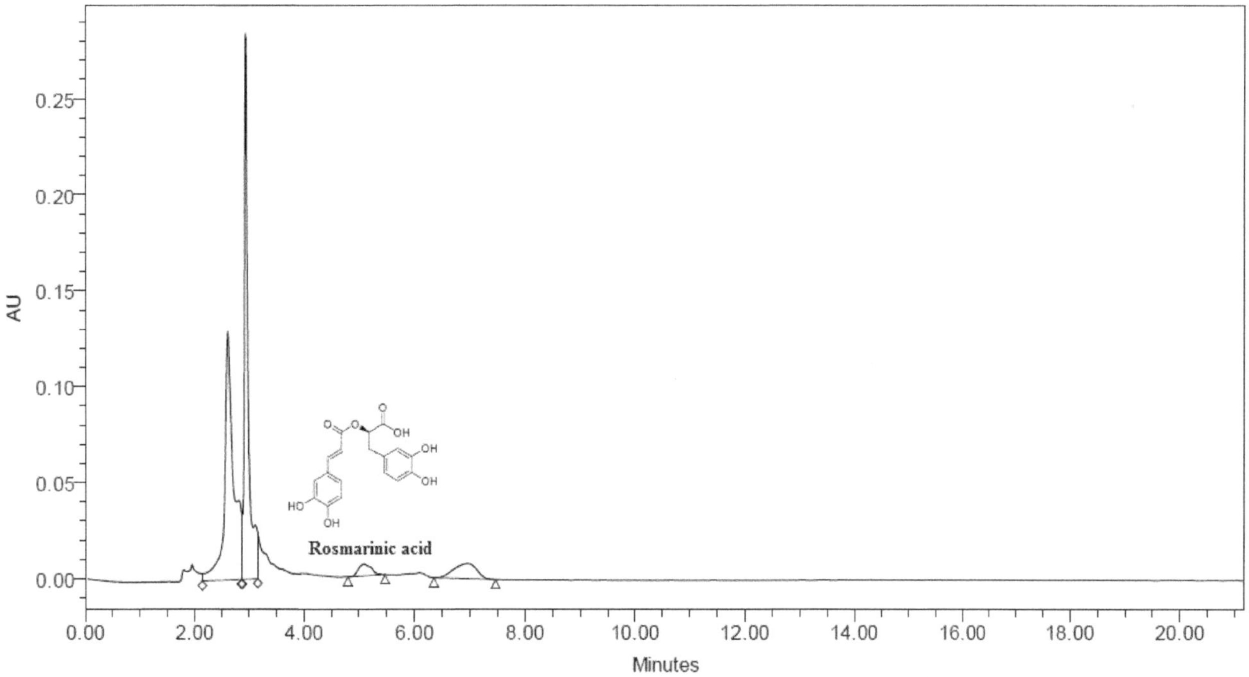

3.5. Assessment of Pharmacological Activity (In Vivo)

There are very few studies that prove pharmacological activities of *B. ramiflora.* The leaf of this plant is known to have hypoglycemic and hypolipidemic properties [8,27]. Hence to add to the pharmacology activity the present study focused on the anti-inflammatory activity of the methanolic fraction of *B. ramiflora leaves.*

3.5.1. Toxicity

BME administration did not exhibit any side effect in survival rate and body weight of the experimental animals. *In vivo* LD_{50} of the BME was estimated to be 1.980 g/kg BW. *In vivo* methods of estimation of toxicity are of critical value in the drug discovery process. They are carried out to evaluate the level to which a chemical entity can cause injurious effects to the cell and to broadly understand the optimum drug concentration to be administered. Determined LD_{50} was used to define the dosage for the study and to the best of our knowledge, this is the first study to report the lethal dose of BME.

3.5.2. Anti-Inflammatory Activity

Carrageenan injection into the rat paw provokes a local, acute inflammation because of arachidonic acid metabolites. The BME extract showed dose dependent inhibition of carrageenan induced edema comparable to the anti-inflammatory drug diclofenac. BME significantly ($p < 0.05$) hindered the incidence of inflammation and a dose dependent reduction of paw edema was seen at 4 h. After 4 h diminution in paw edema by HBME was 62.71%, as compared to edema reduced (71.57%) by the standard drug, diclofenac (10 mg/kg) (Table 1). The BME was found to be highly effective in inhibiting edema. It is evident from our results that phytoconstituents (Rosmarinic acid) would be responsible for the anti-inflammatory activity which might be related to the inhibition of the release or synthesis of cyclooxygenase products. However detailed molecular and chemical level study is required to confirm the same. The results also correlate with our previous studies where phytochemical screening indicates the presence of phenols, flavonoids and terpenoids which might add to the pharmacological activity [1]. Rosmarinic acid is known to possess extraordinary biological anti-inflammatory, anti-mutagen, and anti-microbial activities [28]. The antioxidative potential of rosmarinic acid has been established by various workers [29,30].

Table 1. Paw volume variation due to anti-inflammatory effects of *Baccaurea ramiflora* (BME) [&].

Experimental Animals	Dosages mg/kg	% Increase 1 h	% Increase 2 h	% Increase 4 h
NL	No treatment	0.24 ± 0.12	0.62 ± 0.18	0.36 ± 0.24
IC	No treatment	2.1 ± 0.47 ^{a}	6.2 ± 0.54 ^{a}	9.4 ± 1.76 ^{a}
IC + LBME	100	9.3 ± 0.65 ^{NS}	14.2 ± 2.17 ^{a}	29.6 ± 4.62 ^{b}
IC + HBME	200	13.7 ± 1.75 ^{NS}	31.3 ± 4.13 ^{b}	62.71 ± 4.98 ^{b}
IC + Diclofenac	10	52.3 ± 6.89 ^{b}	60.1 ± 8.90 ^{b}	71.57 ± 7.32 ^{b}

[&] Data represented as mean% ± SEM for number of animals = 6. NL: normal; IC: carrageenan injected control; IC + LBME: control rats treated with 100 mg/mL; IC + HBME: control rats treated with 200 mg/mL; IC + Diclofenac: control rats treated with diclofenac. Different subscripts denotes significant group at $p < 0.05$ *vs.* inflammatory group (IC); [NS] indicates not significant.

The higher antioxidant activity of BME might be due to the synergistic effect of the presence of rosmarinic acid in conjunction with other phyto-constituents [31]. Edema in rat paw induced by carrageenan is used by and large as an effective inflammatory model for new anti-inflammatory drug design [32]. In the first 2 h (phase I), there is liberation of histamine and serotonin in the rat paw after the carrageenan injection. The complementary system increases the activation and release of kinin, prostaglandin, bradykinin, leukotrienes, and polymorphonuclear cells in the damaged tissue in surroundings cells in phase II [33].

3.5.3. Anti-Inflammatory Effect of BME on Cytokines Levels

Cytokines, small glycoproteins, are produced by the body in response to an antigen, and a mediator for regulating the innate and adaptive immune effects. IL-1β and TNF-α are the major cytokines in inflammation [34]. The cytokine level should be suppressed to reduce the severity of the inflammatory reaction. Table 2 shows that the inflammatory reaction induced by carrageenan significantly uplifts the concentration of IL-1β and TNF-α ($p < 0.05$; $p < 0.01$). BME treatment significantly reduced the production of these cytokines. The anti-inflammatory activity of the standard drug diclofenac sodium, normal saline and two doses of BME extract were compared. In this process, HBME showed a partially significant result compared to the standard drug ($p < 0.01$). Our *in vivo* results were in accordance with a previous report [35]. TNF-α and interleukin-1β levels of experimental animals were significantly lowered with HBME induction. HBME was more effective than LBME and comparable with the standard drug.

Earlier reports also indicated the potential role of phenolic compounds in inflammatory conditions [35,36] and their ability to inhibit polymorphonuclear lipoxygenase, an enzyme involved in inflammatory conditions [37].

Table 2. Cytokines levels (interleukin-1β (IL-1β) and tumor necrosis factor-α (TNF-α) of experimental animals ($n = 6$).

Experimental Animals	IL-1β (pg/mg Protein)	TNF-α (ng/μg Protein)
NL	3.3	0.15
IC	10.4	0.58
IC + LBME (100 mg/kg)	6.7 *	0.39 *
IC + HBME (200 mg/kg)	4.4 *	0.21 ***
IC+Diclofenac	4.8 **	0.18 ***

NL: normal; IC: carrageenan induced edema control; IC + LBME: control rats treated with 100 mg/mL; IC + HBME: control rats treated with 200 mg/mL; IC + Diclofenac: control rats treated with Diclofenac. Values are in means ± SEM. * $p < 0.05$ *vs.* inflammatory group; ** $p < 0.01$ *vs.* inflammatory group; *** $p < 0001$ *vs.* inflammatory group.

4. Conclusions

This is reputed to be the first report to provide scientific evidence of the biological activities of the traditional use of *Baccaurea ramiflora*. Rosmarinic acid, detected by an HPLC method is thought to inhibit eicosanoids such as prostaglandin biosynthesis which is the end product of the cyclooxygenase pathway. The phytochemical also has the ability to inhibit the infiltration of neutrophils and its degranulation, thereby decreasing the level of arachidonic acid which is responsible for painful sensation. In conclusion this study supports the mechanism involved in the anti-inflammatory and antioxidant activity of *Baccaurea ramiflora*.

Acknowledgments

The authors would like to express their deepest gratitude to Maharani Lakshmi Ammanni College for Women (MLACW), Bangalore for their help in animal experiments and providing research facilities.

Author Contributions

T.U., S.K.M. and A.K.G. proposed the ideas, analyzed the data and wrote the manuscript; T.U., M.B. and P.L. collected the samples and carried out the chemical analyses. All the authors contributed in planning the experimental methods and reviewing and editing the manuscript. All authors read and approved the final manuscript.

Conflicts of Interest

The authors declare no conflict of interest.

References

1. Goyal, A.K.; Mishra, T.; Sen, A. Antioxidant profiling of Latkan (*Baccaurea ramiflora* Lour.) wine. *Indian J. Biotechnol.* **2013**, *12*, 137–139.
2. Howlader, M.A.; Apu, A.S.; Saha, R.K.; Rizwan, F.; Nasrin, N.; Asaduzzaman, M. Cytotoxic activity of *n*-hexane, chloroform and carbon tetrachloride fractions of the ethanolic extract of leaves and stems of *Baccaurea ramiflora* (Lour.). *Int. J. Pharm. Sci. Res.* **2012**, *3*, 822–825.
3. Lin, Y.F.; Yi, Z.; Zhao, Y.H. *Chinese Dai Medicine Colorful Illustrations*; Yunnan Nationality Press: Kunming, China, 2003; pp. 158–160.
4. Hasan, S.R.; Hossain, M.M.; Akter, R.; Jamila, M.; Mazumder, M.E.H.; Rahman, S. DPPH free radical scavenging activity of some Bangladeshi medicinal plants. *J. Med. Plant Res.* **2009**, *3*, 875–879.
5. Aswal, B.S.; Goel, A.K.; Kulshrestha, D.K.; Mehrotra, B.N.; Patnaik, G.K. Screening of Indian plants for biological activity: Part XV. *Indian J. Exp. Biol.* **1996**, *34*, 444–467.
6. Yang, X.W.; Wang, J.S.; Ma, Y.L.; Xiao, H.T.; Zuo, Y.Q.; Lin, H.; He, H.P.; Li, L.; Hao, X.J. Bioactive Phenols from the Leaves of *Baccaurea ramiflora*. *Planta Med.* **2007**, *73*, 1415–1417.
7. Yang, X.W.; He, H.P.; Ma, Y.L.; Wang, F.; Zuo, Y.Q.; Lin, H.; Li, S.L.; Li, L.; Hao, X.J. Three new vanilloid derivatives from the stems of *Baccaurea ramiflora*. *Planta Med.* **2010**, *76*, 88–90.
8. Obayed Ullah, M.; Urmi, K.F.; Howlader, A.; Hossain, K.; Ahmed, M.T.; Hamid, K. Hypoglycemic, hypolipidemic and antioxidant effects of leaves methanolic extract of *Baccaurea ramiflora*. *Int. J. Pharm. Pharm. Sci.* **2012**, *4*, 266–269.
9. Singh, R.; Singh, M.K.; Chandra, L.R.; Bhat, D.; Arora, M.S.; Nailwal, T.; Pande, V. *In vitro* antioxidant and free radical scavenging activity of *Macrotyloma uniflorum* (Gahat dal.) from Kumauni region. *Int. J. Fund. Appl. Sci.* **2012**, *1*, 7–10.
10. Singleton, V.L.; Rossi, J.A. Colorimetry of total phenolics with phosphomolybdic-phosphotungstic acid reagents. *Am. J. Enol. Vitic.* **1965**, *16*, 144–158.

11. Goyal, A.K.; Middha, S.K.; Sen, A. Evaluation of the DPPH radical scavenging activity, total phenols and antioxidant activities in Indian wild *Bambusa vulgaris* "Vittata" methanolic leaf extract. *J. Nat. Pharm.* **2010**, *1*, 34–39.
12. Zhishen, J.; Mengcheng, T.; Jianming, W. The determination of flavonoid content in mulberry and their scavenging effects on superoxide radicals. *Food Chem.* **1999**, *64*, 555–559.
13. Vuong, Q.V.; Hirun, S.; Roach, P.D.; Bowyer, M.C.; Phillips, P.A.; Scarlett, C.J. Effect of extraction conditions on total phenolic compounds and antioxidant activities of *Carica papaya* leaf aqueous extracts. *J. Herbal Med.* **2013**, *3*, 104–111.
14. Marcocci, L.; Packer, L.; Droy-Lefaix, M.T.; Sekaki, A.; Gardes-Albert, M. Antioxidant action of *Ginkgo biloba* extract (EGb 761). *Meth. Enzymol.* **1994**, *234*, 462–475.
15. Goyal, A.K.; Mishra, T.; Bhattacharya, M.; Kar, P.; Sen, A. Evaluation of phytochemical constituents and antioxidant activity of selected actinorhizal fruits growing in the forests of Northeast India. *J. Biosci.* **2013**, *38*, 797–803.
16. Prieto, P.; Pineda, M.; Aguilar, M. Spectrophotometric quantitation of antioxidant capacity through the formation of a phosphor molybdenum complex: Specific application to the determination of vitamin E. *Ann. Biochem.* **1999**, *269*, 337–341.
17. Middha, S.K.; Usha, T.; Pande, V. HPLC Evaluation of Phenolic Profile, Nutritive Content and Antioxidant Capacity of Extracts Obtained from *Punica granatum* Fruit Peel. *Adv. Pharm. Sci.* **2013**, doi:10.1155/2013/296236.
18. Middha, S.K.; Bhattacharjee, B.; Saini, D.; Baliga, M.S.; Nagaveni, M.B.; Usha, T. Protective role of *Trigonella foenum graceum* extract against oxidative stress in hyperglycemic rats. *Eur. Rev. Med. Pharmacol. Sci.* **2011**, *15*, 427–435.
19. Winter, C.A.; Risley, E.A.; Nuss, G.W. Carrageenan induced oedema in hind paw of rat as an assay for anti-inflammatory drugs. *P. Soc. Exp. Biol. Med.* **1962**, *111*, 544–547.
20. Mohanty, S.K.; Kumaraswamy, M.; Middha, S.K.; Lokesh, P.; Balasubramamanya, S.; Anuradha, M. Analgesic, anti-inflammatory, anti-lipoxygenase activity and characterization of three bioactive compounds in the most active fraction of *Leptadenia reticulata* (Wight & Arn.)—A valuable plant of medicinal importance. *Iran. J. Pharm. Sci.* **2014**, in press.
21. Comalada, M.; Camuesco, D.; Sierra, S.; Ballester, I.; Xaus, J.; Gálvez, J.; Zarzuelo, A. *In vivo* quercitrin anti-inflammatory effect involves release of quercetin, which inhibits inflammation through down-regulation of the NF-κB pathway. *Eur. J. Immunol.* **2005**, *35*, 584–592.
22. Goyal, A.K.; Basistha, B.C.; Sen, A.; Middha, S.K. Antioxidant profiling of *Hippophae salicifolia* growing in sacred forests of Sikkim, India. *Funct. Plant Biol.* **2011**, *38*, 697–701.
23. Goyal, A.K.; Middha, S.K.; Sen, A. *In vitro* antioxidative profiling of different fractions of *Dendrocalamus strictus* (Roxb.) Nees leaf extracts. *Free Rad. Antiox.* **2011**, *1*, 42–48.
24. Usha, T.; Middha, S.K. A new vista an *in vitro* new vista to identify hypoglycemic activity. *Int. J. Fund. Appl. Sci.* **2012**, *1*, 27–29.
25. Hazra, A.; Santana, B.; Nripendranath, M. Antioxidant and free radicals scavenging activity of *Spondias pinnata. BMC Complement. Altern. Med.* **2008**, *8*, 63–66.
26. Petersen, M.; Simmonds, M.S. Rosmarinic acid. *Phytochemistry* **2003**, *62*, 121–125.
27. Usha, T.; Akshya, L.; Kundu, S.; Nair, R.K.; Hussain, I.; Middha, S.K. An updated version of Phytomellitus Database *Int. J. Fund. Appl. Sci.* **2013**, *2*, 29.

28. Lin, L.; Cui, C.; Wen, L.; Yang, B.; Luo, W.; Zhao, M. Assessment of *in vitro* antioxidant capacity of stem and leaf extracts of *Rabdosia serra* (Maxim.) Hara and identification of the major compound. *Food Chem.* **2011**, *126*, 54–59.
29. Chen, J.H.; Ho, C.T. Antioxidant activities of caffeic acid and its related hydroxycinnamic acid compounds. *J. Agric. Food Chem.* **1997**, *45*, 2374–2378.
30. Lu, Y.R.; Foo, L.Y. Antioxidant activities of polyphenols from sage (*Salvia officinalis*). *Food Chem.* **2001**, *75*, 197–202.
31. Rocha, J.P.; Miguel Sepodes, B.; Eduardo-Figueira, M.; Bronze, R.; Duarte, C.; Serra, T.; Silva-Lima, B.; Mota-Filipe, H. Anti-inflammatory effect of Rosmarinic acid on a model of carrageenan-induced paw edema. *FASEB J.* **2009**, *23*, Available online: http://www.fasebj.org/cgi/content/meeting_abstract/23/1_MeetingAbstracts/571.2?sid=1506b73f-6d7a-4e94-b2ad-1e5e915f8d67 (accessed on 8 August 2014).
32. Vinegar, R.; Schreiber, W.; Hugo, R. Biphasic development of carrageenan edema in rats. *J. Pharmacol. Exp. Ther.* **1969**, *166*, 96–103.
33. Miller, N.J.; Rice-Evans, C.; Davies, M.J.; Gopinathan, V.; Milner, A. A novel methods for measuring antioxidant capacity and its application to monitoring the antioxidant status in premature neonates. *Clin. Sci.* **1993**, *84*, 407–412.
34. Loram, L.C.; Fuller, A.; Fick, L.G.; Cartmell, T.; Poole, S.; Mitchell, D. Cytokine profiles during carrageenan-induced inflammatory hyperalgesia in rat muscle and hind paw. *J. Pain.* **2007**, *8*, 127–136.
35. Ravipati, A.S.; Zhang, L.; Koyyalamudi, S.R.; Jeong, S.C.; Reddy, N.; Bartlett, J.; Smith, P.T.; Shanmugam, K.; Munch, G.W.; Ming, J.S.; *et al.* Antioxidant and anti-inflammatory activities of selected Chinese medicinal plants and their relation with antioxidant content. *BMC Complement. Altern. Med.* **2012**, *12*, 1–14.
36. Tohda, A.; Nakayama, N.; Hatanaka, F.; Komatsu, K. Comparison of anti-inflammatory activities of six curcuma rhizomes: A possible curcuminoid-independent pathway mediated by *Curcuma phaeocaulis* extract. *Evid. Based Complement. Altern. Med.* **2006**, *3*, 255–260.
37. Mogana, R.; Teng Jinh, K.; Wiart, C. Anti inflammatory activities of leaf methanol extract of *Cotyledon orbiculata* L. (Crassulaceae). *Evid. Based Complement. Altern. Med.* **2013**, doi:10.1155/2013/734824.

11

Studies on Modulation of Gut Microbiota by Wine Polyphenols: From Isolated Cultures to Omic Approaches

Montserrat Dueñas [1], Carolina Cueva [2], Irene Muñoz-González [2], Ana Jiménez-Girón [2], Fernando Sánchez-Patán [2], Celestino Santos-Buelga [1], M. Victoria Moreno-Arribas [2] and Begoña Bartolomé [2,*]

[1] Grupo de Investigación en Polifenoles, Unidad de Nutrición y Bromatología, Facultad de Farmacia, Universidad de Salamanca, Campus Miguel de Unamuno, Salamanca 37007, Spain; E-Mails: mduenas@usal.es (M.D.); csb@usal.es (C.S.-B.)

[2] Grupo de Biotecnología Enológica Aplicada, Instituto de Investigación en Ciencias de la Alimentación (CIAL), CSIC-UAM, C/Nicolás Cabrera 9, Campus de Cantoblanco, Madrid 28049, Spain; E-Mails: carolina.cueva@csic.es (C.C.); irene.munoz@csic.es (I.M.-G.); a.jimenez.giron@csic.es (A.J.-G.); fspatan@ifi.csic.es (F.S.-P); mvmoreno@ifi.csic.es (M.V.M.-A.)

* Author to whom correspondence should be addressed; E-Mail: b.bartolome@csic.es

Academic Editors: Jeffrey B. Blumberg and C.-Y. Oliver Chen

Abstract: Moderate consumption of wine seems to produce positive health effects derived from the occurrence of bioactive polyphenols. The gut microbiota is involved in the metabolism of phenolic compounds, and these compounds and/or their metabolites may modulate gut microbiota through the stimulation of the growth of beneficial bacteria and the inhibition of pathogenic bacteria. The characterization of bacterial metabolites derived from polyphenols is essential in order to understand their effects, including microbial modulation, and therefore to associate dietary intake with particular health effects. This review aims to summarize the current information about the two-way "wine polyphenols–gut microbiota" interaction, from a perspective based on the experimental and analytical designs used. The availability of advanced methods for monitoring bacterial communities, along with the combination of *in vitro* and *in vivo* models, could help to assess the metabolism of polyphenols in the human body and to monitor total bacterial communities, and, therefore, to elucidate the implications of diet on the modulation of microbiota for delivering health benefits.

Keywords: wine; polyphenols; gut microbiota; probiotics; modulation; batch culture fermentation; gastrointestinal simulators; animal models; human studies

1. Introduction

Wine is considered to be a high bioactive polyphenol content source. Many studies have revealed the key role played by phenolic compounds from grapes and wine on human health; cardiovascular diseases being the pathologies that have received much attention [1,2].

Several studies confirm the importance of the intestinal microbiota to the health of the host, including mental health [3]. Gut bacteria not only help to maximize the absorption of nutrients and energy, but also are essential in the body's defense mechanisms [4]. Although polyphenol metabolism starts in the mouth and proceeds along the gastrointestinal tract, most of the dietary polyphenols reach the colon, where they are subjected to the action of the gut microbiota, thus releasing aglycones that might, to a certain extent, be absorbed and degraded to simpler phenolic derivatives and other metabolites which could present higher activity at a physiological level than the corresponding food precursors [5]. These metabolites could also be absorbed, increasing polyphenol bioavailability [6]. Non-absorbed polyphenols and/or the resultant phenolic metabolites could affect the growth of gut microbiota, thereby modifying their diversity and metabolic activity. In fact, modulation of gut microbiota by polyphenols has been a topic that has gained increasing attention from the scientific community in recent years, as can be seen from several reviews [7–12]. Among the major wine phenolic compounds that may reach the gut, special attention has been given to polymeric flavan-3-ols or proanthocyanidins (also known as condensed tannins) since there is evidence that these polyphenols promote the growth of beneficial bacteria and the inhibition of pathogenic bacteria while they are extensively metabolized by gut microbiota to produce a great range of active metabolites. As relevant reference, an intervention study of cocoa flavan-3-ols in healthy volunteers has shown that they enhance the growth of *Lactobacillus* spp. and *Bifidobacterium* spp. and limit the growth of the *Clostridium histolyticum* group [13]. On the other hand, proanthocyanidins have been found to be largely metabolized into phenylvalerolactones and phenolic acids after cocoa intake in humans and rats [14].

The aim of this review was to summarize the information available on the action of gut microbiota on wine polyphenols (metabolism), as well as the effect of phenolic compounds on the growth of gut bacteria (modulation). This two-way polyphenols–gut microbiota interaction will be assessed from a perspective based on the experimental designs used, from isolated cultures to omic approaches in the case of microbiota analysis, and advanced analytical techniques in the case of metabolite analysis.

2. Phenolic Compounds in Wine

The term "phenolic" describes those compounds that possess a benzenic ring substituted by one or several hydroxyl groups (−OH). Polyphenols are secondary metabolites of plants and play an important role in the plant's defense mechanism against external agents, such as animals or microbial infections, they facilitate pollination and seed dispersion through signals that attract insects and animals, and participate in protection mechanisms against ultraviolet radiation and/or oxidant agents. Phenolic compounds presented in grapes, located in the solid parts of the fruit (skins and seeds), have a wide diversity of chemical structures, including flavonoid compounds (flavan-3-ols (monomers and oligomeric and polymeric proanthocyanidins), anthocyanins, flavonols, and dihydroflavonols) and non-flavonoid compounds (hydroxybenzoic and hydroxycinnamic acids, phenolic alcohols, and stilbenes) (Figure 1). Although the concentration of phenolic compounds in wine is conditioned by several factors related to the grape (variety, soil, geography, climate, *etc.*) and by enological practices, the total polyphenol content is around 50–400 mg/L for white wines, and 900–1400 mg/L for young red wines. Therefore, moderate consumption of wine (250 mL/day) corresponds to an intake of 60 mg of polyphenols for white wines and 210 mg for young red wines [15]. With respect to their distribution by compound groups, acids and hydroxybenzoic derivatives represent approximately 6% of the total; acids and hydroxycinnamic derivatives, 1.1%; stilbenes, 0.5%; alcohols, 3.8%; flavanols, 15%; flavonols, 3.6%; and anthocyanins, 70% in young red wines [15]. Other anthocyanin derivatives such as pyranoanthocyanins present much lower proportions.

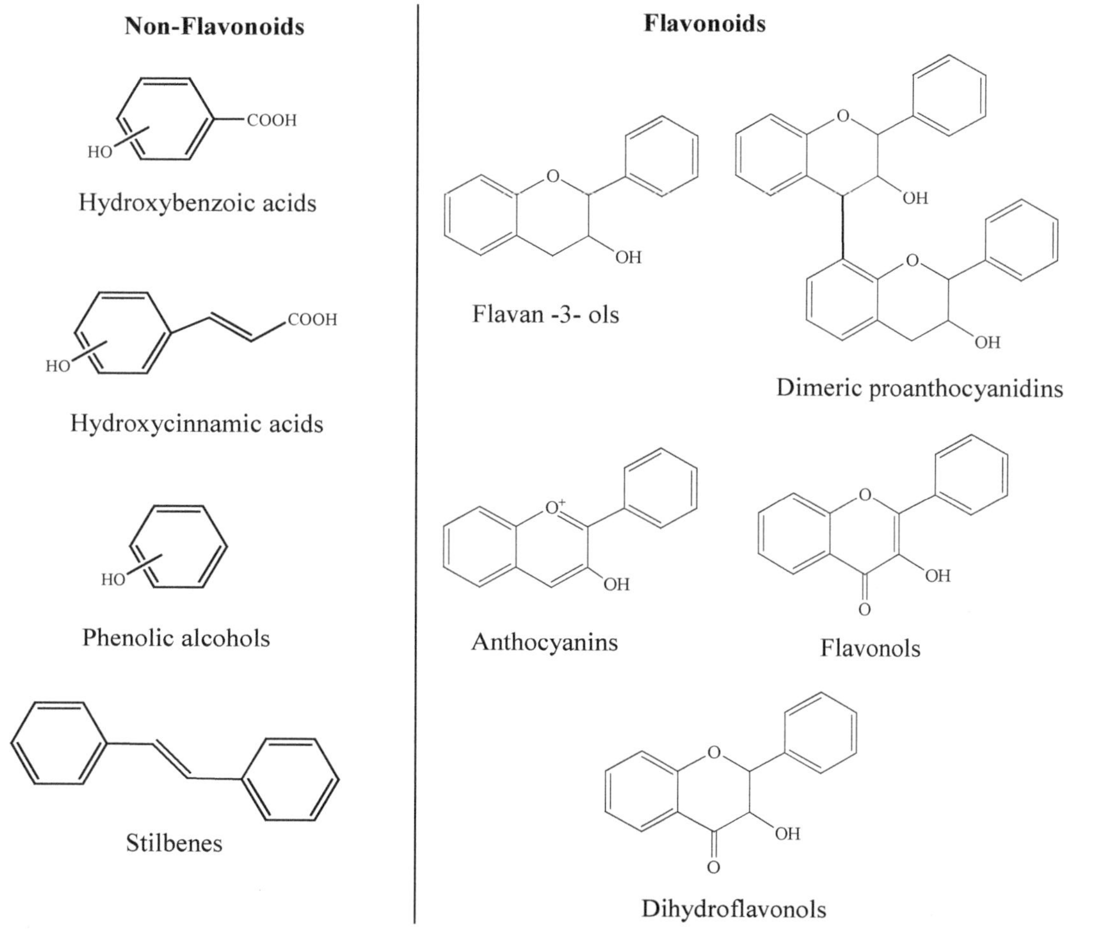

Figure 1. Common phenolic compounds in wine.

Over the last few decades, the role of polyphenols has been the focus of considerable research in the field of nutrition. Several epidemiological studies have shown that the intake of these compounds is inversely associated with the risk of various chronic diseases, such as coronary heart disease, specific cancers, and neurodegenerative disorders [16,17]. Indeed, potential beneficial effects have been demonstrated for different phenolic compounds (especially flavonoids) through *in vitro* assays. In particular, these compounds act as powerful inhibitors of low-density lipoprotein (LDL) oxidation, one of the main mechanisms responsible for the development of atherosclerosis. However, it is currently believed that the physiological activities and mechanisms of these compounds are more diverse and complex. Thus, phenolic compounds are able to inhibit the growth of human cancer cell lines, cholesterol-related processes, and the activity of enzymes, such as telomerase, lipoxygenase and cyclooxygenase involved in inflammatory processes [18,19]. They also interact in different signal transduction pathways, and can affect the cell-cycle regulation, platelet function, and prevent endothelium dysfunction [20,21].

However, the health effects of these compounds depend on their bioavailability, and therefore it is important to understand how they are absorbed, metabolized and eliminated from the body, in order to ascertain their *in vivo* actions.

3. General Metabolism of Polyphenols in the Human Body

Polyphenols are considered as xenobiotics by the human organism and therefore are extensively metabolized and finally eliminated, mainly in the bile but also in the urine. The first step in their metabolism is likely to be deglycosylation before absorption in the small intestine. Hydrolysis of some flavonoid glycosides might have already occurred in the oral cavity, as both saliva and oral microbiota show β-glucosidase activity, giving rise to the corresponding aglycones. The hydrolytic activity begins in the mouth, and continues through the digestive tract into the stomach, where the size of food particles is reduced, which prompts the release of phenolic compounds. It has been estimated that 5%–10% of ingested polyphenols are absorbed in the small intestine, while 90%–95% reach the colon where they are intensively degraded by microbiota into a diversity of bioactive phenolic metabolites, lactones and phenolic acids that are then further absorbed [22].

In particular, the glycosylated polyphenols, such as anthocyanins, flavonol glycosides and glycosides of resveratrol, can be hydrolyzed by intestinal β-glucosidases. In contrast, monomeric flavanols, and dimer procyanidins can be absorbed directly into the small intestine. Once absorbed, the resulting aglycones would enter enterocyte by passive diffusion. Thus, the resulting aglycone is rapidly biotransformed by phase II enzymes into conjugated metabolites (*i.e.*, glucuronides, *O*-methylethers and/or sulfates) within the enterocyte and again in the liver [23]. Other wine polyphenols, mainly oligomeric flavan-3-ols with a degree of polymerization (mDP) >3 and polymeric flavanols (proantocyanidins and condensed tannins), esters of hydroxycinnamic acids, and flavonols conjugated with rhamnose, such as rutin, are not absorbed in their native forms. These compounds reach the colon, where compounds are subjected to the action of the colonic microflora and transformed into various phenolic acids and other metabolites [24]. The methylated, glucuronide and sulfate conjugates (phase II metabolites) can reach the colon via enterohepatic circulation and are also susceptible to degradation by the intestinal microbiota. Finally, the phenolic metabolites are excreted via urine and feces.

4. Gut Microbiota

The orogastrointestinal tract of humans has an abundant microbiota dominated by anaerobic bacteria. The number of bacteria in the oral cavity is about 10^{11} bacteria/g in dental plaque and 10^{8}–10^{9} bacteria/mL in saliva, whereas in feces the corresponding numbers are 10^{11}–10^{12} bacteria/g [25]. More precisely, the gut microbial ecosystem includes native species that permanently colonize the gastrointestinal tract, and a variable number of live microorganisms that temporarily pass through the digestive tract [4]. Native bacteria are mainly acquired at birth and during the first year of life, whereas transient bacteria are continuously being ingested from food, drinks and the environment.

Among the human gastrointestinal microbiota, the majority of the species belong to the phyla Firmicutes, Bacteroides, Actinobacteria and Proteobacteria [26] (Figure 2). The phylogenetic composition of the gut microbiota is considered specific and stable over time for each individual. The species vary greatly between individuals. In fact, interindividual variation in gut microbiota may, in part, reflect differences in dietary intake, although the response of the gut microbiota to dietary change can also differ among individuals. The composition of the individual's microbiota can fluctuate under some circumstances, for instance acute diarrheal illnesses, antibiotic treatment, or to a lesser extent when being induced by dietary interventions, but the individual flora composition patterns usually remain constant [27]. Among the microbiota associated with the esophagus, are included microorganisms belonging to the genera *Streptococcus*, *Prevotella* and *Veillonella*, which also appear in the oral cavity. The density of colonization is increased about eight times from proximal regions of the small intestine (10^{3} bacterias/g) until the colon. In the stomach and duodenum, the number of microorganisms is reduced due to acid, bile and pancreatic secretions; as advances in the small intestine, the acidity decreases due to the dilution of the acid, which facilitates bacterial colonization, reaching 10^{11} CFU (colony forming units)/mL in the colon. The most frequently found bacteria in this area are members of the genus *Bacteroides*, *Bifidobacterium*, *Eubacterium*, *Clostridium*, *Lactobacillus*, and Gram-positive cocci [28], while *Enterococci* and representatives of the *Enterobacteriaceae* family are found to a lesser extent [29].

Currently, there is evidence that confirms the importance of the gut microbiota in host health is associated to bacterial groups that colonize the intestine. The large intestine contains a complex and dynamic microbial ecosystem composed of commensal bacteria, potentially harmful opportunistic bacteria and others that can have both effects [30]. In particular, having a well-balanced gut microbiota composition is essential for human health. Undesirable bacteria include species of the genus *Clostridium*, *Staphylococcus* and *Veillonella*. These species can produce potentially harmful products, such as toxins and carcinogens, which are associated with intestinal disorders such as chronic inflammatory bowel diseases and other immune-related disorders. However, the use of antibiotics can disrupt the ecological balance and allow the overgrowth of species with potential pathogenicity, such as *Clostridium difficile*, associated with pseudomembranous colitis [31]. With regards to beneficial bacteria, among which are mainly included species of the genus *Lactobacillus* and *Bifidobacterium*, these play a key role in nutritional and disease-prevention functions, so they are used as probiotics. The main functions of these bacteria are: decreasing gas production, the production of short-chain fatty acids (SCFA), immunostimulating and antitumor activity [32,33].

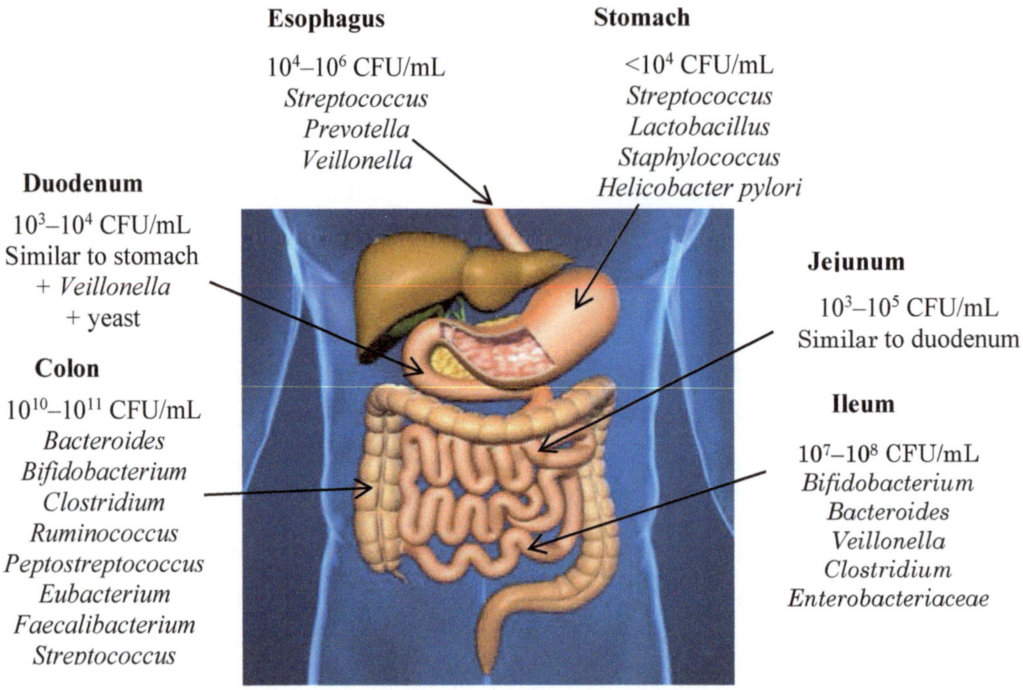

Figure 2. Distribution and composition of bacterial species in the gastrointestinal tract. CFU: colony forming units.

Although the investigations concerning the influence of polyphenols on the gut microbiota and their mechanisms of action in humans are scarce, nevertheless drastic changes in fecal- and mucosa-associated microbiota could be related to determined diseases, such as colon cancer, short bowel syndrome and obesity [34]. It is important, therefore, to investigate gut microbiota composition and the influence of dietary intake on the human microbiome, to elucidate the implications of diet on the modulation of microbiota for delivering health benefits.

The availability and feasibility of advanced methods for monitoring total bacterial communities is a prerequisite to be considered when designing a study to assess gut microbiota. Over recent decades, the introduction of culture-independent techniques like terminal restriction fragment length polymorphism (T-RFLP), fluorescence *in situ* hybridation (FISH), denaturing gradient gel electrophoresis (DGGE), and quantitative polymerase chain reaction (qPCR) have improved the analysis of gut microbiota. At present, next-generation sequencing (NGS) techniques have promoted the emergence of new, high-throughput technologies, such as genomics, metagenomics, transcriptomics, metatranscriptomics, *etc.* The development of these techniques has provided the opportunity to explore the taxonomic, protein-coding gene or expression diversity by applying more comprehensive and less biased measurements to all systems involved (*i.e.*, diet, microbiota and host) [35]. However, the enormous amount of data generated becomes cumbersome to analyze, requiring much dedicated time as well as expertise to manage data in such quantity [36]. The link between high-throughput qPCR and next generation sequencing technologies provides manageable data with valuable quantitative and taxonomic information. NGS platforms involve many different technologies [37] all of which generate large, genome-scale datasets. Examples of current commercial platforms are the 454 (Roche), Solexa (Illumina), SOLiD and Ion Torrent (Life Technologies), and PacBio (Pacific Biosciences) systems.

5. Interaction between Wine Polyphenols and Gut Microbiota

Taking the above-mentioned points into account, it is evident that the gut microbiome is implicated in the metabolism of wine polyphenols, and it also suggests that phenolic compounds and their metabolites may modulate the gut microbiota to a certain extent. The so-called two-way interaction between wine polyphenols and gut microbiota [7] can be subjected to various factors related to the phenolic compounds concentration and structure as well as the characteristics of bacterial strains, microbial environment, food matrix, *etc.* Prior to the specific sections about wine polyphenol metabolism and bacterial modulation by polyphenols, a preliminary section about experimental and analytical approaches used in these studies is included.

5.1. Experimental Designs and Analytical Techniques

5.1.1. *In Vitro* Gastrointestinal Tract Simulators

Studies regarding polyphenol metabolism and microbial modulation have been carried out using different experimental designs, from simple approaches involving fermentation experiments in batch cultures and in continuous simulators, to intervention trials in animals and humans. Although *in vivo* trials are most relevant physiologically, *in vitro* tools have been designed to simulate intestinal conditions and the transformations undergone in foods during transit through the gastrointestinal tract.

Simply, static gut models, also known as batch-type cultures, are generally closed systems using sealed bottles or reactors containing suspensions of fecal material, which are maintained under anaerobic conditions. This model approach is primarily used when assessing the stability of polyphenols in the presence of human-derived gut microbiota and evaluating which environmental conditions favor or limit polyphenol bioconversion.

In contrast to short-duration experiments with static gut models, longer-term experiments are required when the adaptation of the gut microbial community to dietary polyphenols needs to be assessed. To that end, simulators of the gastrointestinal tract comprise stomach and small intestinal sections for the pre-digestion of food as well as vessels stimulating the ascending, transverse and descending sections of the human colon, allowing the assessment of changes in colonic areas that are very challenging to access in a human intervention, as well as the physiological variables (*i.e.*, retention time, pH variations, gastrointestinal fluids, microbiota).

In the past few years, dynamic *in vitro* gut models, such as SHIME (simulator of the human intestinal microbial ecosystem) (University of Ghent, Belgium) [38], the TIM-1 and TIM-2 (developed by TNO, Netherlands Organization for Applied Scientific Research, The Netherlands) [39] and the recent SIMGI (dynamic gastrointestinal tract simulator) (CIAL, CSIC-UAM, Spain), have been developed to simulate physiological conditions that can have an influence on the gut microbiota and their metabolic activity. All these systems contain different compartments representing the stomach, small intestine and colon.

5.1.2. Analytical Approaches to the Analysis of Phenolic Microbial-Derived Metabolites

Rapid, sensitive, and reliable analytical methods are needed for profiling microbe-derived phenolic metabolites in biological fluids in order to determine their contribution to the overall bioavailability of polyphenols and to allow identification of biomarkers of polyphenol exposure that could be further used to correlate dietary intake with particular health effects, including microbial modulation.

Gas chromatography (GC) coupled to mass spectrometry techniques (MS) and liquid chromatography (LC) coupled to a diode array detector (DAD), electrochemical (ECD) and, in particular, tandem mass spectrometry (ESI-MS/MS) are the most widely used analytical techniques for the analysis of microbial phenolic metabolites in biological fluids, which can be found in micromolar concentration. GC methodologies provide higher resolution and sensitivity than LC methodologies but a laborious sample preparation stage is required, since they usually involve the isolation of metabolites by liquid-liquid extraction (LLE) or other extraction procedures followed by further derivatization [40]. In contrast, for LC analysis the sample preparation step is more simple and faster and provides a very sensitive method for the quantification of selected phenolic metabolites when coupled to MS/MS [41]. The efficiency of separation and the sensitivity of LC–MS/MS methodologies can be largely improved by the use of ultra performance liquid chromatography (UPLC), which operates with smaller particle size (<2 μm) sorbent materials and at very high pressures (up to 15,000 psi). This technique coupled to MS detection is considered a good application in the analysis of phenolic compounds in biological fluids and food matrixes [41].

Metabolomics approaches have been used in clinical, pharmaceutical, and toxicological applications and have also recently emerged as fields of increasing importance in food and nutrition sciences [42]. The metabolome can be affected by different external and internal factors. Among them, diet is a very important external factor affecting the urinary metabolome because it produces notable changes in its composition [43]. Metabolomics studies of the polyphenols may lead to the discovery of new phytochemical metabolites and new biomarkers of intake that could allow the intake of dietary phytochemicals to be monitored, and eventually relate them to the expected biological effects [44].

5.2. Biotransformation of Wine Polyphenols by Gut Microbiota

The microbial biotransformation of wine polyphenols is widely influenced by their chemical structure. Oligomers and polymers of flavan-3-ols are the major phenolic compounds present in the wine that reaches the colon. The colonic metabolism of dimeric procyanidins involves C-ring opening, followed by lactonization, decarboxylation, dehydroxylation, and oxidation reactions, among others [45] (Figure 3). In the case of galloylated monomeric flavan-3-ols, the microbial metabolism usually starts with the rapid cleavage of the gallic acid ester moiety by microbial esterases, giving rise to gallic acid which is further decarboxylated into pyrogallol. The C-ring is subsequently opened, giving rise to 1-(3′,4′-dihydroxyphenyl)-3-(2″,4″,6″-trihydroxyphenyl)-propan-2-ol, which is later converted into 5-(3′,4′-dihydroxyphenyl)-γ-valerolactone in the case of (epi)catechin or 5-(3′,4′,5′-trihydroxyphenyl)-γ-valerolactone in the case of (epi)gallocatechin. The valerolactone ring later breaks, giving rise to 5-(3′,4′-dihydroxyphenyl)valeric acid and/or 4-hydroxy-5-(3′,4′-dihydroxyphenyl) valeric acid. Subsequent biotransformations of these valeric acids give rise to hydroxyphenylpropionic and hydroxybenzoic acids by successive loss of carbon atoms from the side chain through β-oxidation (Figure 3).

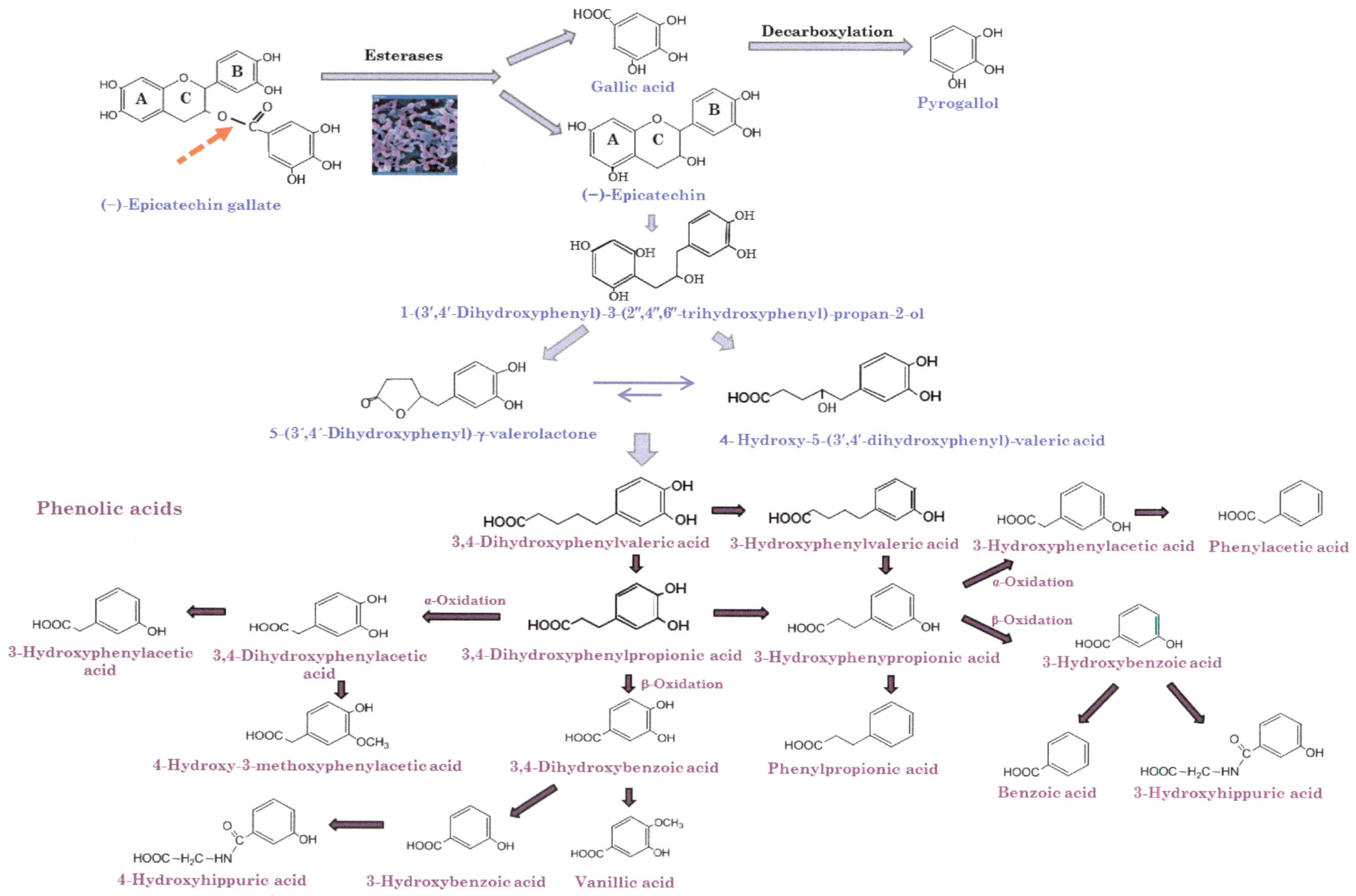

Figure 3. Catabolism of monomeric flavan-3-ols by gut microbiota.

With regard to microbial catabolism of flavonols, other flavonoid compounds present in the wine are directly transformed into 3,4- or 3,5-dihydroxylated phenylacetic acids [46]. In the case of anthocyanins, they are converted into 3,4-dihydroxy-, 4-hydroxy-, 3,4-dimethoxy- or 3-methoxyl-4-hydroxyl benzoic acids according to the substitution pattern of the B-ring of the precursor anthocyanin molecule [46]. However, in spite of the fact that anthocyanins are abundant in wine, their circulating levels in plasma are very low, which has been attributed to anthocyanin instability under neutral pH [47], their extensive metabolism *in vivo* [48], and their probable catabolism by intestinal microbiota [5]. Thus, it has been described that less than 1% of the ingested anthocyanins are typically absorbed and excreted in the urine [49], which might be found in low concentrations in serum, feces, urine (in the nanomolar concentrations) [50].

In the case of non-flavonoid compounds present in wine, hydroxycinnamic esters (*i.e.*, caffeic acid derivatives) are mainly transformed into 3-hydroxyphenylpropionic acid, benzoic acid and 4-ethylcatechol. Once absorbed, the microbial metabolites are mainly metabolized in the liver by phase II enzymes as conjugated metabolites (glucuronides and sulfates). Other reactions that can take place are glycination, dehydrogenation, hydroxylation and methylation. In fact, different studies have proved the presence of phase II and microbial-derived phenolic metabolites in plasma and urine after ingestion of wine and wine polyphenols (see for review [51,52]). Some previous studies have confirmed that a moderate intake of red wine effectively led to a significant increase of phenolic metabolites in human feces [53,54].

Despite the advances recently made in the knowledge of the identification of phenolic metabolites, the specific bacterial species able to metabolize most wine polyphenols in the gastrointestinal tract and the anaerobic degradation pathways have still not been identified. So far, only a few bacterial species, some of them belonging to the class *Clostridiales*, such as *Eubacterium* sp. and *Flavonifractor* sp., and *Eggerthella* spp., have been reported to be able to initiate the metabolism of flavanol-3-ols [55–57].

5.3. Modulation of Gut Microbiota by Wine

Tables 1 and 2 report different studies of modulation of gut microbiota by wine and grape polyphenols using different experimental approaches: batch culture fermentations, human gastrointestinal simulators, animal model studies and human interventions. For all of them, details about experimental conditions and/or study design, microbial techniques used and main effects on bacteria groups (growth enhancement, growth inhibition or no effect) are included.

Table 1. Studies regarding modulation of gut microbiota by wine polyphenols using batch culture fermentations and gastrointestinal tract simulator.

Studies Using Batch Culture Fermentation								
Reference	**Fecal Concentration**	**Phenolic Compound/Food**	**Dose**	**Time of Incubation**	**Microbial Technique**	**Growth Enhancement**	**Growth Inhibition**	**No Effect**
[58]	10%, w/v	(+)-Catechin	150 mg/L, 1000 mg/L	<48 h	FISH	*Lactobacillus–Enterococcus* spp.; *Bifidobacterium* spp.; *C. coccoides–E. Rectale* group *E. coli*	*C. histolyticum* group	
[59]	10%, w/v	Malvidin-3-*O*-glucoside Anthocyanidins mixture	20 mg/L and 200 mg/L 4850 mg/L and 48,500 mg/L	<24 h	FISH	*Lactobacillus–Enterococcus* spp.; *Bifidobacterium* spp.; *C. coccoides–E. rectale* group		
[60]	10%, w/v	Grape seed extract fractions	300–450 mg/L	<48 h	FISH	*Lactobacillus–Enterococcus* spp.	*C. histolyticum* group	
[61]	1% w/v	Red wine extract	600 mg/L	48 h	FISH		*C. histolyticum* group	*Lactobacillus–Enterococcus spp.*
[62]		Red wine extract	500 mg/L	48 h	qPCR	*Lactobacillus* spp.; *Bifidobacterium* spp.; *Bacteroides* spp.; *Ruminococcus* spp.		
[63]	20% w/v	Red wine/ grape extract	500–1000 mg/L	72 h	HITChip			

Studies Using a Gastrointestinal Simulator								
Reference	**Simulator**	**Phenolic Compound/Food**	**Dose**	**Time**	**Microbial Technique**	**Population Increase**	**Population Decrease**	**No Effect**
[64]	Twin-SHIME	Red wine-grape extract	3 × daily dosing (1000 mg polyphenols as total daily dose)	2 weeks	Plate count qPCR PCR-DGGE; Pyrosequencing	*Klebsiella* spp.; *Alistipes* spp.; *Cloacibacillus* spp.; *Victivallis* spp.; *Akkermansia* spp.	Bifidobacteria; *Blautia coccoides* group; *Anaeroglobus* spp.; *Subdoligranulum* spp. *Bacteroides*	

Table 2. Studies regarding modulation of gut microbiota by wine polyphenols in studies with animals and humans.

Animal Model Studies								
Reference	**Simulator**	**Phenolic Compound/Food**	**Dose**	**Time**	**Microbial Technique**	**Population Increase**	**Population Decrease**	**No Effect**
[65]	Rats	Red wine polyphenols powder	50 mg/kg	16 weeks	Plate count	Lactobacilli; Bifidobacteria	Propionibacteria; *Bacteroides*; Clostridia	
[66]	Broiler chicks	Grape seed extract (GSE)	7.2 g/kg diet (GSE) (free access)	21 days	Plate count T-RFLP	*E. coli; Enterococcus* spp.; *Lactobacillus* spp.		
[67]	Pigs	Grape seed extract	1% (free access)	4 weeks	qPCR		*Streptococcus* spp.; *Clostridium* Cluster XIVa	*Lactobacillus* spp; *Bifidobacterium* spp.
[68]	Pigs	Grape seed extract	1% w/w	6 days	Ilumina MiSeq platform	*Lachnospiraceae, Clostridales, Lactobacillu, Ruminococcacceae*		

Human Intervention Studies								
Reference	**Volunteer Numbers**	**Phenolic Compound/Food**	**Dose**	**Treatment Duration**	**Microbial Technique**	**Population Increase**	**Population Decrease**	**No Effect**
[69]	9	Proantocyanidin-rich extract from grape seeds	0.5 g/day	6 weeks	Plate count	*Bifidobacterium* spp.	*Enterobacteriaceae*	
[70]	10	Red wine	272 mL/day	20 days	qPCR	*Enterococcus* spp.; *Prevotella* spp.; *Bacteroides Bifidobacterium* spp.; *Bacteroides uniformis* *Eggerthella lenta Blautia coccoides–E. rectale* group	*Clostridium* spp.; *C. histolyticum* group	Actinobacteria

5.3.1. Studies Using Batch Culture Fermentations

One of the first relevant experiments using batch culture fermentation and a standard compound present in grape seeds and wine was carried out by Tzounis *et al.* [58] who found that the flavan-3-ol monomers ((−)-epicatechin and (+)-catechin) promoted the growth of the *Clostridium coccoides–Eubacterium rectale* group, which is known to produce large amounts of butyrate, a short-chain fatty acid (SCFA) with anti-inflammatory and antineoplastic properties (Table 1). Compared to (−)-epicatechin, (+)-catechin exposure resulted in a greater modification of the growth of the bacterial groups; (+)-catechin also increased the growth of *Lactobacillus–Enterococcus* spp., *Bifidobacterium* spp. and *Escherichia coli* but decreased the growth of *Clostridium histolyticum*. In contrast, the effect of (−)-epicatechin only significantly increased the growth of the *C. coccoides–E. rectale* group. Also using standard compounds of anthocyanins (*i.e.*, malvidin-3-glucoside and a mixture of anthocyanins), Hidalgo *et al.* [59] found a significant increase in the growth of *Lactobacillus–Enterococcus* spp. and *Bifidobacterium* spp.

Recently, our research group carried out several batch culture fermentations of two flavan-3-ol fractions with different degrees of polymerization from grape seed extract and of wine polyphenols, with fecal microbiota from three healthy volunteers [60,61]. Both flavan-3-ol fractions induced the growth of *Lactobacillus/Enterococcus* spp. and inhibited the *C. histolyticum* group during fermentation, although the effects were only statistically significant with the less polymerized fraction. However, wine polyphenols only showed a slight inhibition in the *C. histolyticum* group, probably due to their lower content in flavan-3-ols and the time of exposure. Additionally, this type of fermentation has also been used to assess the contribution of certain probiotic strains to the colonic metabolism of polyphenols. With this in mind, Barroso *et al.* [62] carried out fermentations of a red wine extract inoculated with human microbiota obtained from the colonic compartments of a dynamic simulator, in the presence and absence of the probiotic strain *Lactobacillus plantarum* IFPL935. Microbial analysis by qPCR indicated that red wine polyphenols induced greater variations among *in vitro* batches harboring different colon-region (ascending colon, descending colon and effluent) microbiota than those found when *L. plantarum* IFPL935 was added. Batches inoculated with microbiota from the ascending colon were shown to harbor the major proportion of saccharolytic bacteria (*Bacteroides*, *Bifidobacterium*, *Prevotella*) whereas *Clostridium* groups were found in major numbers in the batches inoculated with microbiota simulating the distal regions [62].

Gross *et al.* [63] carried out *in vitro* fermentation of a red wine/grape and black tea polyphenols with fecal samples from 10 human volunteers. Microbial composition of fecal samples using HITChip analysis indicated different bacterial populations present in the individual fecal microbial communities. This inter-individual variability was related to the production of different metabolites, suggesting that each individual generates a specific microbial metabolome.

5.3.2. Studies Using Human Gastrointestinal Simulators

As an example of the versatility and potential of human gastrointestinal simulators, Table 1 also reports a study concerning the modulation of gut microbiota by polyphenols using the SHIME [38]. This validated model comprises stomach and small intestinal sections for predigestion of food as well as vessels stimulating the ascending, transverse and descending parts of the human colon, allowing assessment of changes in the different colonic areas that are very challenging to access in a human intervention.

The use of the twin-SHIME to investigate the effects of a red wine extract on the colonic microbiota was the aim of a study recently carried out by Kemperman *et al.* [64]. These authors characterized microbial community changes using a combination of analyses including cultivation, PCR-denaturing gradient gel electrophoresis (DGGE), quantitative PCR and high-throughput pyrosequencing of the 16S ribosomal RNA gene. They observed that the continuous administration of red wine extracts for two weeks could modulate select members of the human gut microbiota, revealing novel targets potentially involved in polyphenol metabolism and/or resistant microbes.

5.3.3. Animal Model Studies

It is widely known that preliminary evidence should be warranted in animal models before human intervention trials. Animal models contribute to a better understanding of the mechanisms and biological effects that could be likely to happen in the human body. Caution is required in extrapolating results to humans because culture-independent comparisons have revealed that most bacterial genera and species found in mice are not seen in humans, although the distal gut microbiota of mice and humans harbor the same bacterial phyla. Nevertheless, the use of germ-free mice inoculated with human fecal microbiota is of great relevance. Such humanized mice provide a model system for controlling host genotype, gut microbiota composition, diet, and housing conditions. To date, no studies have followed the composition of the intestinal microbiota in humanized rodents fed a diet rich in wine polyphenols [71].

In this section, studies performed on animals in order to assess the effects of wine polyphenols on the modulation of intestinal microbiota are summarized (Table 2).

Dolara *et al.* [65] showed that treatment with wine polyphenols in carcinogen-treated F344 rats was associated with a strong variation in the colonic microbiota, compared to the control-fed rats. Although the total bacterial counts and anaerobe/aerobe ratio of microorganisms in the feces from polyphenol-treated rats were similar to those from control rats, Propionibacteria, *Bacteroides* and *Clostridia* decreased while Lactobacilli and Bifidobacteria increased. Based on additional experiments, these authors concluded that reduction of oxidative damage, modulation of colonic flora and variation in gene expression may all be connected in the action of wine polyphenols on the intestinal function and carcinogenesis.

Another animal experiment was conducted to study the effect of the inclusion of grape seed extracts in the diet of broiler chicks [66] on intestinal microbiota. It was observed that, for the cecum, birds fed grape extracts had higher populations of *E. coli*, and *Lactobacillus* and *Enterococcus* species than birds in control groups. These authors concluded that polyphenol-rich grape products modified the gut morphology and intestinal microbiota and increased the degree of biodiversity in intestinal bacteria of broiler chicks.

Animal studies performed in pigs [67,68] demonstrated that grape seed extract administration caused an ecological shift in the microbiome. Thus, the administration of grape seed extracts produced lower counts of *Streptococcus* spp and *Clostridium* cluster XIVa [67], and increased *Lachnospiraceae*, *Clostridales*, *Lactobacillus* and *Ruminococcus* during the intervention period [68].

5.3.4. Human Intervention Studies

Investigations carried out with humans potentially provide the best models for studying the interactions of food components (e.g., polyphenols) with microbiota, although *in vivo* intervention trials hold inevitable practical and ethical limitations [8]. The use of crossover designs, where volunteers serve as their own control, permit multilevel analysis schemes that increase power, but as with other human intervention studies, they require a relevant number of volunteers to allow for statistically significant multivariate models [72]. Up to now, only a few studies have examined the *in vivo* impact of dietary polyphenols on the human gut microbiota, and most of them were focused on single polyphenol molecules and selected bacterial populations (Table 2).

In a study with a reduced number of subjects (n = 9), Yamakoski *et al.* [69] reported that administration of a proanthocyanidin-rich extract from grape seed significantly increased the fecal number of *Bifidobacterium* spp., whereas a reduction in the level of putrefactive bacteria, such as *Enterobacteriaceae* was observed.

Queipo-Ortuño *et al.* [70] performed a randomized, crossover and controlled trial (n = 10), which studied the effect of the intake of red wine, de-alcoholized red wine and gin over three consecutive periods. After the red wine period, the bacterial concentrations of *Proteobacteria*, *Fusobacteria*, *Firmicutes* and *Bacterioidetes* markedly increased compared with the washout period; significant increases in the number of *Bifidobacterium* spp. and *Prevotella* spp. were also observed. In contrast, *Clostridium* spp. and *C. histolyticum* group concentrations decreased after the red wine period.

6. Conclusions

Assessment of microbial metabolism of wine polyphenols is necessary to clarify the potential physiological effects on human health associated with moderate consumption of wine. There is scientific evidence that polyphenols may contribute to the maintenance of gut health by the modulation of the gut microbial balance through the stimulation of the growth of beneficial bacteria and the inhibition of pathogen bacteria, exerting prebiotic-like effects. Although genetic and environmental factors are the main determinants of gut microbiota composition, it is well established that diet influences microbial fermentation and total bacteria in the intestine. In fact, inter-individual variation in gut microbiota may, in part, reflect differences in dietary intake, although the response of the gut microbiota to dietary change can also differ among individuals. Therefore, the availability of advanced methods for monitoring bacterial communities could provide a better understanding of the underlying mechanisms in the polyphenols–microbiota–host triangle, and elucidate the implications of polyphenols on host health.

Acknowledgments

Authors of this review were funded by the Spanish MINECO through different projects (AGL2012-04172-C02-01, AGL2010-17499, BFU2012-35228), the CONSOLIDER INGENIO 2010 Program (project FUN-C-FOOD, CSD2007-063), as well by the CSIC ("Intramural" project 201270E065) and the "Comunidad de Madrid" (project ALIBIRD P2009/AGR-1469). Author M.D. thanks the Spanish "Ramón y Cajal" Programme for a contract.

Conflicts of Interest

The authors declare no conflict of interest.

References

1. Pozo-Bayón, M.A.; Monagas, M.; Bartolome, B.; Moreno-Arribas, M.V. Wine features related to safety and consumer health: An integrated perspective. *Crit. Rev. Food Sci. Nutr.* **2012**, *52*, 31–54.
2. Arranz, S.; Chiva-Blanch, G.; Valderas-Martínez, P.; Medina-Remón, A.; Lamuela-Raventós, R.M.; Estruch, R. Wine, beer, alcohol and polyphenols on cardiovascular disease and cancer. *Nutrients* **2012**, *4*, 759–781.
3. Berthoud, H.R.; Shin, A.C.; Zheng, H. Obesity surgery and gut-brain communication. *Physiol. Behav.* **2011**, *105*, 106–119.
4. Guarner, F.; Malagelada, J.R. Gut flora in health and disease. *Lancet* **2003**, *361*, 512–519.
5. Williamson, G.; Clifford, M.N. Colonic metabolites of berry polyphenols: The missing link to biological activity? *Br. J. Nutr.* **2010**, *104*, S48–S66.
6. Selma, M.V.; Espín, J.C.; Tomás-Barberán, F.A. Interaction between phenolics and gut microbiota: Role in human health. *J. Agric. Food Chem.* **2009**, *57*, 6485–6501.
7. Requena, T.; Monagas, M.; Pozo-Bayón, M.A.; Martín-Alvárez, P.J.; Bartolomé, B.; del Campo, R.; Ávila, M.; Martínez-Cuesta, M.C.; Peláez, C.; Moreno-Arribas, M.V. Perspectives of the potential implications of wine polyphenols on human oral and gut microbiota. *Trends Food Sci. Tech.* **2010**, *21*, 332–344.
8. Hervert-Hernández, D.; Goñi, I. Dietary polyphenols and human gut microbiota: A review. *Food Rev. Int.* **2011**, *27*, 154–169.
9. Tuohy, K.M.; Conterno, L.; Gasperotti, M.; Viola, R. Up-regulating the human intestinal microbiome using whole plant foods, polyphenols, and/or fiber. *J. Agric. Food Chem.* **2012**, *60*, 8776–8782.
10. Etxebarria, U.; Fernández-Quintela, A.; Milagro, F.I.; Aguirre, L.; Martinez, J.A.; Portillo, M.P. Impact of polyphenols and polyphenol-rich dietary sources on gut microbiota composition. *J. Agric. Food Chem.* **2013**, *61*, 9517–9533.
11. Van Duynhoven, J.; Vaughan, E.E.; van Drosten, F.; Gomez-Roldan,V.; de Vos, R.; Vervoort, J.; van der Hooft, J.J.J.; Roger, L.; Draijer, R.; Jacobs, D.M. Interactions of black tea polyphenols with human gut microbiota: Implications for gut and cardiovascular health. *Am. J. Clin. Nutr.* **2013**, *98*, 1631S–1641S.
12. Cardona, F.; Andrés-Lacueva, C.; Tulipani, S.; Tinahones, F.J.; Queipo-Ortuño, M.I. Benefits of polyphenols on gut microbiota and implications in human health. *J. Nutr. Biochem.* **2013**, *24*, 1415–1422.
13. Tzounis, X.; Rodriguez-Mateos, A.; Vulevic, J.; Gibson, G.R.; Kwik-Uribe, C.; Spencer, J.P.E. Prebiotic evaluation of cocoa-derived flavanols in healthy humans by using a randomized, controlled, double-blind, crossover intervention study. *Am. J. Clin. Nutr.* **2011**, *93*, 62–72.

14. Urpi-Sarda, M.; Monagas, M.; Khan, N.; Lamuela-Raventos, R.M.; Santos-Buelga, C.; Sacanella, E.; Castell, M.; Permanyer, J.; Andrés-Lacueva, C. Epicatechin, procyanidins, and phenolic microbial metabolites after cocoa intake in humans and rats. *Anal. Bioanal. Chem.* **2009**, *394*, 1545–1556.
15. García-Ruiz, A.; Bartolomé, B.; Martínez-Rodríguez, A.J.; Pueyo, E.; Martín-Alvárez, P.J.; Moreno-Arribas, M.V. Potential of phenolic compounds for controlling lactic acid bacteria growth in wine. *Food Control.* **2008**, *19*, 835–841.
16. Hertog, M.G.L.; Feskens, E.J.M.; Hollman, P.C.H.; Katan, M.B.; Kromhout, D. Dietary antioxidant flavonoids and risk of coronary heart disease: The Zutphen Elderly Study. *Lancet* **1993**, *342*, 1007–1011.
17. Del Rio, D.; Rodriguez-Mateos, A.; Spencer, J.P.E.; Tognolini, M.; Borges, G.; Crozier, A. Dietary (poly)phenolics in human health: structures, bioavailability, and evidence of protective effects against chronic diseases. *Antioxid. Redox Sign*. **2013**, *18*, 1818–1892.
18. Vison, J.A.; Teufel, K.; Wu, N. Red wine, dealcoholized red wine, and especially grape juice, inhibit atherosclerosis in a harmster model. *Atherosclerosis* **2001**, *156*, 67–72.
19. Khan, N.; Adhami, V.M.; Mukhtar, H. Apoptosis by dietary agents for prevention and treatment of prostate cancer. *Endocr. Relat. Cancer* **2010**, *17*, R39–R52.
20. Russo, P.; Tedesco, I.; Russo, M.; Russo, G.L.; Venezia, A.; Cicala, C. Effects of de-alcoholated red wine and its phenolic fractions on platelet aggregation. *Nutr. Metab. Cardiovasc. Dis.* **2001**, *11*, 25–29.
21. Rechner, A.R.; Kroner, C. Anthocyanins and colonic metabolites of dietary polyphenols inhibit platelet function. *Thromb. Res.* **2005**, *116*, 327–334.
22. Clifford, M.N. Diet-derived phenols in plasma and tissues and their implications for health. *Planta Medica* **2004**, *70*, 1103–1114.
23. Scalbert, A.; Williamson, G. Dietary intake and bioavailability of polyphenols. *J. Nutr.* **2000**, *130*, 2073S–2085S.
24. Manach, C.; Scalbert, A.; Morand, C.; Rémésy, C.; Jiménez, L. Polyphenols: Food sources and bioavailability. *Am. J. Clin. Nutr.* **2004**, *79*, 727–747.
25. Maukonen, J.; Motto, J.; Suihko, M.L.; Saarela, M. Intra-individual diversity and similarity of salivary and faecal microbiota. *J. Med. Microbiol.* **2008**, *57*, 1560–1568.
26. Gerritsen, J.; Smidt, H.; Rijkers, G.T.; de Vos, W.M. Intestinal microbiota in human health and disease: the impact of probiotics. *Genes Nutr*. **2011**, *6*, 209–240.
27. Moore, W.E.; Moore, L.H. Intestinal floras of populations that have a high risk of colon cancer. *Appl. Environ. Microbiol.* **1995**, *61*, 3200–3207.
28. Gibson, G.R.; Roberfroid, M.B. Dietary modulation of the human colonic microbiota-Introducing the concept of prebiotics. *J. Nutr.* **1995**, *125*, 1401–1412.
29. Guarner, F. Enteric flora in health and disease. *Digestion* **2006**, *73*, 5–12.
30. Gibson, G.R. Dietary modulation of the human gut microflora using prebiotics. *Br. J. Nutr.* **1998**, *80*, S209–S212.
31. Van der Waaij, L.A.; Limburg, P.C.; Mesander, G.; van der Waaij, D. *In vivo* IgA coating of anaerobic bacteria in human faeces. *Gut* **1996**, *38*, 348–354.

32. Salminen, S.; Wright, A.; Morelli, L.; Marteau, P.; Brassart, D.; Vos, W.M.; Fondem, R.; Saxelin, M.; Collins, K.; Mogensen, G.; *et al.* Demostration of safety of probiotic—A review. *Int. J. Food Microbiol.* **1998**, *44*, 93–106.
33. Rastall, R.A.; Gibson, G.R.; Gill, H.S.; Guarner, F.; Klaenhammer, T.R.; Pot, B.; Reid, G.; Rowland, I.R.; Sanders, M.E. Modulation of the microbial ecology of the human colon by probiotic, prebiotics and synbiotics to enhance human health: An overview of enabling science and potential applications. *FEMS Microbiol. Ecol.* **2005**, *52*, 145–152.
34. Joly, F.; Mayeur, C.; Bruneau, A.; Noordine, M.L.; Meylheuc, T.; Langella, P.; Messing, B.; Duée, P.H.; Cherbuy, C.; Thomas, M. Drastic changes in fecal and mucosa-associated microbiota in adult patients with short bowel syndrome. *Biochimie* **2010**, *92*, 753–761.
35. Kemperman, R.A.; Bolca, S.; Roger, L.C.; Vaughan, E.E. Novel approaches for analysing gut microbes and dietary polyphenols: Challeges and opportunities. *Microbiology* **2010**, *156*, 3224–3231.
36. Fraher, M.H.; O'Toole, P.W.; Quigley, E.M.M. Techniques used to characterize the gut microbiota: A guide for the clinician. *Nat. Rev. Gastroentero* **2012**, *9*, 312–322.
37. Glenn, T.C. Field guide to next-generation DNA sequencers. *Mol. Ecol. Resources.* **2011**, *11*, 759–769.
38. De Boever, P.; Deplancke, B.; Verstraete, W. Fermentation by gut microbiota cultured in a simulator of the human intestinal microbial ecosystem is improved by supplementing a soygerm powder. *J. Nutr.* **2000**, *130*, 2599–2606.
39. Molly, K.; Vande Woestyne, M.; Verstraete, W. Development of a 5 step multi-chamber reactor as a simulation of the human intestinal microbial ecosystem. *Appl. Microbiol. Biotechnol.* **1993**, *39*, 254–258.
40. Muñoz-González, C.; Moreno-Arribas, M.V.; Rodríguez-Bencomo, J.J.; Cueva, C.; Martín-Álvarez, P.J.; Pozo-Bayón, M.A. Feasibility and application of liquid-liquid extraction combined with gas chromatography-mass spectrometry for the analysis of phenolic acids from grape polyphenols degraded by human faecal microbiota. *Food Chem.* **2012**, *133*, 526–535.
41. Sánchez-Patán, F.; Monagas, M.; Moreno-Arribas, M.V.; Bartolomé, B. Determination of microbial phenolic acids in human faeces by UPLC-ESI-TQ MS. *J. Agric. Food Chem.* **2011**, *59*, 2241–2247.
42. García-Cañas, V.; Simó, C.; León, C.; Cifuentes, A. Advances in nutrigenomics research: Novel and future analytical approaches to invesigate the biological activity of natural compounds and food functions. *J. Pharmaceut. Biomed.* **2010**, *51*, 290–304.
43. Brennan, L. Metabolomics in nutrition research: current status and perspectives. *Biochem. Soc. Trans.* **2013**, *41*, 670–673.
44. Llorach, R.; Garrido, I.; Monagas, M.; Urpi-Sarda, M.; Tulipani, S.; Bartolomé B.; Andrés-Lacueva, C. Metabolomics study of human urinary metabolome modifications after intake of almond (*Prunus. dulcis* (Mill.) D.A. Webb) skin polyphenols. *J. Proteome Res.* **2010**, *9*, 5859–5867.
45. Monagas, M.; Urpi-Sarda, M.; Sánchez-Patán, F.; Llorach, R.; Garrido, I.; Gómez-Cordovés, C.; Andrés-Lacueva, C.; Bartolomé, B. Insights into the metabolism and microbial biotransformation of dietary flavan-3-ols and the bioactivity of their metabolites. *Food Funct.* **2010**, *1*, 233–253.

46. Aura, A.M. Microbial metabolism of dietary phenolic compounds in the colon. *Phytochem. Rev.* **2008**, *7*, 407–429.
47. Fossen, T.; Cabrita, L.; Andersen, O.M. Colour and stability of pure anthocyanins influenced by pH including the alkaline region. *Food Chem.* **1998**, *63*, 435–440.
48. Mullen, W.; Edwards, C.A.; Serafini, M.; Crozier, A. Bioavailability of pelargonidin-3-*O*-glucoside and its metabolites in humans following the ingestion of strawberries with and without cream. *J. Agric. Food Chem.* **2008**, *56*, 713–719.
49. McGhie, T.K.; Walton, M.C. The bioavailability and absorption of anthocyanins: Towards a better understanding. *Mol. Nutr. Food Res.* **2007**, *51*, 702–713.
50. De Ferrars, R.M.; Czank, C.; Saha, S.; Needs, P.W.; Zhang, Q.; Raheem, K.S.; Botting, N.P.; Kroon, P.A.; Kay, C.D. Methods for isolating, identifying, and quantifying anthocyanin metabolites in clinical samples. *Anal. Chem.* **2014**, *86*, 10052–10058.
51. Stockley, C.; Teissedre, P.L.; Boban, M.; di Lorenzo, C.; Restani, P. Bioavailability of wine-derived phenolic compounds in humans: A review. *Food Funct.* **2012**, *3*, 995–1007.
52. Boto-Ordoñez, M.; Rothwell, J.A.; Andres-Lacueva, C.; Manach, C.; Scalbert, A.; Urpi-Sarda, M. Prediction of the wine polyphenol metabolic space: An application of the Phenol-Explorer database. *Mol. Nutr. Food Res.* **2014**, *3*, 466–477.
53. Jiménez-Girón, A.; Queipo-Ortuno, M.I.; Boto-Ordoñez, M.; Muñoz-González, I.; Sánchez-Patán, F.; Monagas, M.; Martín-Álvarez, P.J.; Murri, M.; Tinahones, F.J.; Andres-Lacueva, C.; *et al.* More comparative study of microbial-derived phenolic metabolites in human feces after intake of gin, red wine, and dealcoholized red wine. *J. Agric. Food Chem.* **2012**, *61*, 3909–3915.
54. Muñoz-González, I.; Jiménez-Girón, A.; Martín-Álvarez, P.J.; Bartolomé, B.; Moreno-Arribas, M.V. Profiling of microbial-derived phenolic metabolites in human feces after moderate red wine intake. *J. Agric. Food Chem.* **2013**, *61*, 9470–9479.
55. Wang, L.Q.; Meselhy, M.R.; Li, Y.; Nakamura, N.; Min, B.S.; Qin, G.W.; Hattori, M. The heterocyclic ring fission and dehydroxylation of catechins and related compounds by *Eubacterium* sp. strain SDG-2, a human intestinal. *Chem. Pharm. Bull.* **2001**, *49*, 1640–1643.
56. Kutschera, M.; Engst, W.; Blaut, M.; Braune, A. Isolation of catechin-converting human intestinal bacteria. *J. Appl. Microbiol.* **2011**, *111*, 165–175.
57. Jin, J.S.; Hattori, M. Isolation and characterization of a human intestinal bacterium *Eggerthella* sp. CAT-1 capable of cleaving the C-Ring of (+)-catechin and (−)-epicatechin, followed by *p*-dehydroxylation of the B-Ring. *Biol. Pharm. Bull.* **2012**, *35*, 2252–2256.
58. Tzounis, X.; Vulevic, J.; Kuhnle, G.G.C.; George, T.; Leonczak, J.; Gibson, G.R.; Kwik-Uribe, C.; Spencer, J.P.E. Flavanol monomer-induced changes to the human faecal microflora. *Br. J. Nutr.* **2008**, *99*, 782–792.
59. Hidalgo, M.; Oruna-Concha, M.J.; Kolida, S.; Walton, G.E.; Kallithraka, S.; Spencer, S.P.E.; Gibson, G.R.; Pascual-Teresa, S. Metabolism of antocyanins by human gut microflora and their influence on gut bacterial growth. *J. Agric. Food Chem.* **2012**, *60*, 3882–3890.

60. Cueva, C.; Sánchez-Patán, F.; Monagas, M.; Walton, G.E.; Gibson, G.R.; Martin-Alvarez, P.J.; Bartolome, B.; Moreno-Arribas, M.V. *In vitro* fermentation of grape seed flavan-3-ol fractions by human faecal microbiota: Changes in microbial groups and phenolic metabolites. *FEMS Microbiol. Ecol.* **2013**, *83*, 729–805.
61. Sánchez-Patán, F.; Cueva, C.; Monagas, M.; Walton, G.E.; Gibson, G.R.; Quintanilla-Lopez, J.E.; Lebron-Aguilar, R.; Martin-Alvarez, P.J.; Moreno-Arribas, M.V.; Bartolome, B. *In vitro* fermentation of a red wine extract by human gut microbiota; changes in microbial groups and formation and phenolic metabolites. *J. Agric. Food Chem.* **2012**, *60*, 2136–2147.
62. Barroso, E.; Sánchez-Patán, F.; Martin-Alvarez, P.J.; Bartolome, B.; Moreno-Arribas, M.V.; Pelaez, C.; Requena, T.; van de Wiele, T.; Martinez-Cuesta, M.C. *Lactobacillus plantarum* IFPL935 favors the initial metabolism of red wine polyphenols when added to a colonic microbiota. *J. Agric. Food Chem.* **2013**, *61*, 10163–10172.
63. Gross, G.; Jacobs, D.M.; Peters, S.; Possemiers, S.; van Duynhoven, J.; Vaughan, E.E.; van de Wiele, T. *In vitro* bioconversion of polyphenols from black tea and red wine/grape juice by human intestinal microbiota displays strong interindividual variability. *J. Agric. Food Chem.* **2010**, *58*, 10236–10246.
64. Kemperman, R.A.; Gross, G.; Mondot, S.; Possemiers, S.; Marzorati, M.; van de Wiele, T.; Dore, J.; Vaughan, E.E. Impact of polyphenols from black tea and red wine/grape juice on a gut model microbiome. *Food Res. Int.* **2013**, *53*, 659–669.
65. Dolara, P.; Luceri, C.; de Filippo, C.; Femia, A.P.; Giovannelli, L.; Caderni, G.; Cecchini, C.; Silvi, S.; Orpianesi, C.; Cresci, A. Red wine polyphenols influence carcinogenesis, intetinalmicroflora, oxidative damage and gene expression profile of colonic mucosa in F344 rats. *Mutat. Res.* **2005**, *591*, 237–246.
66. Viveros, A.; Chamorro, S.; Pizarro, M.; Arija, I.; Centeno, C.; Brenes, A. Effects of dietary polyphenols-rich grape productus on intestinal microflora and gut morphology in broiler chicks. *Poultry Sci.* **2011**, *90*, 566–578.
67. Fiesel, A.; Gessner, D.K.; Most, E.; Eder, K. Effects of dietary polyphenol-rich plant products from grape or hop on pro-inflammatory gene expression in the intestine, nutrient digestibility and faecal microbiota of weaned pigs. *BMC Vet. Res.* **2014**, *10*, 196.
68. Choy, Y.Y.; Quifer-Rada, F.; Holstege, D.M.; Frese, S.A.; Calvert, C.C.; Mills, D.A.; Lamuela-Raventos, R.M.; Waterhouse, A.L. Phenolic metabolites and substantial microbiome changes in pig feces by ingesting grape seed proantocyanidins. *Food Funct.* **2014**, *5*, 2298–2308.
69. Yamakoshi, J.; Tokutake, S.; Kikuchi, M.; Kubota, Y.; Konishi, H.; Mitsuoka, T. Effect of proantocyanidin-rich extract from grape seed on human fecal flora and fecal odor. *Microb. Ecol. Health Dis.* **2001**, *13*, 25–31.
70. Queipo-Ortuño, M.I.; Boto-Ordoñez, M.; Murri, M.; Gómez-Zumaquero, J.M.; Clemente-Postigo, M.; Estruch, R.; Diaz, F.C.; Andres-Lacueva, C.; Tinahones, F.J. Influence of red wine polyphenols and ethanol on the gut microbiota ecology and biochemical biomarkers. *Am. J. Clin. Nutr.* **2012**, *95*, 1323–1334.
71. Van Duynhoven, J.; Vaughan, E.E.; Jacobs, D.M.; Kemperman, R.A.; van Velzen, E.J.J.; Gross, G.; Roger, L.C.; Possemiers, S.; Smilde, A.K.; Dore, J.; *et al.* Metabolic fate of polyphenols in the human superorganism. *Proc. Natl. Acad. Sci. USA* **2011**, *108*, 4531–4538.

72. Van Velzen, E.J.J.; Wasterhuis, J.A.; van Duynhoven, J.P.M.; van Drosten, F.A.; Hoefsloot, H.C.J.; Jacobs, D.M.; Smit, S.; Draijer, R.; Kroner, C.I.; Smilde, A.K. Multilevel data analysis of a crossover designed human nutritional intervention study. *J. Proteome Res.* **2008**, *7*, 4483–4491.

12

Antioxidant, Biomolecule Oxidation Protective Activities of *Nardostachys jatamansi* DC and Its Phytochemical Analysis by RP-HPLC and GC-MS

Sakina Razack [1], Kandikattu Hemanth Kumar [1], Ilaiyaraja Nallamuthu [1], Mahadeva Naika [2] and Farhath Khanum [1,*]

[1] Biochemistry and Nanosciences Division, Defence Food Research Laboratory, Mysore-570011, India; E-Mails: razacksakina@yahoo.co.in (S.R.); kandikattu.hemanth@gmail.com (K.H.K.); nilaiyaraja@gmail.com (I.N.)

[2] Applied Nutrition Division, Defence Food Research Laboratory, Mysore-570011, India; E-Mail: mnaika@rediffmail.com

* Author to whom correspondence should be addressed; E-Mail: farhathkhanum@gmail.com

Academic Editors: Antonio Segura-Carretero and David Arráez-Román

---

Abstract: The study aimed at analyzing the metabolite profile of *Nardostachys jatamansi* using RP-HPLC, GC-MS and also its antioxidant, biomolecule protective and cytoprotective properties. The 70% ethanolic extract of *Nardostachys jatamansi* (NJE) showed the presence of polyphenols and flavonoids (gallic acid, catechin, chlorogenic acid, homovanillin, epicatechin, rutin hydrate and quercetin-3-rhamnoside) analyzed by RP-HPLC, whereas hexane extract revealed an array of metabolites (fatty acids, sesquiterpenes, alkane hydrocarbons and esters) by GC-MS analysis. The antioxidant assays showed the enhanced potency of NJE with a half maximal inhibitory concentration (IC_{50}) value of 222.22 ± 7.4 μg/mL for 2,2-diphenyl-1-picrylhydrazyl (DPPH), 13.90 ± 0.5 μg/mL for 2,2′-azino-bis(3-ethyl benzothiazoline-6-sulfonic acid) diammonium salt (ABTS), 113.81 ± 4.2 μg/mL for superoxide, 948 ± 21.1 μg/mL for metal chelating and 12.3 ± 0.43 mg $FeSO_4$ equivalent/g of extract for ferric reducing antioxidant power assays and was more potent than hexane extract. NJE effectively inhibited 2,2′-azobis(2-methylpropionamidine) dihydrochloride (AAPH)-induced oxidation of biomolecules analyzed by pBR322 plasmid DNA damage, protein oxidation of bovine serum albumin and lipid peroxidation assays. The observed effects might be due to the high content of polyphenols, 53.06 ± 2.2 mg gallic acid equivalents/g, and flavonoids, 25.303 ± 0.9 mg catechin equivalents/g, of NJE compared to the hexane fraction. Additionally, the extract abrogated the protein, carbonyl, and ROS formation, and NJE showed cytotoxicity in SH-SY5Y neuronal cells above 75 μg/mL. Thus, the study suggests that the herb unequivocally is a potential source of antioxidants and could aid in alleviating oxidative stress-mediated disorders.

Keywords: antioxidant activity; DNA damage protective; GC-MS; *Nardostachys jatamansi*; phytochemical analysis; RP-HPLC

1. Introduction

Reactive oxygen species (ROS) comprise free radicals, such as superoxide anion (O_2^-) and hydroxyl ($HO^{\cdot}$) radicals, and non-free radical species, such as H_2O_2 and singlet oxygen ($^1O_2$), and these comprise different forms of activated oxygen [1–3]. ROS are known to damage cellular membranes by inducing lipid peroxidation, damage DNA, proteins and lipids [4]. Thus, excessive production of free radicals/ROS leads to oxidation of biomolecules, and this has been implicated in the etiology of several human diseases, such as atherosclerosis, ischemia reperfusion injury, ageing and liver-related diseases [5,6]. To protect the cells and organ systems of the body from free radicals/ROS, the human body has evolved a highly sophisticated and complex antioxidant protection system. Antioxidants are capable of stabilizing, or deactivating, free radicals before they damage cells. The body's antioxidant defense system comprises endogenous antioxidant enzymes, *viz.* superoxide dismutase, glutathione peroxidase, glutathione reductase, catalase, *etc.*, that catalyze free radical quenching reactions. In addition, dietary antioxidants, like ascorbic acid (vitamin C), tocopherols and tocotrienols (vitamin E), carotenoids and other low molecular weight compounds, such as glutathione and lipoic acid, also aid in stabilizing free radicals [7].

Plant-derived substances, collectively called phytonutrients or phytochemicals, are known for their antioxidant activity [7]. Consumption of natural antioxidants is associated with reduced risks of cancer, cardiovascular disease, diabetes and other diseases associated with ageing [8,9]. In recent years, consumption of the natural phytochemicals present in berry crops, teas, herbs, oilseeds, beans, fruits and vegetables have markedly increased [10,11]. The oxidative DNA damage plays a significant role in mutagenesis, cancer, aging and other human pathologies [12,13]. The best strategy to counter oxidative stress is supplementation of food rich in herbal bioactives [14].

Nardostachys jatamansi DC is a member of the plant family, *Valerianaceae*, is a perennial herb growing to about 10–60 m in height and is commonly referred to as Indian spikenard. It is found in the alpine Himalayas from Punjab to Sikkim and Bhutan at an altitude of 3000–5000 m. The plant has been reported to possess antianxiety, hepatoprotective, cardioprotective, anticonvulsant, antiparkinsons, antidepressant and antimicrobial activities [15–23].

Nardostachys jatamansi has been attributed with exorbitant medicinal properties, and it was our interest to analyze the 70% ethanolic and hexane extracts for their *in vitro* antioxidant activities, with the former extract comprising both polar and non-polar metabolites and the latter exclusively non-polar. Scarce literature is available with regards to the antioxidant and biomolecules protective properties of the 70% ethanolic extract and also with regard to the hexane extract. Hence, the rationale of the present study was to characterize the 70% ethanol and hexane extracts of *N. jatamansi* by RP-HPLC and GC-MS analysis and to assess the extract for its antioxidant potency, quantification of polyphenols and flavonoids and, further, to verify the ability of the extract to inhibit the oxidation of biomolecules to understand the possible role of the antioxidant activity of the herb.

2. Experimental Section

2.1. Chemicals and Reagents

2,2′-azino-bis(3-ethyl benzothiazoline-6-sulfonic acid) diammonium salt (ABTS), 2,2′-azobis(2-methylpropionamidine) dihydrochloride (AAPH), agarose, ethidium bromide, bovine serum albumin, gallic acid, catechin, homovanillic acid, epicatechin, chlorogenic acid, rutin hydrate and quercetin-3-rhamnoside were purchased from Sigma-Aldrich, St. Louis, MO, USA. pBR322 plasmid DNA was purchased from Genetix, Bangalore. India. The other chemicals and reagents were of analytical grade and were procured from Merck, Bangalore, India.

2.2. Plant Material and Preparation of the Extract

The root material was procured from a local supplier at Mysore, India. The plant was identified by Dr. K. Madhava Chetty, Botanist, Department of Botany, Sri Venkateswara University, Tirupati, India. The voucher specimen (Herbarium Accession Number 1911) was deposited in the herbarium, Department of Botany, Sri Venkateswara University, Tirupati, India. The roots were washed thoroughly with distilled water to remove the adhering sand particles and shade dried. The thoroughly dried roots were powdered and extracted by shaking with 70% ethanol and hexane with a plant material:solvent ratio of 1:10, filtered through Whatmann filter paper No. 1 (Sigma-Aldrich, St. Louis, MO, USA), and were evaporated to dryness under vacuum on a rotary evaporator (Heidolph rotacool, Germany) into a thick residue. The 70% ethanol fraction was subsequently lyophilized (Lyolab, Hyderabad, India), and the powder was used for analysis. The yield of the 70% ethanol fraction (NJE) was 7.4% and the hexane fraction (NJH) was 4.2%.

2.3. Metabolite Profile of N. jatamansi

2.3.1. Reversed Phase-HPLC Analysis of the 70% Ethanol Fraction

Phenolic compounds of NJE were identified using a diode array detector (JASCO Pu-1580 HPLC system, JASCO Inc., Easton, MD, USA) on a reverse phase C_{18} column (150 × 4.5 mm, JASCO Inc.). The mobile phase was with two solvents: 0.1% formic acid in water (A) and 100% methanol (B). The total run time was 60 min. The eluting compounds were detected by monitoring at 270 nm. The phenolic compounds were identified by comparing the retention time (RT) of the unknowns with the standards [24]. The polyphenols were partially extracted from the sample using the solid-phase extraction method on a C_{18} reverse phase sep-pak column (Catologue Number: WAT094226, HLB3CC, Waters India Private Limited, Bengaluru, Karnataka, India), and the eluant obtained was injected into reversed phase-HPLC for analysis. The lyophilized plant extract (1 mg/mL) was redissolved in 0.1%

formic acid in water, and 20 μL of the sample and the standards were injected for analysis. The standards injected include gallic acid (RT-9.98), catechin (RT-18.17), chlorogenic acid (RT-22.02), homovanillin (RT-23.12), epicatechin (RT-24.09), rutin hydrate (RT-37.33) and quercetin-3-rhamnoside (RT-40.61).

2.3.2. Gas Chromatography and Mass Spectrometry Analysis (GC-MS-HT-TOF) of the Hexane Fraction

The metabolite profiling of NJH was carried out by GC–MS analysis on an Agilent 7890 gas chromatograph (Agilent Technologies, Santa Clara, CA, US) coupled with LECO Corporation (St. Joseph, MI, USA) with the following conditions: the column used was 29.3 m × 0.7 m, 320 μm, capillary column −29.300 m, operating in an electron ionization mode at 70 eV; with helium as a carrier gas at a constant flow of 1.50 mL/min and an injection volume of 1.0 μL; injector temperature 250 °C; ion source temperature 200 °C. The oven temperature was initially 70 °C for 1 min, increased to 160 °C at a rate of 4 °C/min and increased finally to 320 °C for 15 min at a rate of 3 °C/min. The mass spectrometer was operated at an acquisition rate of 10 spectra/s, and the scan range was set between 50 and 600 m/z. The detector voltage was set to 1450 V and the electron energy to −70 V.

2.4. Phytochemical Screening of the Plant Extract

2.4.1. Determination of Total Phenolic Content

The total polyphenols were estimated by the Folin–Ciocalteu method [25]. Briefly, to different concentrations of the plant extract (100–1000 μg/mL), 2.5 mL of Folin-Ciocalteu reagent (1:10) and 2 mL of 7% Na_2CO_3 were added and incubated at room temperature for 90 min, and the absorbance was measured spectrophotometrically at 765 nm (Shimadzu, Kyoto, Japan). Gallic acid (10–100 μg/mL) was used as a standard, and the amount of total polyphenolic content was expressed as mg gallic acid equivalent per gram extract (mg GAE/g).

2.4.2. Determination of Total Flavonoid Content

Briefly, to different concentrations of plant extracts (100–1000 μg/mL), 75 μL of 5% $NaNO_2$ and 150 μL of 10% $AlCl_3$ were added and incubated at room temperature for 5 min. Then, 500 μL of 1 M NaOH followed by 775 μL of water were added, vortexed and the absorbance measured at 510 nm [26]. Catechin (10–100 μg/mL) was used as the standard, and the amount of total flavonoids was expressed as mg of catechin equivalents per gram extract (mg CE/g).

2.5. In Vitro Free Radical Scavenging and Antioxidant Activities

2.5.1. DPPH Radical Scavenging Activity

The free radical scavenging activity of NJE and NJH were measured by the 2,2-diphenyl-1-picrylhydrazyl (DPPH) assay [27]. Briefly, 500 μL of DPPH (0.5 mM in methanol) solution was mixed with different amounts of sample (100–1000 μg/mL), and the volume was made to 3 mL with methanol. The mixture was incubated in the dark for 45 min at room temperature, and the absorbance was recorded at 515 nm.

Butylated hydroxyanisole (BHA; 10–100 μg/mL) was used as the standard antioxidant compound. Methanol was used as the blank. The percent inhibition was calculated as follows:

$$\text{Percentage inhibition (\%)} = [A_C - A_S/A_C] \times 100 \quad (1)$$

where A_C is the absorbance of positive control and A_S is the absorbance of the sample.

2.5.2. ABTS Radical Cation Decolorization Assay

The ABTS (2,2′-azino-bis(3-ethyl benzothiazoline-6-sulfonic acid) diammonium salt) radical cation scavenging activity of NJE/NJH was assayed [28]. The working solution was prepared by mixing two stock solutions of 14 mM ABTS solution and 4.9 mM potassium persulfate solution in equal amounts and allowed to react for 12–16 h at room temperature in the dark. This solution (5 mL) was diluted with absolute ethanol until the OD value reached 0.70 ± 0.02 at 734 nm. The working solution was thoroughly mixed at room temperature, and 1 mL of this solution was mixed with different concentrations of plant extracts (10–100 μg/mL). Additionally, the scavenging capacity of the extract for $ABTS^+$ was calculated and expressed as the percent inhibition. The IC_{50} value was calculated and compared with ascorbic acid (5–50 μg/mL), used as a standard.

2.5.3. Metal Chelating

Briefly, to different concentrations of the extracts (100–2000 μg/mL), 50 μL of $FeCl_2$ (2 mM) were added. The reaction was initiated by the addition of 200 μL of ferrozine (2 mM); the mixture was vortexed and left at room temperature for 10 min. Absorbance was then measured spectrophotometrically at 562 nm [29]. The % of inhibition of the ferrozine-Fe^{2+} complex was calculated by:

$$\text{\% Inhibition} = [(A_0 - A_1)/A_0] \times 100 \quad (2)$$

where A_0 is the absorbance of the control and A_1 is the absorbance of the samples and standards. The ability of the extracts to chelate ferrous ion was compared with EDTA (100 μg/mL), used as a standard.

2.5.4. Superoxide Radical Scavenging Activity

The superoxide radicals were generated in 16 mM Tris-HCl buffer (pH 8.0) containing 50 μM nitro blue tetrazolium (NBT), 78 μM NADH and different concentrations of plant extract (100–1000 μg/mL). The reaction was started by adding 10 μM PMS to the mixture. The reaction mixture was incubated at 25 °C for 5 min, and the absorbance was measured at 560 nm [30]. L-ascorbic acid (10–100 μg/mL) was used as the control. The percent inhibition of superoxide anion generation was calculated by using the following formula:

$$\text{\% Inhibition} = [(A_0 - A_1)/A_0] \times 100 \quad (3)$$

where A_0 is the absorbance of the control and A_1 is the absorbance of sample extract/standard.

2.5.5. Ferric-Reducing Antioxidant Power Assay

The antioxidant capacity was determined by the ferric-reducing antioxidant power (FRAP) assay [31]. The stock solutions included 300 mM acetate buffer (3.1 g of CH_3COONa and 16 mL of CH_3OOH), pH 3.6, 10 mM TPTZ (2,4,6-tripyridyl-*s*-triazine) solution in 40 mM HCl and 20 mM $FeCl_3 \cdot 6H_2O$ solution. The fresh working solution was prepared by mixing 25 mL acetate buffer, 2.5 mL TPTZ and 2.5 mL $FeCl_3 \cdot 6H_2O$. The temperature of the solution was raised to 37 °C before the assay. Plant extract (10–100 μg/mL) was allowed to react with 2850 μL of the FRAP solution for 30 min in the dark. Readings of the colored product (ferrous tripyridyltriazine complex) were measured at 593 nm.

2.6. Biomolecules Oxidation Assays

2.6.1. AAPH-Induced Plasmid Nick Assay

Conversion of the supercoiled form of plasmid DNA to open circular and further linear form has been used as an index of DNA damage. The DNA damage protective activity of the NJE was assessed using pBR322 plasmid DNA [32]. Different concentrations of NJE (5, 10, 15 μg from a stock of 1 mg/mL) and plasmid DNA (0.5 μg) were incubated at room temperature followed by the addition of AAPH (200 mM). The reaction mixture was made up to 20 μL and incubated for one hour at 37 °C. The DNA samples were electrophoresed on 1% agarose gel, and the band intensities were analyzed using Easy win 32 software from Herolab (Herolab GmbH Laborgeraete, Wiesloch, Germany).

2.6.2. Protein Oxidation

The oxidation of BSA (5 μg) in phosphate buffer was initiated by 200 mM AAPH, and the inhibitory properties of NJE were measured at a fixed concentration (1 mg/mL). After incubation for one hour at 37 °C, BHT (0.02%) was added to prevent further formation of peroxyl radicals, and the samples were assayed by SDS-PAGE [33].

2.6.3. Thiobarbituric Acid-Reactive Substances Assay

Male Wistar albino rats weighing 200–220 g were housed under conventional conditions and were allowed free access to food and water. The experiments were carried out according to Committee for the Purpose of Control and Supervision of Experiments on Animals (CPCSEA) guidelines for the care and use of experiment animals approved by the Institutional Animal Ethics Committee. The rats were anaesthetized, and the liver and brain were removed and homogenized (10% w/v) in sodium phosphate buffer (pH, 7.4). Homogenates were then centrifuged at 5000× *g* for 15 min at 4 °C. Oxidative stress was induced using AAPH [34]. The reaction mixture was composed of liver/brain homogenate (15 mg protein), 120 mM AAPH and different concentrations of NJE (100–1000 μg/mL). The reaction mixture was incubated at 37 °C for 2 h, and the extent of lipid peroxidation of the liver homogenate in the presence and absence of the plant extract was evaluated by measuring the product of thiobarbituric acid reactive substances. After incubation, the reaction was terminated by adding 2% BHT followed by the addition of 1 mL of TCA (20% w/v) to the mixture. After centrifugation at 3000× *g* for 15 min, the supernatant was incubated with 1 mL of thiobarbituric acid (TBA, 0.67%) at 100 °C for 15 min, and the absorbance was measured at 532 nm. The data were expressed in terms of the percent inhibition.

$$\% \text{ Inhibition} = [A\text{c} - A\text{s}/A\text{c}] \times 100 \quad (4)$$

where $A\text{c}$ is the absorbance of the control, and $A\text{s}$ is the absorbance of the sample.

2.6.4. Estimation of Reactive Oxygen Species

The extent of inhibition of reactive oxygen species (ROS) was evaluated using 2′,7′-dichlorodihydro fluorescein diacetate (DCHF-DA). The brain and liver homogenates (10% w/v) were incubated with different concentrations of NJE (10–100 µg/mL) and gallic acid (1 mg/mL) as a standard. After incubation for half an hour, AAPH (120 mM), a peroxyl radical inducer, was added to all of the wells, except the control, and incubated for 30 min. Later, 10 µM of DCFH-DA was added and incubated for 60 min. The fluorescence of DCF (dichlorofluorescein) was measured at 492-nm excitation and 520-nm emission. Results were expressed as the percent of fluorescence [35].

2.6.5. Estimation of Protein Carbonyls

The protein carbonyl content (PCC) of the rat liver homogenates in the presence and absence of NJE (10–100 µg/mL)/AAPH (120 mM) was evaluated [36]. One milliliter of 10 mM DNPH (dinitrophenyl hydrazine) in 2 N HCl was added to the reaction mixture (2 mg of protein), and samples were incubated for 1 h at room temperature. Then, 1 mL of trichloroacetic acid 10% was added to each reaction mixture and centrifuged at 3000× *g* for 10 min. The protein pellet was washed thrice with 2 mL of ethanol/ethyl acetate (1:1, v/v) and dissolved in 1 mL of guanidine hydrochloride (6 M, pH 2.3) and incubated for 10 min at 37 °C. The carbonyl content was calculated based on the molar extinction coefficient of DNPH ($\varepsilon = 2.2 \times 104\ \text{cm}^{-1}\ \text{M}^{-1}$). The data were expressed in terms of the percent of inhibition.

2.7. Cytoprotective Effects

LDH Leakage Assay

The human neuroblastoma SH-SY5Y cells were procured from the National Center for Cell Sciences, Pune, India. Cells were seeded into 25-cm^2 flasks in DMEM/F-12, 1:1 mixture supplemented with 10% fetal bovine serum, 2 mM L-glutamine, antibiotic and antimycotic solution and maintained in a humid atmosphere with 5% CO_2 and 95% air at 37 °C. Media were changed every alternate day, and the experiment was carried out in serum-free media. Once the cells reached confluence, they were treated with NJE at different concentrations (10, 25, 50, 75, 100, 150, 200, 300, 400, 500 and 1000 µg/mL) for 24 h. After 24 h, to the untreated cells, 10 µL of cell lysis solution (2% Triton X-100 in PBS) were added; this accounted for the total lactate dehydrogenase (LDH) activity. Cells were centrifuged at 2000× *g* for 5 min at 4 °C, and 100 µL of the supernatant were collected. To this, 900 µL of the LDH estimation (Agappe-11407002, Kerala, India) reaction mixture was added and measured spectrophotometrically at 340 nm for 3 min.

2.8. Statistical Analysis

All of the analyses were carried out in triplicate. The values of the *in vitro* antioxidant assays are expressed as the mean ± standard deviation. The other tests were analyzed by one-way ANOVA followed by Tukey's *post hoc* test. $p < 0.05$ was considered significant.

3. Results and Discussion

3.1. Metabolite Profile of N. jatamansi

3.1.1. Reversed Phase-HPLC Analysis of Phenolic Compounds

The reversed phase-HPLC analysis aided in the identification (Figure 1) and quantification (Table 1) of polyphenols present in NJE. The identified polyphenols include gallic acid (0.18 mg/g), catechin (4.37 mg/g), chlorogenic acid (19.90 mg/g), homovanillic acid (32.02 mg/g), epicatechin (4.23 mg/g), rutin hydrate (0.08 mg/g) and quercetin-3-rhamnoside (7.13 mg/g). However, a few of the polyphenols could not be identified, due to a lack of standards. The relative concentrations of each identified metabolite was estimated on the basis of the relative standard calibration curve.

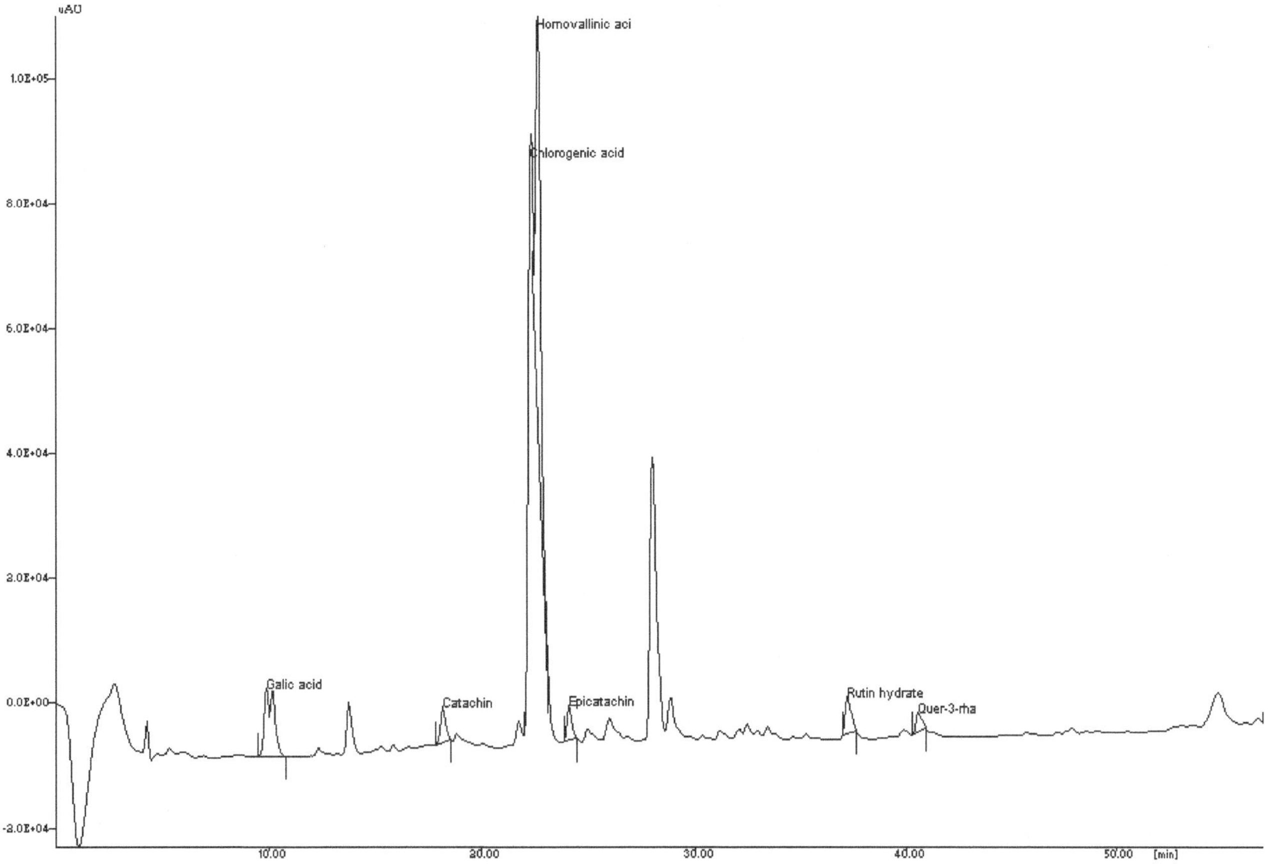

Figure 1. RP-HPLC chromatogram of the identified polyphenols in the ethanolic extract of *N. jatamansi*.

Table 1. Quantification of the identified polyphenols by RP-HPLC.

No.	Polyphenols	RT (min)	Concentration (mg/g)
1	Gallic acid	9.88	0.18
2	Catechin	18.16	4.37
3	Chlorogenic acid	22.27	19.90
4	Homovanillin	22.60	32.02
5	Epicatechin	24.09	4.23
6	Rutin hydrate	37.17	0.08
7	Quercetin-3-rhamnoside	40.56	7.13

3.1.2. GC-MS Analysis

The total ion chromatogram (TIC) of the hexane fraction of *N. jatamansi* is shown in Figure 2. The chemical formulae, retention time and mass of the identified compounds are listed in Table 2.

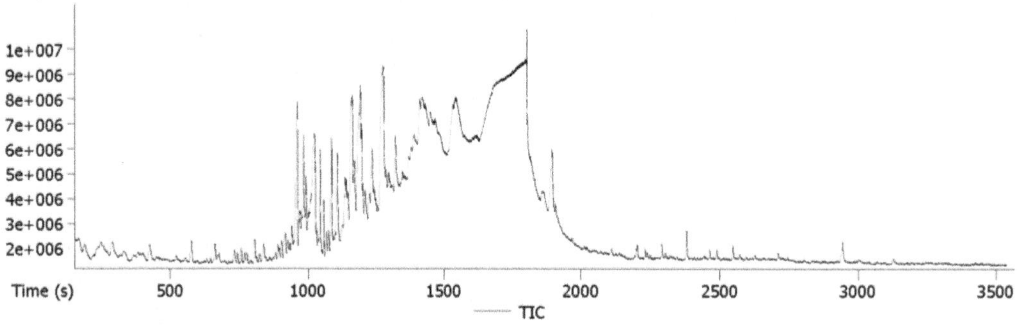

Figure 2. Total ion chromatogram of the hexane fraction of *N. jatamansi*.

Table 2. Phytochemical analysis of the *N. jatamansi* hexane extract.

No.	RT (min)	Compound Name	Chemical Formula	Mass	Area (%)	Hit
1	9.62	Dodecane	$C_{12}H_{26}$	170.204	1.615	1
2	11.22	2-Furanmethanol, tetrahydro-5-methyl-*trans*	$C_6H_{12}O$	116.084	0.031	1
3	13.98	Linalool	$C_{10}H_{18}O$	154.135	0.362	1
4	15.30	Benzenemethanol, a-methyl-propanoate	$C_{11}H_{14}O_2$	178.099	0.705	1
5	15.68	1H-Cyclopropa(a)naphthalene, 1a,2,3,5,6,7,7a,7b-octahydro-1,1,7,7a-tetramethyl-(1aR-(1aa,7a,7aa,7ba))	$C_{15}H_{24}$	204.188	1.061	1
6	16.02	a-Muurolene	$C_{15}H_{24}$	204.188	14.071	1
7	16.39	1H-Cycloprop(e)azulene,1a,2,3,4,4a,5,6,7b-octahydro-1,1,4,7-tetramethyl-(1aR-(1aa,4a,4aa,7ba))	$C_{15}H_{24}$	204.188	9.184	5
8	16.69	L-calamenene	$C_{15}H_{22}$	202.172	0.065	1
9	17.11	A-ionone	$C_{13}H_{20}O$	192.151	0.643	1
10	17.41	Bicyclo(3.3.1)nonan-2-one,1-methyl-9-(1-methylethylidene)	$C_{13}H_{20}O$	192.151	8.329	1
11	17.63	Neoisolongifolene,8,9-dehydro	$C_{15}H_{22}$	202.172	2.002	1

Table 2. *Cont.*

12	17.81	(−)-à-Panasinsen	$C_{15}H_{24}$	204.188	1.716	1
13	18.13	Neoisolongifolene,8,9-dehydro	$C_{15}H_{22}$	202.172	10.86	1
14	18.96	Bicyclo(2.2.2)octa-2,5-diene, 1,2,3,6-tetramethyl	$C_{12}H_{18}$	162.141	3.643	1
15	19.15	Isolongifolene,4,5,9,10-dehydro	$C_{15}H_{20}$	200.157	1.24	1
16	19.76	*trans*-Nerolidol	$C_{15}H_{26}O$	222.198	4.781	1
17	19.88	para-methoxyphenylpiperazine	$C_{11}H_{16}N_2O$	192.1263	5.576	1
18	20.2	Cyclolongifolene oxide, dehydro	$C_{15}H_{22}O$	218.167	1.368	1
19	21.51	1(2H)-Naphthalenone,octahydro-4a,8a-dimethyl-7-(1-ethylethyl)-, (4aR-(4aa,7a,8aa))	$C_{15}H_{26}O$	222.198	0.329	1
20	22.05	2-tetradecenal	$C_{14}H_{26}O$	210.1984	4.873	1
21	30.1	Palmitic acid	$C_{16}H_{32}O_2$	256.24	1.537	1
22	31.64	Oleic acid	$C_{18}H_{34}O_2$	282.256	5.499	1
23	32.86	2,8,9-Trioxa-5-aza-1-silabicyclo(3.3.3)undecane,1-methoxy	$C_7H_{15}NO_4Si$	205.077	0.165	1
24	37.22	Tridecanoic acid, methyl ester	$C_{14}H_{28}O_2$	228.209	0.625	1
25	39.71	Heptacosane	$C_{27}H_{56}$	380.438	2.664	1

3.2. In Vitro Antioxidant and Free Radical Scavenging Activities

The DPPH free radical scavenging assay results represent a strong effect of the extract to scavenge free radicals. The percent scavenging activity of the extracts showed an increase with the subsequent increase in the concentration of the extract. The 70% ethanol extract was more potent than the hexane fraction, and the IC_{50} values for DPPH, $ABTS^+$, superoxide radical scavenging, metal chelation and ferric reducing antioxidant activities were found to be 222.22 ± 7.4 μg/mL, 13.90 ± 0.5 μg/mL, 113.81 ± 4.2 μg/mL, 948 ± 21.1 μg/mL and 12.3 ± 0.43 mg $FeSO_4E$/g of extract, respectively. However, the hexane fraction also exhibited antioxidant activity with IC_{50} values of 432.68 ± 13.7 μg/mL for DPPH, 23.57 ± 1.4 μg/mL for $ABTS^+$, 255.72 ± 9.7 μg/mL for superoxide radical scavenging, 1211 ± 27.8 μg/mL for metal chelation and 45.62 ± 1.34 mg $FeSO_4E$/g of extract for the ferric-reducing antioxidant assay (Table 3).

Table 3. Polyphenol and flavonoid content, antioxidant and free radical scavenging activities of ethanolic extract of *Nardostachys jatamansi* (NJE) and the hexane fraction (NJH). Values are represented as the means ± SD of the triplicate determination. ABTS, 2,2′-azino-bis(3-ethyl benzothiazoline-6-sulfonic acid) diammonium salt. CE, catechin equivalents.

Assay	Ethanolic Extract (NJE)	Hexane Extract (NJH)
Total polyphenolic content	53.06 ± 2.2 mg GAE/g of extract	13.87 ± 1.3 mg GAE/g of extract
Total flavonoids	25.303 ± 0.9 mg CE/g of extract	4.58 ± 0.3 mg GAE/g of extract
DPPH radical scavenging assay (IC_{50})	222.22 ± 7.4 μg/mL	432.68 ± 13.7 μg/mL
Metal chelation (IC_{50})	948 ± 21.1 μg/mL	1211 ± 27.8 μg/mL
ABTS (IC_{50})	13.90 ± 0.5 μg/mL	23.57 ± 1.4 μg/mL
Superoxide (IC_{50})	113.81 ± 4.2 μg/mL	255.72 ± 9.7 μg/mL
Anti-lipid peroxidation (IC_{50})	465.11 ± 14.3 μg/mL (brain) 539.08 ± 18.9 μg/mL (liver)	587.53 ± 17.6 μg/mL (brain) 685.15 ± 13.4 μg/mL (liver)
Ferric-reducing antioxidant power	12.3 ± 0.43 mg $FeSO_4E$/g of extract	45.62 ± 1.34 mg $FeSO_4E$/g of extract

The 70% ethanol extract exhibited strong antioxidant capacity in comparison to that of the hexane extract. Hence, the 70% ethanol extract was further used to analyze the protective effect against AAPH-induced oxidative damage of biomolecules.

Polyphenol and Flavonoid Contents

NJE showed a high content of polyphenols, 53.06 ± 2.2 mg GAE/g of the extract (gallic acid equivalent), and flavonoids, 25.303 ± 0.9 mg CE/g of the extract (catechin equivalents), whereas NJH showed 13.87 ± 1.3 mg GAE/g of the extract of polyphenols and 4.58 ± 0.3 mg GAE/g of the extract of flavonoids (Table 3).

3.3. Protective Properties against Biomolecules Oxidation

3.3.1. NJE Inhibits AAPH-Induced DNA Damage

AAPH is a water-soluble initiator, which decomposes at physiological temperature, producing alkyl radicals, which react with oxygen to form alkyl peroxyl radicals to initiate protein oxidation [37]. The protective potential of the plant extract against DNA damage induced by AAPH (200 mM) was measured using pBR322 DNA. As shown in Figure 3, the plasmid DNA was mainly of the supercoiled form (bottom band) and open circular form (top band) in the absence of AAPH (Lane 1). With the addition of 10 mM AAPH, the supercoiled form decreased and converted into the open circular form (Lane 2). The addition of the extract significantly inhibited the formation of the open circular form as compared to the positive control (85%; 15 μg/mL extract). Gallic acid at 5 μg/mL also showed significant protection against DNA damage (95%).

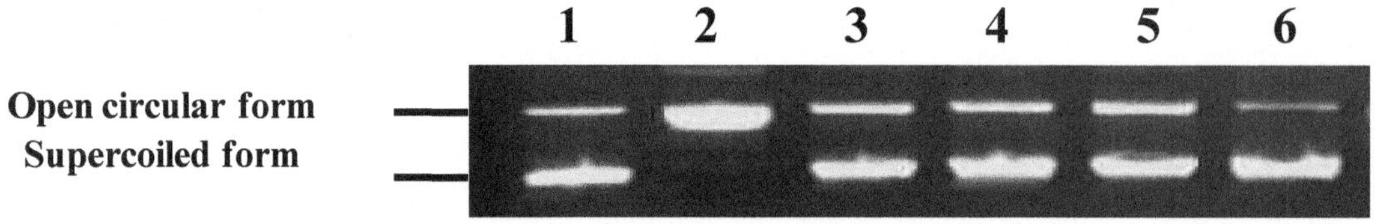

Figure 3. Inhibition of DNA damage by NJE. Lane 1, plasmid DNA (pBR322); Lane 2, DNA + 2,2′-azobis(2-methylpropionamidine) dihydrochloride (AAPH); Lane 3, DNA + AAPH + 5 μg extract; Lane 4, DNA + AAPH + 10 μg extract; Lane 5, DNA + AAPH + 15 μg extract; Lane 6, DNA + AAPH + 10 μg GA.

3.3.2. NJE Inhibits Protein Oxidation

The protective property of the plant extract against protein oxidation was determined by the oxidation of BSA initiated by AAPH. After incubation for 2 h at 37 °C, the samples were analyzed using SDS-PAGE. The results demonstrated that the plant extract showed 50% protection against BSA oxidation (Lanes 3, 4, 5, 6, 7) in a dose-dependent manner. In Lane 2 (positive control), in the presence of the radical inducer AAPH (200 mM), BSA was completely degraded, whereas gallic acid at 10 μg/mL showed a complete protective effect (Figure 4).

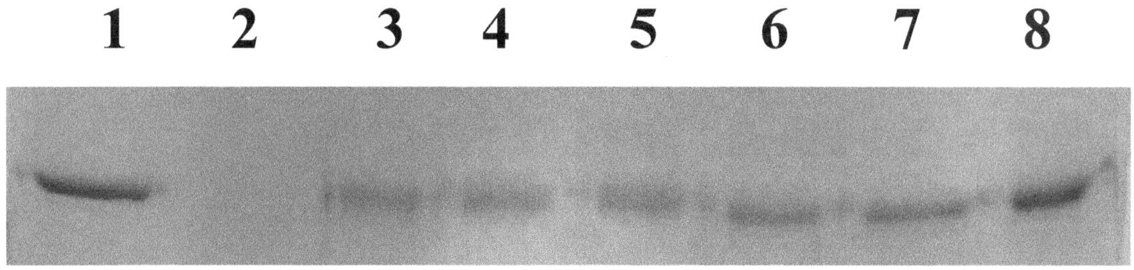

Figure 4. Inhibition of protein oxidation by NJE. Lane 1, BSA; Lane 2, BSA + AAPH; Lane 3, BSA + AAPH + 5 μg extract; Lane 4, BSA + AAPH + 10 μg extract; Lane 5, BSA + AAPH + 15 μg extract; Lane 6, BSA + AAPH + 20 μg extract; Lane 7, BSA + AAPH + 25 μg extract; Lane 8, BSA + AAPH + 10 μg GA.

3.3.3. Inhibition of Lipid Peroxidation

Further, the potential of 70% ethanol extract to inhibit lipid peroxidation induced by AAPH (a peroxyl radical inducer) in rat liver and brain homogenate was analyzed by the TBARS assay. The IC_{50} values of NJE for inhibiting lipid peroxides in the brain and liver are 465.11 ± 14.3 μg/mL, 539.08 ± 18.9 μg/mL, respectively, and for NJH are 587.53 ± 17.6 μg/mL (brain) and 685.15 ± 13.4 μg/mL (liver) (Table 3).

3.3.4. Inhibition of Reactive Oxygen Species

This assay was performed to check the protective efficacy of the plant extract against intracellular ROS generation. AAPH caused a significant increase in ROS production in comparison with that of the control. NJE at 100 μg/mL was effective at combating ROS and showed 58% and 55% ROS inhibition in the liver and brain homogenates (Figure 5).

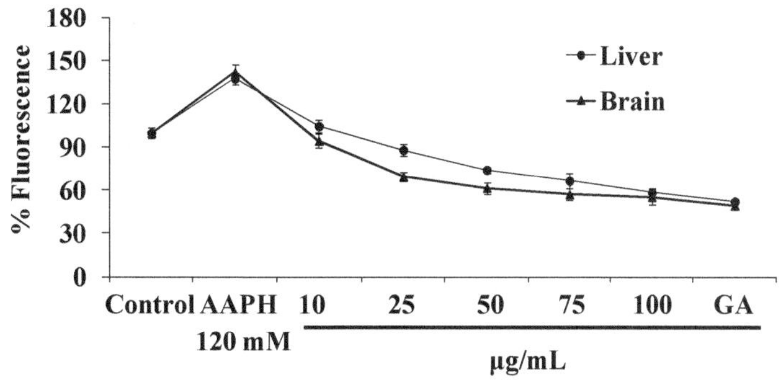

Figure 5. Antioxidant effects of NJE and gallic acid on AAPH-induced ROS generation.

3.3.5. NJE Inhibits Protein Carbonyl Content

The protein carbonyl formation increases drastically under various pathological conditions of oxidative stress [38]. AAPH significantly increased the formation of protein carbonyls in comparison to that of the control. The increasing concentration of the extract was effective at inhibiting the protein carbonyls formed with 100 μg/mL of NJE, showing a 61% and 54% inhibitory effect on protein carbonyls formed in the liver and brain homogenates (Figure 6).

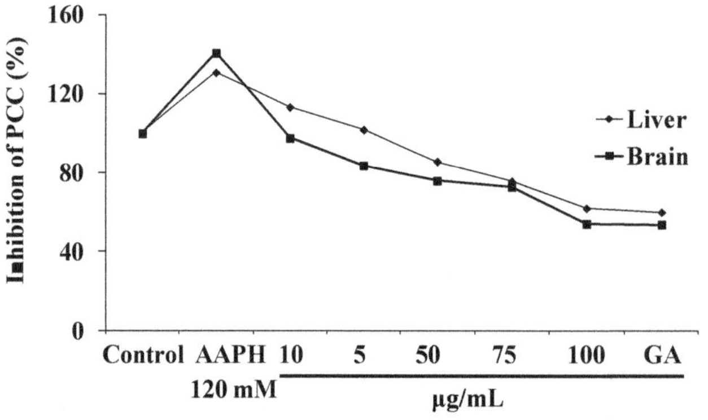

Figure 6. Inhibitory effects of NJE and gallic acid on AAPH-induced protein carbonyl formation.

3.4. Cytotoxicity of NJE by the Plasma Membrane Leakage Assay

Cell damage was evaluated by measuring the leakage of intracellular LDH into the medium. The extract was not toxic up to 75 μg/mL. However, pretreatment of the cells above 100 μg/mL exhibited cell damage, and an increased LDH release was observed (Figure 7).

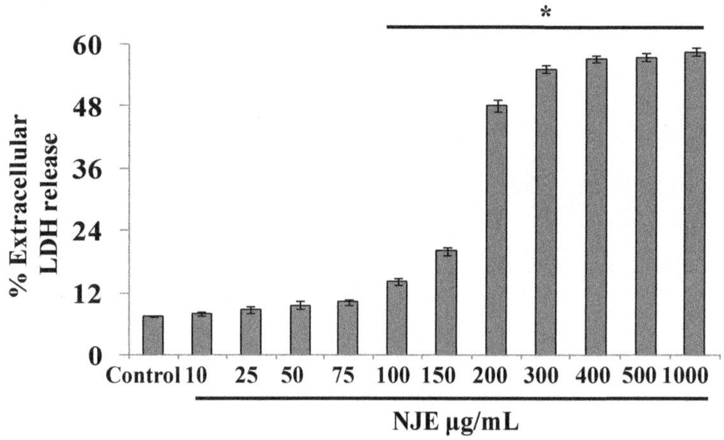

Figure 7. Effects of NJE on LDH release in SH-SY5Y neuronal cells (* $p < 0.05$ *vs.* control).

4. Discussion

Modern lifestyle, including the fast pace of life and reduced physical activity, seems to have a direct, as well as indirect effect on human health, especially mental and physical stress. The first and foremost manifestation is anxiety. Excessive anxiety can prove to be disabling and could lead to other neuropsychiatric disorders. There are several reports demonstrating that physical/mental stress leads to oxidative stress, which, in turn, becomes a cause for more serious illness [39,40].

In the present investigation, preliminarily, we analyzed the phytochemicals present in *N. jatamansi*. Reversed phase-HPLC analysis aided in quantifying the polyphenols present in NJE. The identified polyphenols contribute to the enhanced antioxidant potency of the herb and help in scavenging ROS. In plants, polyphenols and flavonoids serve as protectors against a wide variety of environmental stresses, while, in humans, flavonoids appear to function as "biological response modifiers". The broad therapeutic effects of polyphenols and flavonoids can be largely attributed to their antioxidant properties [7]. Polyphenols and flavonoids are naturally-occurring antioxidant substances, which exhibit myriad pharmacological actions on the central nervous system and are effective at managing anxiety and depression [41]. Reports demonstrated that polyphenols, such as gallic acid, catechin, chlorogenic acid and quercetin-3-rhamnoside, exhibit anxiolytic effects [42]. GC-MS analysis aided in the identification of an array phytochemicals, such as fatty acids, sesquiterpenes, alkane hydrocarbons and esters. Among the identified compounds, alpha-muurolene, a sesquiterpene, was found to be abundant. Sesquiterpenes have been ascribed as potent antioxidants [43]. Another identified metabolite, linalool, a monoterpene, has been used as a sleep aid and is effective in the treatment of psychosis and anxiety [44]. Omega 3-fatty acids and long-chain polyunsaturated fatty acids have been implicated with diminishing inflammation [45] and reduce the risk of thrombosis [46].

Polyphenols and flavonoids are gaining importance for their antioxidant properties, as they are effective hydrogen donors [47]. These natural antioxidants provide health benefits associated with preventing damage due to biological degeneration and any oxidative onslaught [34]. Previous studies also reported that the dried roots of *N. jatamansi* contain sesquiterpenes, like jatamansone, spirojatamol, patchouli alcohol, norseychelanone, jatamol A and B, lignans and neolignans, like virolin, pinoresinol, jatamansic acid, and terpenic coumarins, like oroselol and jatamansin. The essential oil is composed of sesquiterpenoids and coumarins. Jatamansone is the principle sesquiterpenoid and is attributed with rendering majority of the biological activity [48,49].

The free radical scavenging activity of *N. jatamansi* was evaluated by an array of antioxidant assays. The DPPH free radical scavenging assay is extensively employed to determine the antioxidant activities of different samples in a short time. This suggests that the plant extract contains compounds that are capable of donating hydrogen to a free radical in order to remove the odd electron, which is responsible for the radical's reactivity [50]. The $ABTS^{+}$ radical scavenging assay reflects the capacity of an antioxidant species to donate electrons or hydrogen atoms to inactivate this radical cation. The reducing power of a compound serves as a significant indicator of its potential antioxidant activity [51]. The ferric-reducing antioxidant power assay is described as a rapid and novel method for assessing the antioxidant power of various samples. Fe^{3+} reduction is often used as an indicator of electron donating activity, which is an important mechanism of phenolic antioxidant action and can be strongly correlated with other antioxidant properties [52]. The superoxide anion radical is one of the strongest reactive oxygen species generated among the free radicals. Thus, the antioxidant activity is attributed to various mechanisms, like the prevention of chain initiation, binding of transition metal ion catalysts, prevention of continued hydrogen abstraction, reductive capacity, radical scavenging activity and decomposition of peroxides [53,54]. In the present study, the results suggest that 70% ethanol extract is a potent scavenger of free radicals compared to the hexane extract. This might be due to the diverse content of phytochemicals in the 70% ethanol fraction of *N. jatamansi*.

The excessive production of free radicals, collectively called ROS, is known to cause oxidation of biomolecules and is observed in degenerative diseases and ageing. Thus, the inhibition of oxidation of these biomolecules may be vital for the alleviation of an array of diseases. Damage to DNA is known to alter replication and transcription, causing cell death or mutations. Oxidative modifications of DNA have been suggested to contribute to aging and various diseases, including cancer and chronic inflammation [55]. Lipid peroxidation is a free radical chain reaction, which arises from the oxidative conversion of polyunsaturated fatty acids by $HO^{\bullet}$ to lipid peroxides, which, in turn, can damage biological membranes. The malondialdehyde level is widely utilized as a marker of lipid peroxidation in states of elevated oxidative stress. Thus, it can be correlated to the identified metabolites serving as antioxidants, being mainly capable of scavenging these free radicals and thereby preventing DNA damage, protein oxidation and lipid peroxidation. The results are suggestive of the fact that the plant extract is effective at the prevention of lipid peroxidation, a common causative agent in cell membrane disruption and cell damage [56].

The protective effect of antioxidants on cells is thought to reside in their ability to reduce the level of intracellular ROS generation [57]. The fluorescent probe 2′,7′-dichlorodihydro fluorescein diacetate (DCFH-DA) is used to measure ROS that crosses cell membranes and that are enzymatically hydrolysed to nonfluorescent DCFH. In the presence of ROS, DCFH is oxidized to a highly fluorescent compound, 2′,7′-dichlorofluorescein (DCF), and measured to quantify the ROS generation [58]. AAPH, a peroxyl radical generator, caused an increase in ROS production that was alleviated by NJE dose dependently.

The oxidative damage to the proteins caused by free radicals has been shown to play a significant role in aging and several pathological events [59]. The major molecular mechanism leading to structural modifications in proteins is free radical-mediated protein oxidation characterized by carbonyl formation. The measurement of protein carbonyls is used to assess the oxidative damage of proteins [60]. As observed, NJE inhibited protein carbonyl (PCO) formation, confirming its antioxidative role and demonstrating that it could be a good source of antioxidants and could protect against oxidative insults. The cytotoxic effect of the extract assessed by the LDH assay on SH-SY5Y cells showed that the extract is non-toxic up to the dose of 75 g/mL and could be used to further verify the antioxidant effects at the cellular level.

5. Conclusions

Oxidative stress causes cellular damage and subsequent cell death. The results obtained suggests that *Nardostachys jatamansi* rhizome contains good amounts of polyphenols, flavonoids and sesquiterpenes and thereby exhibits high antioxidant and free radical scavenging activities. The extract was found to chelate iron, scavenge superoxide radical and has a high reducing power. It inhibited protein oxidation, DNA damage and decreased the malondialdehyde level. The extract effectively scavenged ROS and inhibited protein carbonyls. Hence, these *in vitro* assays support the idea that *Nardostachys jatamansi* can be a good source of natural antioxidants to alleviate oxidative stress-mediated disorders.

Acknowledgments

This work has been carried out for Project No. DFR 297, funded by the Defence Research and Development Organization, India. The authors acknowledge the constant support and encouragement given by the Director, Defence Food Research Laboratory (DFRL), Mysore, India. The authors are grateful to Prasanth Bitla, School of Life Sciences, University of Hyderabad, Hyderabad, for the GC-MS-HT-TOF analysis.

Author Contributions

Sakina Razack, Kandikattu Hemanth Kumar, Ilaiyaraja Nallamuthu and Farhath Khanum conceived of and designed the experiments. Sakina Razack performed the experiments and wrote the paper. Kandikattu Hemanth Kumar participated in GC-MS and cell culture studies. Sakina Razack and Ilaiyaraja Nallamuthu analyzed the data. Mahadeva Naika facilitated the reversed phase-HPLC analysis. All of the authors have read and approved the final manuscript.

Conflicts of Interest

The authors declare no conflict of interest.

References

1. Halliwell, B.; Gutteridge, J.M. *Free Radicals in Biology and Medicine*; Oxford University Press: Oxford, UK, 1999.
2. Yildirim, A.; Mavi, A.; Oktay, M.; Kara, A.A.; Algur, O.F.; Bilaloglu, V. Comparison of antioxidant and antimicrobial activities of Tilia (*Tilia argenta* Desf Ex DC), sage (*Salvia triloba* L.) and black tea (*Camellia sinensis*) extracts. *J. Agric. Food Chem.* **2000**, *48*, 5030–5034.
3. Gulcin, I.; Oktay, M.O.; Kufrevioglu, O.I.; Aslan, A. Determination of antioxidant activity of lichen *Cetraria islandica* (L.). *J. Ethnopharmacol.* **2002**, *79*, 325–329.
4. Mittova, V.; Volokitam, M.; Guy, M.; Tal, M. Activities of SOD and the ascorbate-glutathione cycle enzymes in subcellular compartments in leaves and roots of the cultivated tomato and its wild salt-tolerant relative *Lycopersicon pennellii*. *Physiol. Plant.* **2000**, *110*, 45.
5. Dean, R.T.; Stocker, R.; Davies, M.J. Biochemistry and pathology of radical-mediated protein. *J. Biochem.* **1997**, *324*, 1–18.
6. Aruoma, O.I. Free radicals, oxidative stress and antioxidants in human health and disease. *J. Am. Oil Chem. Soc.* **1998**, *75*, 199–212.
7. Percival, M. Antioxidants. *Clin. Nutr. Insight* **1998**, *31*, 1–4.
8. Ashokkumar, D.; Mazumder, U.K.; Gupta, M.; Senthilkumar, G.P.; Selvan, V.T. Evaluation of antioxidant and free radical scavenging activities of *Oxystelma esculentum* in various *in vitro* models. *J. Compliment. Integr. Med.* **2008**, *5*, 1553–3840.
9. Veerapur, V.P.; Prabhakar, K.R.; Parihar, V.P.; Kandadi, M.R.; Ramakrishana, S.; Mishra, B.; Satish Rao, B.S.; Srinivasan, K.K.; Priyadarsini, K.I.; Unnikrishnan, M.K. *Ficus racemosa* stem bark extract: A potent antioxidant and a probable natural radioprotector. *Evid. Based Complement. Alternat Med.* **2009**, *6*, 317–324.
10. Kitts, D.D.; Yuan, Y.V.; Wijewickreme, A.N.; Hu, C. Antioxidant properties of a North American gingseng extract. *Mol. Cell. Biochem.* **2000**, *203*, 1–10.
11. Muselík, J.; García-Alonso, M.; Martín-López, M.P.; Želmièka, M.; Rivas-Gonzalo, J.C. Measurement of antioxidant activity of wine catechins, procyanidins, antocyanins and piranoantocyanins. *Int. J. Mol. Sci.* **2007**, *8*, 797–809.
12. Aviram, M. Review of human studies on oxidative damage and antioxidant protection related to cardiovascular diseases. *Free Radic. Res.* **2000**, *33*, 85–97.

13. Baynes, J.W. From life to death-the struggle between chemistry and biology during aging: The Maillard reaction as an amplifier of genomic damage. *Biogerontology* **2000**, *1*, 235–246.
14. Halliwell, B. Oxidative stress, nutrition and health. Experimental strategies for optimization of nutritional antioxidant intake in humans. *Free Radic. Res.* **1996**, *25*, 57–74.
15. Rao, V.S.; Rao, A.; Karanth, K.S. Anticonvulsant and neurotoxicity profile of *Nardostachys jatamansi* in rats. *J. Ethnopharmacol.* **2005**, *102*, 351–356.
16. Ahmad, M.; Yousuf, S.; Khan, M.B.; Hoda, M.N.; Ahmad, A.S.; Ansari, M.A.; Ishrata, T.; Agrawalb, A.K.; Islama, F. Attenuation by *Nardostachys jatamansi* of 6-hydroxydopamine-induced parkinsonism in rats: Behavioral, neurochemical, and immunohistochemical studies. *Pharmacol. Biochem. Behav.* **2006**, *83*, 150–160.
17. Subashini, R.; Yogeeta, S.; Gnanapragasam, A.; Devaki, T. Protective effect of *Nardostachys jatamansi* on oxidative injury and cellular abnormalities during doxorubicin-induced cardiac damage in rats. *J. Pharm. Pharmacol.* **2006**, *58*, 257–262.
18. Joshi, H.; Parle, M. *Nardostachys jatamansi* improves learning and memory in mice. *J. Med. Food* **2006**, *9*, 113–118.
19. Kumar, V.P.; Chauhan, N.S.; Padh, H.; Rajani, M. Search for antibacterial and antifungal agents from selected Indian medicinal plants. *J. Ethnopharmacol.* **2006**, *107*, 182–188.
20. Dhingra, D.; Goyal, P.K. Inhibition of MAO and GABA: Probable mechanisms for antidepressant-like activity of *Nardostachys jatamansi* DC. in mice. *Indian J. Exp. Biol.* **2008**, *46*, 212–218.
21. Dandagi, P.M.; Patil, M.B.; Mastiholimath, V.S.; Gadad, A.P.; Dhumansure, R.H. Development and evaluation of hepatoprotective polyherbal formulation containing some indigenous medicinal plants. *Indian J. Pharm. Sci.* **2008**, *70*, 265.
22. Jadhav, V.M.; Thorat, R.M.; Kadam, V.J.; Kamble, S.S. Herbal anxiolyte: *Nardostachys jatamansi*. *J. Pharm. Res.* **2009**, *2*, 1208–1211.
23. Khan, M.B.; Hoda, M.N.; Ishrat, T.; Ahmad, S.; Khan, M.M.; Ahmad, A.; Yusuf, S.; Islama, F. Neuroprotective efficacy of *Nardostachys jatamansi* and crocetin in conjunction with selenium in cognitive impairment. *Neurol. Sci.* **2012**, *33*, 1011–1020.
24. Ross, K.A.; Beta, T.; Arntfield, S.D. A comparative study on the phenolic acids identified and quantified in dry beans using HPLC as affected by different extraction and hydrolysis methods. *Food Chem.* **2009**, *113*, 336–344.
25. Singleton, V.L.; Rossi, J.A. Colorimetry of total phenolics with phosphomolybdic acid-phosphotungstic acid reagents. *Am. J. Enol. Vitic.* **1965**, *16*, 144–158.
26. Sakanaka, S.; Tachibana, Y.; Okada, Y. Preparation and antioxidant properties of extracts of Japanese persimmon leaf tea (*kakinoha-cha*). *Food Chem.* **2005**, *89*, 569–575.
27. Eberhardt, M.V.; Lee, C.Y.; Liu, R.H. Antioxidant activity of fresh apple. *Nature* **2000**, *405*, 903–904.
28. Re, R.; Pellegrini, N.; Proteggente, A.; Pannala, A.; Yang, M.; Rice-Evans, C. Antioxidant activity applying an improved ABTS radical cation decolorization assay. *Free Radic. Biol. Med.* **1999**, *26*, 1231–1237.

29. Dinis, T.C.P.; Madeira, V.M.C.; Almeida, L.M. Action of phenolic derivatives (acetoaminophen, Salicylate and 5-aminosalicylate) as inhibitora of membrane lipid peroxidation and as peroxyl radical scavengers. *Arch. Biochem. Biophys.* **1994**, *315*, 161–169.
30. Ye, X.Y.; Wang, H.X.; Liu, F.; Ng, T.B. Ribonuclease, cell free transalation inhibitory and superoxide radical scavenging activities of the iron-binding protein lactoferrin from bovine mill. *Int. J. Biochem. Cell Biol.* **2000**, *32*, 235–241.
31. Benzie, I.E.F.; Strain, J.J. The ferric reducing ability of plasma (FRAP) as a measure of "antioxidant power": The FRAP assay. *Anal. Biochem.* **1996**, *239*, 70–76.
32. Lee, J.C.; Kim, H.R.; Kim, T.; Jang, Y.S. Antioxidant property of an ethanol extract of the stem of *Opuntia ficus-indica* Var. Saboten. *J. Agric. Food Chem.* **2002**, *50*, 6490–6496.
33. Kwon, H.Y.; Choi, S.Y.; Won, M.H.; Kang, T.C.; Kang, J.H. Oxidative modification and inactivation of Cu, Zn-superoxide dismutase by 2, 2′-azobis(2-amidinopropane) dihydrochloride. *Biochim. Biophys. Acta* **2000**, *1543*, 69–76.
34. Bahramikia, S.; Ardestani, A.; Yazdanparast, R. Protective effects of four Iranian plants against free radical-mediated protein oxidation. *Food Chem.* **2009**, *115*, 37–42.
35. Kobayashi, D.; Kondo, K.; Uehara, N.; Otokozawa, S.; Tsuji, N.; Yagihashi, A.; Watanabe, N. Endogenous reactive oxygen species is an important endogenous reactive oxygen species is an important mediator of miconazole antifungal effect. *Phytomedicine* **2008**, *15*, 391–399.
36. Reznick, A.Z.; Packer, L. Oxidative damage to proteins: Spectrophotometric method for carbonyl assay. *Methods Enzymol.* **1994**, *233*, 357–363.
37. Cai, Y.J.; Fang, J.G.; Ma, L.P.; Li, Y.; Liu, Z.L. Inhibition of free radical induced peroxidation of rat liver microsomes by resveratrol and its analogues. *Biochim. Biophys. Acta* **2003**, *1637*, 31–38.
38. Dalle-Donnea, I.; Rossib, R.; Giustarinib, D.; Milzania, A.; Colomboa, R. Protein carbonyl groups as biomarkers of oxidative stress. *Clin. Chim. Acta* **2003**, *329*, 23–38.
39. Leeuwenburgh, C.; Heinecke, J.W. Oxidative stress and antioxidants in exercise. *Curr. Med. Chem.* **2001**, *8*, 829–838.
40. Liu, J.K.; Mori, A. Stress, aging, brain oxidative damage. *Neurochem Res* **1999**, *24*, 1479–1497.
41. Bouayed, J. Polyphenols: A potential new strategy for the prevention and treatment of anxiety and depression. *Curr. Nutr. Food Sci.* **2010**, *6*, 13–18.
42. Bouayed, J.; Rammal, H.; Dicko, A.; Younos, C.; Soulimani, R. Chlorogenic acid, a polyphenol from *Prunus domestica* (Mirabelle), with coupled anxiolytic and antioxidant effects. *J. Neurol. Sci.* **2007**, *262*, 77–84.
43. Neganova, M.E.; Afanas'eva, S.V.; Klochkov, S.G.; Shevtsovs, E.F. Mechanisms of antioxidant effect of natural sesquiterpene lactone and alkaloid derivatives. *Bull. Exp. Biol. Med.* **2012**, *152*, 720–722.
44. Linck, V.M.; da Silva, A.L.; Figueiro, M.; Caramao, E.B.; Morena, P.R.H.; Elisabetsky, E. Effects of inhaled linalool in anxiety, social interaction and aggressive behaviour in mice. *Phytomedicine* **2010**, *17*, 679–683.
45. Richard, D.; Kefi, K.; Barbe, U.; Bausero, P.; Visioli, F. Polyunsaturated fatty acids as antioxidants. *Pharmacol. Res.* **2008**, *57*, 451–455.
46. Kalmijn, S.; Feskens, E.J.M.; Launer, L.J.; Kromhout, D. Polyunsaturated fatty acids, antioxidants, and cognitive function in very old men. *Am. J. Epidemiol.* **1997**, *145*, 33–41.

47. Rice-Evans, C.A.; Miller, N.J.; Bramley, P.M.; Pridham, J.B. The relative antioxidant activity of plant derived polyphenolic flavonoids. *Free Radic. Res.* **1995**, *22*, 375–383.
48. Hoerster, H.; Ruecker, G.; Tautges, J. Valeranone content in the roots of *Nardostachys jatamansi* and *Valeriana officinalis*. *Phytochem* **1977**, *1*, 1070–1071.
49. Rucker, G.; Tautges, J.; Sleck, A.; Wenzl, H.; Graf, E. Isolation and pharmacological activity of the sesquiterpene valeranone from *Nardostachys jatamansi* DC (in German). *Arzneimittelforschung* **1978**, *28*, 7–13.
50. Aiyegoro, O.A.; Okoh, A.I. Prelimnary phytochemical screening and invitro antioxidant activities of the aqueous extract of *Helichrysum longifolium* DC. *BMC Complement. Altern. Med.* **2010**, *10*, 21.
51. Gordon, M.H. The mechanism of antioxidant action *in vitro*. In *Food Antioxidants*; Hudson, B.J.F., Ed.; Elsevier Applied Science: London, UK, 1990; pp. 1–18.
52. Dorman, H.J.D.; Peltoketo, A.; Hiltunen, R.; Tikkanen, M.J. Characterisation of the antioxidant properties of deodorized aqueous extracts from selected Lamiaceae herbs. *Food Chem.* **2003**, *83*, 255–262.
53. Kumar, R.S.; Rajkapoor, B.; Perumal, P. Antioxidant activities of *Indigofera cassioides* Rottl. Ex. DC. using various *in vitro* assay models. *Asian Pac. J. Trop. Biomed.* **2012**, *2*, 256–261.
54. Hemanth Kumar, K.; Razack, S.; Nallamuthu, I.; Khanum, F. Phytochemical analysis and biological properties of *Cyperus rotundus* L. *Ind. Crops Prod.* **2014**, *52*, 815–826.
55. Yang, J.L.; Weissman, L.; Bohr, V.A.; Mattson, M.P. Mitochondrial DNA damage and repair in neurodegenerative disorders. *DNA Repair* **2008**, *7*, 1110–1120.
56. Barrera, G.; Pizzimenti, S.; Dianzani, M.U. Lipid peroxidation: Control of cell proliferation cell differentiation and cell death. *Mol. Aspects Med.* **2008**, *29*, 1–8.
57. Roche, M.; Tarnus, E.; Rondeau, P.; Bourdon, E. Effects of nutritional antioxidants on AAPH- or AGEs-induced oxidative stress in human SW872 liposarcoma cells. *Cell Biol. Toxicol.* **2009**, *25*, 635–644.
58. LeBel, C.P.; Ischiropoulos, H.; Bondys, S.C. Evaluation of the probe 2′,7′-dichiorofluorescin as an indicator of reactive oxygen species formation and oxidative stress. *Chem. Res. Toxicol.* **1992**, *5*, 227–231.
59. Stadtman, E.R.; Levin, R.L. Protein oxidation. *Ann. N. Y. Acad. Sci.* **2000**, *899*, 191–208.
60. Ardestani, A.; Yazdanparast, R. Antioxidant and free radical scavenging potential of *Achillea santolina* extracts. *Food Chem.* **2007**, *104*, 21–29.

13

Silymarin as a Natural Antioxidant: An Overview of the Current Evidence and Perspectives

Peter F. Surai [1,2,3,4]

[1] Department of Microbiology and Biochemistry, Faculty of Veterinary Medicine, Trakia University, Stara Zagora 6000, Bulgaria; E-Mail: psurai@feedfood.co.uk

[2] Department of Animal Nutrition, Faculty of Agricultural and Environmental Sciences, Szent Istvan University, Gödöllo H-2103, Hungary

[3] Department of Veterinary Expertise and Microbiology, Faculty of Veterinary Medicine, Sumy National Agrarian University, Sumy 40021, Ukraine

[4] Odessa National Academy of Food Technology, Odessa 65039, Ukraine

Academic Editor: Ehab Abourashed

Abstract: Silymarin (SM), an extract from the *Silybum marianum* (milk thistle) plant containing various flavonolignans (with silybin being the major one), has received a tremendous amount of attention over the last decade as a herbal remedy for liver treatment. In many cases, the antioxidant properties of SM are considered to be responsible for its protective actions. Possible antioxidant mechanisms of SM are evaluated in this review. (1) Direct scavenging free radicals and chelating free Fe and Cu are mainly effective in the gut. (2) Preventing free radical formation by inhibiting specific ROS-producing enzymes, or improving an integrity of mitochondria in stress conditions, are of great importance. (3) Maintaining an optimal redox balance in the cell by activating a range of antioxidant enzymes and non-enzymatic antioxidants, mainly via Nrf2 activation is probably the main driving force of antioxidant (AO) action of SM. (4) Decreasing inflammatory responses by inhibiting NF-κB pathways is an emerging mechanism of SM protective effects in liver toxicity and various liver diseases. (5) Activating vitagenes, responsible for synthesis of protective molecules, including heat shock proteins (HSPs), thioredoxin and sirtuins and providing additional protection in stress conditions deserves more attention. (6) Affecting the microenvironment of the gut, including SM-bacteria interactions, awaits future investigations. (7) In animal nutrition and disease prevention strategy, SM alone, or in combination with other hepatho-active compounds (carnitine, betaine, vitamin B_{12}, *etc.*), might have similar hepatoprotective effects as described in human nutrition.

Keywords: silymarin; silybin; silibinin; antioxidant; Nrf2; NF-κB; vitagenes; gut

1. Introduction

Silymarin (SM) is a C25 containing flavonoid mixture, extracted from the *Silybum marianum* (milk thistle) plant. Today's standardized (according to its silibinin, often called silybin, content) SM extract contains approximately 65% to 80% flavonolignans (silybin A and silybin B, isosilybin A, isosilybin B, silychristin and silydianin), with small amounts of flavonoids, and approximately 20% to 35% of fatty acids and polyphenolic compounds possessing a range of metabolic regulatory effects [1]. Silybin was discovered as the first member of a new family of natural compounds called flavonolignans in 1959 [2] and it is known as the predominant and primary active ingredient in SM [3,4]. That is why compounds containing milk thistle ingredients showing silybin content and silybin antioxidant, as well as other activities in various model systems, are used to explain the biological activity of SM. In particular, SM has been the gold standard drug to treat liver disorders of different etiologies and milk thistle extracts have been used as traditional herbal remedies ("liver tonics") for almost 2000 years. Therefore, SM is most well known for its antioxidant and chemoprotective effects on the liver [5–11] and it is often prescribed and self-prescribed as a complementary and alternative hepatoprotective medicine [12]. SM is being studied as a hepato-, neuro-, nephro- and cardio-protective ingredient due to its strong antioxidant and tissue regenerative properties [12–17]. There is a range of recent comprehensive reviews covering various routes and mechanisms of action of SM in animal models and human trials [13–17] very often referring to its antioxidant properties. However, it seems likely that direct antioxidant (AO) activity of polyphenols does not contribute directly to the antioxidant defence of the body [18,19] and only limited work has been carried out to explore SM/silybin impact on the induction of cellular antioxidant defence via the modulation of various transcription factors, including Nrf2 and NF-κB and respective gene and protein expressions. Potential molecular proliferative signaling targets for anti-cancer activity of silibinin include the receptor tyrosine kinase, STAT, androgen receptor and NF-κB pathways [20], however, anti-cancer activity of SM is beyond the scope of the present review. Therefore, this review focuses on evaluating recent studies on SM (silibinin) antioxidant effects in various *in vitro* and *in vivo* model systems in the context of its contribution to the antioxidant systems regulation and participation in cell signaling.

2. Absorption and Metabolism of Silibinin

SM and its main constituent silibinin sources, metabolism and bioavailability have already been reviewed extensively [3,21,22]. It has been shown that after oral consumption silibinin is characterised by comparatively low availability, e.g., in rats it is only about 0.95% [23]. In fact, after the oral administration of the standardized milk thistle extract Legalon, flavonolignans were rapidly absorbed and eliminated [22] with a half-life for silibinin of 6 h [24–26]. The main biotransformation route of silybin and derivatives was identified to be conjugation [3]. It is interesting to note that silibinin, similar to other flavonoids is recognized by the body as a foreign matter and quickly metabolized via phase II enzymes. Indeed, oral consumption of silibinin was associated with a significant increase in both glutathione *S*-transferase (GST) and quinone reductase (QR) activities in liver, lung, stomach, skin and small bowel in a dose- and time-dependent manner [27]. Silibinin present in the systemic circulation was found mainly in conjugated form [28,29]. In fact, after oral SM administration to healthy volunteers, only 10% [30] to 17% [31] of the total silibinin in the plasma was found in the free unconjugated form. Indeed, mono-, di-, and sulpho-glucuronides are shown to be formed, and 31 metabolites have been identified [32]. Indeed, silibinin in humans and rats shows fast elimination of both the free and conjugated forms with the mean elimination half-life being 6.32 h [24]. Therefore, similar to other flavonoids, after oral consumption silibinin, the main constituent of SM, is characterised by comparatively low availability, fast metabolism and its concentration in plasma is mainly in nano-molar range and only in some cases reaching micro-molar concentrations.

3. Antioxidant Systems of the Body

During evolution, living organisms have developed specific antioxidant protective mechanisms to deal with reactive oxygen species (ROS) and reactive nitrogen species (RNS) [33]. Therefore, it is only the presence of natural antioxidants in living organisms which enable them to survive in an oxygen-rich environment. The general term "antioxidant systems" describes these mechanisms, which are diverse and responsible for the protection of cells from the actions of free radicals. The protective antioxidant compounds are located in organelles, subcellular compartments, or the extracellular space enabling maximum cellular protection to occur. Thus, antioxidant systems of the living cell include three major levels of defence [33–37].

The first level of defence is responsible for prevention of free radical formation by removing precursors of free radicals or by inactivating catalysts and consists of three antioxidant enzymes, namely superoxide dismutase (SOD), glutathione peroxidase (GSH-Px), and catalase (CAT) plus metal-binding proteins. Since the superoxide radical is the main free radical produced in physiological conditions in the cell, SOD (EC 1.15.1.1) is considered to be the main element of the first level of antioxidant defence in the cell. This enzyme dismutates the superoxide radical in the following reaction:

$$2\ O_2^{-} + 2\ H^+ \rightarrow H_2O_2 + O_2$$

The hydrogen peroxide formed by SOD action can be detoxified by GSH-Px or CAT which reduce it to water as follows:

$$H_2O_2 + 2\ GSH \rightarrow GSSG + 2\ H_2O$$
$$2\ H_2O_2 \rightarrow 2\ H_2O + O_2$$

Transition metal ions also accelerate the decomposition of lipid hydroperoxides into cytotoxic products such as aldehydes, alkoxyl radicals, and peroxyl radicals. Therefore, metal-binding proteins (transferrin, lactoferrin, haptoglobin, hemopexin, metallothionein, ceruloplasmin, ferritin, albumin, myoglobin, *etc.*) also belong to the first level of defence. Unfortunately, the first level of antioxidant defence in the cell is not sufficient to completely prevent free radical formation and some radicals do escape through this level, initiating lipid peroxidation and causing damage to polyunsaturated fatty acids

(PUFAs), DNA and proteins. Therefore, the second level of defence consists of chain-breaking antioxidants—vitamin E, ubiquinol, carotenoids, vitamin A, ascorbic acid, uric acid, and some other antioxidants. Glutathione (GSH) and thioredoxin systems also have a substantial role in the second level of antioxidant defence. Chain-breaking antioxidants inhibit peroxidation by keeping the chain length of the propagation reaction as small as possible. Therefore, they prevent the propagation step of lipid peroxidation by scavenging peroxyl radical intermediates in the chain reaction.

However, even the second level of antioxidant defence in the cell is not able to prevent lipid and protein oxidation and some biological molecules are damaged. In this case, the third level of defence is based on systems that eliminate damaged molecules or repair them. This level of antioxidant defence includes lipolitic (lipases), proteolytic (peptidases or proteases) and other enzymes (DNA repair enzymes, ligases, nucleases, polymerases, proteinases, phospholipases, various transferases, *etc.*). All these antioxidants are operating in the body in association with each other forming an integrated antioxidant system. The cooperative interactions between antioxidants in the cell are vital for maximum protection from the deleterious effects of free radicals and toxic products of their metabolism.

Therefore, the antioxidant defences include several options [33–37]:

- Decrease localized oxygen concentration
- Decrease activity of pro-oxidant enzymes and improve efficiency of electron chain in the mitochondria and decreasing electron leakage leading to superoxide production
- Prevention of first-chain initiation by scavenging initial radicals by inducing various transcription factors (e.g., Nrf2 and NF-κB) and ARE-related synthesis of AO enzymes (SOD, GSH-Px, CAT, glutathione reductase (GR), GST, *etc.*)
- Vita-gene activation and synthesis and increased expression of protective molecules (GSH, Thioredoxins, heat shock proteins (HSPs), sirtuins, *etc.*)
- Binding metal ions (metal-binding proteins) and metal chelating
- Decomposition of peroxides by converting them to non-radical, nontoxic products (Se-GSH-Px);
- Chain breaking by scavenging intermediate radicals such as peroxyl and alkoxyl radicals (vitamins E, C, GSH, uric acid, ubiquinol, bilirubin, *etc.*)
- Repair and removal of damaged molecules (methionine sulfoxide reductase (Msr), DNA-repair enzymes, chaperons, *etc.*)

4. Antioxidant Properties of Silymarin (SM)

It should be noted that SM can contribute to the antioxidant defences in different ways. Firstly, by direct free radical scavenging. Secondly, by preventing free radical formation by inhibiting specific enzymes responsible for free radical production, or by maintaining the integrity of electron-transport chain of mitochondria in stress conditions. Thirdly, by participating in the maintenance of optimal redox status of the cell by activating a range of antioxidant enzymes and non-enzymatic antioxidants, mainly via transcription factors, including Nrf2 and NF-κB. Finally, by activating an array of vitagenes, responsible for the synthesis of protective molecules, including HSP, thioredoxin (Trx), sirtuins, *etc.*, and providing additional protection in stress conditions. In most studies pure silybin, as the main component of SM, was used, however, in some cases SM also showed antioxidant action *in vivo*.

4.1. Direct Free Radical Scavenging

The effects of silibinin on the formation of ROS and eicosanoids by human platelets, white blood and endothelial cells were studied [32]. Silibinin showed to be a strong scavenger of HOCI (IC_{50} 7 μM), but not of O_2^- (IC_{50} > 200 μM) produced by human granulocytes. Similarly, production of O_2^- and NO in isolated rat Kupffer cells were inhibited by silibinin in a dose-dependent manner, with IC_{50} 80 μM [38,39]. Indeed, silybin has no superoxide anion scavenging capability but was able to significantly decrease (at 20 μM) hem-mediated low dencity lipoprotein (LDL) oxidation and showed slight inhibition of hydroxyl radical formation [40]. The rate constants of silybin with OH radical (1.8×10^{10} dm/mol/s) is diffusion controlled, suggesting that silybin is a potent free radical scavenger [41]. Indeed, silibinin (2.5 μM) significantly decreased the concentration of H_2O_2 in Ab_{1-42}-stressed neurons and prevented oxidative-related injuries [42]. Furthermore, silybin (10 μM and higher) is shown to have protective activity in ameliorating DNA damage in a model system [41]. Treatment *in vitro* with silibinin significantly inhibited spontaneous O_2^- and H_2O_2 release and TNF-α production by monocytes from pre-eclamptic women. The main effect of silibinin was obtained at 50 μM concentration [43]. Thus, the authors concluded that silibinin exerts anti-oxidative and anti-inflammatory effects on monocytes from pre-eclamptic pregnant women by inhibiting the *in vitro* endogenous release of ROS and TNF-α production.

Antioxidant action of SM (equal to 62 μM of silybin) has been shown in various model systems. For example, IC_{50} for H_2O_2 was 38 μM, while IC_{50} for NO was 266 μM [44]. It was shown that SM inhibited 82.7% lipid peroxidation of linoleic acid emulsion; while BHA, BHT, alpha-tocopherol and Trolox indicated inhibition of 83.3, 82.1, 68.1 and 81.3% respectively on peroxidation of linoleic acid emulsion at the same concentration [45]. In another *in vitro* study free radical scavenging activity and antioxidant properties of SM (>200 μM) were showed by four different assays [46]. It is important to mention that the free radical scavenging activity of pure individual compounds of the SM is reported to vary considerably, with silydianin and silychristin being 2-10-fold more active than the silibinin and on a mass basis, SM is shown to be about 8-fold more potent than silibinin as a free radical scavenger [47]. Indeed, SM and silybin (silibinin) are not the same compounds and their AO activities could differ substantially depending on their concentrations in the target tissues. Therefore, in the following sections a distinction will be made between AO activities of SM and its active constituent silybin.

It seems likely, that SM/silibinin AO effects could be on the cells further upstream of the ROS and TNFa release. In fact, SM has been shown to alter trafficking within cells [48,49] and affect the energy status of the cell [50], so the ROS scavenging effects, if any, could be secondary. Furthermore, many of the effects of SM/silibinin described *in vitro* occur at concentrations not currently achievable in humans or animals. Therefore, a direct scavenging ROS by silibinin in biological systems (except the gut) is not likely to substantially contribute to the antioxidant protection.

4.2. Protective Effects of Silybin on Mitochondria, a Main Source of Free Radical Production in the Cell

Mitochondria are the primary cellular consumers of oxygen and contain numerous redox enzymes capable of transferring single electrons to oxygen, generating the ROS superoxide (O_2^-). It is well documented that mitochondrial enzymes known to generate ROS include the tricarboxylic acid (TCA) cycle enzymes aconitase and α-ketoglutarate dehydrogenase; the electron-transport chain (ETC) complexes I, II and III; pyruvate dehydrogenase and glycerol-3-phosphate dehydrogenase; dihydroorotate dehydrogenase; the monoamine oxidases (MAO) A and B; and cytochrome b5 reductase [51]. Furthermore, mitochondrial insults, including oxidative damage itself, can cause an imbalance between ROS production and removal, resulting in net ROS production. For example, ROS can induce protein modifications, lipid peroxidation and mitochondrial DNA damage, which ultimately results in mitochondrial dysfunction [52]. Many studies have focused on the detrimental effects of ROS, but it is now clear that mitochondrially generated ROS are also involved in regulating intracellular signal transduction pathways that result in cell adaptation to stress [53].

One of the mechanisms responsible for the decrease in oxidative stress is the protective effect of SM/silibinin on mitochondrial structure and function. Indeed SM protects mitochondria from pathological events by triggering pro-survival cell signaling. For example, silibinin supplementation is shown to optimize electron-transport chain, decreasing electron leakage and ROS formation and directly reducing activities of ROS-producing enzymes in the mitochondria. In rats subjected to ischemia/reperfusion (I/R), compared with the control group, a severe impairment of mitochondrial bioenergetics was observed. SM prevented the most significant changes that occurred in mitochondria during I/R (decreased ATP levels, membrane potential and state 3 respiration) and the associated cell dysfunction [54].

Silibinin (100–500 μM), was evaluated for its protective effect against beta-adrenergic agonist isoproterenol-induced injury in cultured rat neonatal cardiac myocytes [55]. It was shown that silibinin addition was associated with increased SOD activity and upregulated mitochondrial membrane potential and with a prevention of mitochondrial dysfunction and cell injury [54]. Silibinin, at a concentration as low as 10 μM, fully mitigated the rise in metabolic flow-driven ROS formation in perfused rat hepatocytes. In addition, studies on isolated liver mitochondria revealed that this low dose of silibinin depressed ROS production linked to the electron transfer chain activity [56]. It has been shown that cold preservation and warm reperfusion of the rat liver were associated with increased lipid peroxidation and superoxide anion generation, as well as with decreased GSH, mitochondrial ATP content and respiratory control ratio (RCR). However, preservation conducted in presence of silibinin (100 μM) improved parameters affected by preservation and reperfusion. Indeed, silibinin promoted an increase of ATP and RCR by 39 and 16% respectively and decreased oxidative stress to values observed in livers never preserved nor perfused [57]. It has been suggested that the uncoupler-like activity of dehydrosilybin could be the basis of its ROS modulation effect in various experimental systems. In fact, dehydrosilybin uncoupled the respiration of isolated rat heart mitochondria with a very high potency in suppressing ROS formation in isolated rat heart mitochondria with IC(50) = 0.15 μM [58]. It is interesting to note that silibinin in mitochondria was far more effective than its effect in a purely chemical system generating superoxide or in cells capable of oxidative burst, where the IC(50) for dehydrosilybin exceeds 50 μM. Changes in mitochondrial respiratory complexes in fatty hepatocytes were also attenuated by silibinin-vitamin E complex (15 mg vitamin E and 47 mg silybin) fed to rats with a major protective effect on Complex II subunit CII-30 [59]. Similarly, silybin (0.4 g/kg) in complex with phospholipid (SILIPHOS) was effective in decreasing severe oxidative stress and preserving hepatic mitochondrial bioenergetics and mitochondrial proton leak and ATP reduction in nonalcoholic steatohepatitis induced by the methionine- and choline-deficient (MCD) diet [60]. Silymarin oil (10 mL/kg/BW) significantly increased levels of membrane fluidity and membrane potential of liver mitochondria [61]. It has been suggested that protective mechanism of action of silibinin (50–200 μM) in intrastriatal MPP+-injected rats may be linked to maintenance of mitochondrial bioenergetics and integrity [62]. An *in vitro* study demonstrated that silibinin inhibits the activity of ROS-generating monoamine oxidase (MAO) that catalyzes the oxidative deamination of monoamines [63]. Similarly, silymarin oil (10 mL/kg/BW) decreased MAO activity in mice livers [61]. The formation of leukotrienes via the 5-lipoxygenase pathway was indicated to be strongly inhibited by silibinin. In particular, in human granulocytes IC50-values of 15 μM and 14.5 μM silibinin were detected for LTB4 and LTC4/D4/E4/F4 formation respectively. However, much higher silibinin concentrations (45–69 μM) were necessary to inhibit the cyclooxygenase pathway [38]. Rats exposed to a carcinogen (1,2-di-methylhydrazine DMH) showed increased activities of phase I enzymes (cytochrome b5, cytochrome b5 reductase, cytochromeP450, cytochromeP450 reductase, cytochromP4502E1) in the liver and colonic mucosa as compared to control rats. Silibinin supplementation (50 mg/kg/BW) modulates the xenobiotic metabolizing enzymes including decreasing activity of ROS-producing cytochrome b5 reductase [64]. Although energy generation in the mitochondrion is an essential and extremely important process for cell survival, excessive mitochondrial ROS production also has detrimental consequences for the cell and the whole body.

The different behaviour of silymarin/silibinin in normal and cancerous cells should be mentioned. In particular, SM is shown to have a protective effect against diabetes-induced cardiomyocyte apoptosis [65] as well as apoptosis caused by various toxicants (see Section 4.4), while it causes apoptosis in cancerous cells. For example, SM effectively suppressed cell growth in a dose- and time-dependent manner, and arrested cell cycle progression at G1/S phase in human ovarian cancer line A2780s and PA-1 cells via up-regulation of p53, p21, and p27 protein expression, and down-regulation of CDK2 protein expression [66]. Furthermore, in the aforementioned study the silymarin treatment significantly induced apoptosis in A2780s and PA-1 cells by increasing Bax and decreasing Bcl-2 protein expression, and activation of caspase-9 and caspase-3. Indeed, anti-cancer properties of SM are well characterised but they are beyond the scope of the review.

4.3. Inhibition of Free-Radical Producing Enzymes by Silybin

4.3.1. Xanthine Oxidase

Xanthine oxidase (XO) catalyzes the terminal two steps of purine degradation, converting hypoxanthine to uric acid and is considered to be a critical source of both O_2^- and H_2O_2 in inflammatory diseases [67]. In particular, XO in its oxidase form is considered to be a main source of oxidative stress and destructive free radicals in ischemia reperfusion injury associated with heart attacks, stroke and spinal cord injury, as well as being a destructive force in myocardial or renal hypoxia and infarctions [68]. It seems likely that silibinin can affect XO activity, and inhibition of XO by silibinin in the steady-state appears to be independent of the silibinin concentration. Furthermore, in the presence of 50 μM silibinin, the production of superoxide by XO was decreased by approximately 20% relative to the control at 100 μM xanthine [69]. When O_2^- production was induced in xanthine/xanthine oxidase system, silybin inhibited uric acid formation with an IC_{50} of 32.2 μM [68]. Indeed, inhibition in uric acid formation occurred parallel with inhibition of cytochrome C reduction (e.g., with O_2^- formation) suggesting that the observed effects are due to the inhibition of xanthine oxidase activity rather than the scavenging of O_2^- radicals.

4.3.2. NADPH Oxidase

ROS generation by NADPH oxidase enzyme complex plays a critical role in several physiological processes including host defense, posttranslational modification of proteins, cell differentiation and regulation of gene expression [70]. Silybin and its derivatives inhibited the NADPH oxidase activity in PMA-stimulated cell lysate in a concentration dependent manner. The most distinct inhibition was observed when cell lysate was pre-incubated with 5,7,4″-trimethylsilybin possessing high lipid solubility. Indeed, 50% inhibition of NADPH oxidase activity was achieved when using 5 μM of 5,7,4″-trimethylsilybin and 10 μM of silybin at final concentrations [68]. It is interesting to note that As-intoxicated rats showed a significant ($p < 0.05$) up-regulation of myocardial NADPH (NOX) oxidase sub units such as NOX2 and NOX4 as well as Keap-1, and down-regulation of Nrf2 and heme oxygenase (HO-1) protein expressions. Pre-administration of silibinin (75 mg/kg/BW) remarkably recovered all those altered parameters to near normalcy in As-induced cardiotoxic rat [71]. Co-administration of silibinin (75 mg/kg/BW) along with As-resulted in a reversal of As-induced biochemical changes in the kidney, accompanied by a significant decrease in lipid peroxidation and an increase in the level of renal antioxidant defense system [72]. It also decreased the NADPH oxidase, iNOS and NF-κB over expression by As and upregulated the Nrf2 expression in the renal tissue. The effect of silybin in cultures of mouse podocytes and in the OVE26 mouse, a model of type 1 diabetes mellitus and diabetic nephropathy was investigated. Exposure of podocytes to high glucose increased intracellular superoxide production by 60%, NADPH oxidase activity by 90%, NOX4 expression by 100%, and the number of apoptotic cells by 150%, effects that were completely blocked by 10 μM silybin. These *in vitro* observations were confirmed by similar *in vivo* findings. The kidney cortex of vehicle-treated control OVE26 mice displayed greater NOX4 expression and twice as much superoxide production than cortex of silybin-treated mice [73]. Therefore, the *in vitro* data demonstrated that silybin decreased superoxide generation in cultured podocytes and the *in vivo* study showed a similar effect in the kidney cortex. This is not surprising considering that oxidative stress and enhanced generation of ROS is the underlying mechanism of injury to multiple cell types.

4.4. Antioxidant Protective Properties of SM (Silibinin) in Prevention of Toxic Effects of Various Chemicals

The liver is a primary organ involved in biotransformation of food toxicants and drugs. Toxins absorbed from the intestinal tract first enter the liver resulting in a variety of liver disorders. Hepatic disorders are mainly caused by toxic chemicals (alcohol), xenobiotics (carbon tetrachloride, chlorinated hydrocarbons), anticancer (doxorubicin, cisplatin), immunosuppressant (cyclosporine), analgesic anti-inflammatory (paracetamol, thioacetamide), anti-tubercular (isoniazid, rifampicin) drugs, radiations (gamma radiations), heavy metals (cadmium, arsenic), mycotoxins (aflatoxin, T-2 toxin, fumonisins, ochratoxins), galactosamine, lipopolysaccharides, *etc.* [74]. Increasing evidence supports the important role of oxidative stress as a key mechanism of hepatotoxicity. Therefore, the aforementioned toxicants were used in various experimental models to assess antioxidant protective properties of SM. These are especially important, since SM is primarily used in clinical practice to protect, support and repair the liver.

4.4.1. Arsenic

As-induced toxicity is associated with increased production of ROS, elevation in lipid peroxidation, protein carbonylation and DNA oxidation, and disruption of the cell cycle that results in apoptosis [75]. Indeed, the toxic effect of arsenic was indicated by significantly decreased activities of AO enzymes (SOD, CAT, GSH-Px, GST, GR and glucose 6-phosphate dehydrogenase (G6PDH)) along with non-enzymatic antioxidants (GSH, vitamins C and E) in the rat liver; [76] and heart [71] and decreased Nrf2 and HO-1 activities in the heart [71]. Similar As-induced changes were seen in the rat kidney, including decreased activity of AO enzymes (SOD, CAT, GSH-Px, GSH, *etc.*), GSH, ascorbic acid and vitamin E, increased thiobarbituric acid reactive substances (TBARS), lipid peroxides and protein carbonyls, overexpression of NADPH oxidase, iNOS and NF-κB [72]. Silibinin was effective in counteracting all the parameters of oxidative stress. Indeed, administration of silibinin (50–75 mg/kg/BW) exhibited a significant preservation of antioxidant defences by maintaining activity of AO enzymes and non-enzymatic antioxidants in As-treated rats [76–78]. All the changes were supported by reduction of DNA damage in hepatocytes and histopathological observations of the liver.

4.4.2. Carbone Tetrachloride

Carbone tetrachloride (CCl4) is considered as an important hepatotoxin due to its severe oxidative effect on the liver. Indeed, metabolism of CCl4 via CYP2E1 to highly reactive ROS plays a major role in the mode of action of the toxicant [79]. It is well established that CCl4 inhibits AO enzymes (SOD, GSH-Px and CAT) and GSH in liver samples [80], increases the secretion of ALT, aspartate transaminase (AST) and ALP due to hepatic injuries caused by ROS [81] and enhances TBARS in the liver [82] and serum [83]. It has been shown that practically all elevated indexes of the oxidative stress caused by CCl4 were restored almost to the initial physiological levels by SM treatment (25–100 mg/kg/BW) [80–87] and hepatic injuries were significantly decreased. It is important to mention that in a model system based on CCl4-induced liver fibrosis in mice, microarray analysis showed that SM downregulated the expression levels of cytoskeleton organization genes and mitochondrion electron-transfer chain genes, such as cytochrome c oxidase, Cox6a2, Cox7a1, and Cox8b genes [88]. Furthermore, SM (0.1–1.0 g/kg diet) had a protective effect against CCl4-induced liver damage and AO system defences in common carp [89].

4.4.3. Mycotoxins

It has been shown that major mycotoxins (aflatoxins, AF; ochratoxin A, OTA; T-2 toxin and fumonisins, FM) cause oxidative stress and apoptosis in the cell [90–92]. Indeed, mycotoxins generate ROS, which induce lipid and protein oxidation leading to changes in membrane integrity, cellular redox signaling, and in the antioxidant status of the cells. Silibinin is shown to be a potent protective compound against apoptosis and cytotoxicity caused by OTA [93]. In fact, silibinin in doses from 130 to 260 μM prevented chromatin condensation, caspase-3 activation, apoptotic DNA fragmentation and lipid peroxidation that were induced by OTA, H_2O_2 and ActD/TNF-α, respectively. Furthermore, silibinin (0.04–26.0 μM) had hepatoprotective effects against OTA- or LPS-mediated TNF-alpha release and also reduced the cytotoxicity of both compounds [94]. Pre-treatment of the rats with SM (100 mg/kg/BW) prior to AFB1 was found to show significant protection against AFB1-induced liver damage, as evidenced by a significant lowering of the activity of the serum enzymes glutamic-oxaloacetic transaminase (GOT) and glutamic-pyruvic transaminase (GPT) and enhanced hepatic reduced GSH status [95]. SM (100 mg/kg/BW) also ameliorated toxic liver damage caused by FB_1 in BALB/c mice by preventing elevation of the levels of caspase-8 and tumor necrosis factor-alpha mediators [96].

4.4.4. Thioacetamide

Thiacetamide (TAA) is a potent hepatotoxicant which requires metabolic activation by the mixed-function oxidases. In the development of TAA-induced toxicity, free radicals are thought to play a critical role [97] and TAA-induced cirrhosis resulted in oxidative stress in the liver, as observed by a high level of lipid peroxidation, accompanied by distorted status of antioxidants [98]. SM is shown to have a protective effect against oxidative stress caused by TAA in rats. Liver fibrosis was induced in male Sprague Dawley rats by TAA administration (0.03% w/v) in drinking water for a period of 12 weeks. SM (50 mg/kg/BW) was shown to significantly increase hepatic antioxidant enzymes (SOD, CAT and GSH-Px) activity in the TAA-treated rats [99]. Indeed, a significant decrease in TAA-induced liver damage was observed in SM-pre-treated rats indicated by a reduction in serum GPT, GOT and alkaline phosphatase (ALP), restoration of the antioxidant system (SOD, CAT, GR, GSH-Px and GST) and decreased lipid peroxidation [100]. Increases in hepatic levels of malondialdehyde (MDA) and protein carbonyls in rats associated with TAA administration were partially blocked by SM consumption [101]. Furthermore, GSH depletion and heat shock protein-47 gene expressions were also decreased in response to SM administration.

4.4.5. Cisplatin

Cisplatin (CDDP) is a chemotherapeutic drug widely used against a variety of cancers and its nephrotoxicity is mainly due to ROS production and oxidative stress [102]. It was shown that CDDP caused decreased activities of AO enzymes (SOD and GSH-Px) and GSH, increased MDA in rat liver [103,104] and significantly elevated serum activities of lactate dehydrogenase (LDH) and creatine kinase (CK) [104]. SM (100 mg/kg/BW) significantly prevented the cisplatin-evoked disturbances in the above-mentioned antioxidant indexes [103,104]. Furthermore, *in vitro* pre-treatment with 25–200 μM of SM significantly protected against cisplatin-induced cell death in a dose-dependent manner [105], inhibited apoptotic cascade and increased cell viability in the HEI-OC1 cells [106].

4.4.6. Mn Toxicity

It is well known that manganese toxicity in animals is associated with increased oxidative stress, apoptosis and inflammation [107]. In fact, animals exposed to manganese chloride were characterised by a significant increase in TBARS levels associated with a decrease of enzymatic (SOD, CAT, GSH-Px) and non-enzymatic (GSH, Vitamin C) antioxidants [108,109], increased lipid and protein oxidation [110], DNA fragmentation and urinary hydrogen peroxide [111]. Co-administration of SM (100 mg/kg/BW) to Mn-treated rats significantly restored antioxidant defences and attenuated oxidative damages observed in the liver, kidney and brain [108–111].

4.4.7. Benzo[a]pyrene

In experiments with benzo(a)pyrene (B(a)P), an environmental contaminant causing oxidative stress, the protective effect of SM/silibinin (400–500 μM) was shown. It was associated with decreasing DNA damage and apoptosis [112], preventing protein thiol oxidation and AO enzyme (SOD, GSH-Px, CAT, GR, GST) inhibition in the hemolysate [113], restoring redox status, modulating glutathione metabolizing enzymes, decreasing formation of protein oxidation products and changing the level of antioxidant enzymes (5 mM; [114]). SM (50–250 mg/kg/BW) showed substantial protective effects against B(a)P-induced damages by inhibiting phase I detoxification enzyme CYP1A1 and modulating phase II conjugating enzymes (glutathione-*S*-transferase, epoxide hydroxylases, uridine diphosphate glucuronosyltransferases, NAD(P)H: quinone oxidoreductase 1, sulfotransferases) in rat liver [115].

4.4.8. Doxorubicin

Doxorubicin (DOX, also called adriamycin) is an effective chemotherapeutic agent.However, its side effect is cardiotoxicity, caused by oxidative stress, mitochondrial dysfunction and apoptosis [116]. It has been shown that DOX-induced biochemical and histopathological alterations in rat testis could be prevented and/or protected by SM (50 mg/kg/BW). The SM protective and preventive effects which result in the reduction of DOX-induced carbonyl stress and DNA damage [117], decrease in the plasma creatine phosphokinase (CPK), LDH, creatinine and urea, myocardial MDA contents and renal tissue contents of MDA and GSH have also been shown [118]. Similarly, SM (16 mg/kg/BW) significantly decreased an oxidative stress in DOX-treated rats by decreasing MDA and DNA fragmentation in the liver [119]. A protective influence of SM (60 mg/kg/BW) on the heart and liver tissue against toxicity induced by DOX in rats was associated with a prevention of an increase in AST and CK serum activity [120].

4.4.9. Ethanol

Oxidative stress plays an important role in the pathogenesis of alcoholic liver damage. During ethanol metabolism, ROS and RNS are formed causing oxidative stress [121]. Supplementation with a standardized SM (200 mg/kg/BW) attenuated an oxidative stress caused by acute ethanol administration, including prevention of elevation of serum ALT activity, decrease in hepatic GSH and enhanced lipid peroxidation, and increased hepatic TNF production [122]. Similarly, SM (100–200 mg/kg/BW) effectively protected liver from alcohol-induced oxidative stress as evidenced by reducing ALT and AST activities in serum, increasing SOD and GSH-Px activities and decreasing MDA content in the liver [123]. Furthermore silybin (250 mg/kg/BW) normalized alcohol-altered AO parameters in the whole blood hemolysate of mice, including AO enzymes (SOD, CAT, GR, GSH-Px), GSH and TBARS [124].

4.4.10. Other Toxicants

The protective effects of similar doses of SM/silybin on the antioxidant systems of the body are also shown in pyrogallol toxicity [125–127], bleomicyn-injected mice [128], acrolein toxicity in mice [129], diethylnitrosamine-induced oxidative stress in rat liver [130], sodium fluoride [131] and sodium nitrite [132] toxicity.

Indeed, in animal experiments and *in vitro* studies SM/silibinin blocked the oxidative toxicities of As, CCl4, various mycotoxins, thioacetamide, cisplatin, Mn overload, benzo[a]pyrene, doxorubicin, ethanol and some other toxicants by decreasing an abnormally high production of ROS in living cells and restoring signal transduction and gene expression. In the aforementioned studies the free radical scavenging and antioxidant properties of SM and silibinin are demonstrated by: (a) restoration of the endogenous AO enzymes (SOD, CAT, GSH-Px, GR and GST) and non-enzymatic antioxidants (vitamins E and C) in the liver and other tissues of stressed animals; (b) increased intracellular concentration of GSH in liver and other tissues; (c) decreased lipid and protein oxidation, detected as reduced MDA/TBARS and carbonyl content; (d) decreased DNA fragmentation/damage and apoptosis; (e) reduced secretion of ALT, AST, ALP from the liver into the plasma due to hepatic injuries caused by ROS; (f) restored Nrf2 and HO-1 activities; (g) reduced NF-κB expression and concentration of pro-inflammatory cytokines, including tumor necrosis factor. It seems likely that the use of SM before or soon after a toxic insult is more protective than in chronic liver disease [133]. A number of *in vitro* and *in vivo* studies have also shown an anticancer protective effect of SM and silibinin in various *in vitro* and *in vivo* model systems and they were comprehensively reviewed [124–138]. However, because of limitation of space in this review they will not be included in the analysis.

4.5. Is Effective Silybin Concentration in Vitro Achievable in Vivo?

As mentioned above, because of their poor absorption and rapid elimination, plasma values of silybin will rarely exceed micromole concentrations. Therefore, the SM concentration used in most of the aforementioned studies are not achievable in the *in vivo* target tissues, where the silybin concentration is usually at least 10-100-fold lower. Indeed, the active free silibinin concentration in plasma after oral consumption of SM, depending on dose of supplementation, could be in the range 0.2–2.0 μM. For example, in healthy volunteers, after an oral administration of SM (equivalent to 120 mg silibinin), total (unconjugated + conjugated) silibinin concentration in plasma was 1.1–1.3 μg/mL [139]. Lower bioavailability of silibinin was shown in another study with higher dose of SM consumption (equivalent to 240 mg silibinin), where maximum silibinin concentration was about 0.7 μM [24]. Comparable silibinin concentrations (0.3–1.9 μM) were reported in healthy volunteers receiving about 240 mg silibinin [140]. More recently, silybin A maximal concentration in human plasma was shown to be 0.2–0.6 μM and silybin B concentrations comprised about 30% of those shown for silybin A [22]. The concentrations of free and conjugated silibinin in the target tissues of mice have been shown to reach their maximum levels within 1 h after 50 mg/kg/BW silibinin administration. The C_{max} of free/conjugated silibinin in the tissues were 18.3/11.8; 8.9/5.8; 255/560; 2.9/8.9; 5.3/12.6; 12.0/21.9 μM for liver, lung, stomach, skin, prostate and pancreas, respectively [27]. Indeed, the average concentration of most plant polyphenols in plasma rarely exceeds 1–2 μM in healthy subjects [141–143]. Furthermore, the effect of SM in physiologically relevant concentrations would be relatively limited when taking into account concentrations of other antioxidants in plasma. For example, vitamin C (26.1 to 84.6 μM), vitamin E (20–30 μM), urate (150–200 μM), and albumin (several hundred μM) are found in plasma [19,144] and polyphenol contribution to plasma TAC will certainly be less than 2% [19]. A similar situation could be found in animals (chickens) where antioxidant concentrations in plasma are in the same range—vitamin C: 61–66 μM; [145–147]; vitamin E: 13–15 μM [34]; and total antioxidant value in plasma could vary from 336.9 μM [147] up to 740–830 μM, [148]. Indeed, the general concentration of antioxidants in human plasma is about 1000 μM [149], therefore 20–50 μM additional antioxidant from dietary sources would be required to make a significant contribution to systemic antioxidant capacity [150]. While SM displays potent antioxidant activity *in vitro*, the bioactive forms of flavonoids *in vivo* are not those forms found in plants (*i.e.*, flavonoid glycosides) due to their extensive biotransformation in the small intestine and hepatic metabolism upon absorption. Therefore, metabolic modifications occurring *in vivo* may substantially influence the antioxidant activity of dietary flavonoids. Silibinin, similar to other flavonoids, is recognized by the body as foreign matter and quickly metabolized via phase II enzymes.

The aforementioned data clearly indicates that SM protects cells against the damaging effects of oxidative/nitrosative stress and inflammation. It also prevents unwanted cell proliferation in cancer cells [66] as well as pathogen infection [151].

5. Oxidative Stress and Transcription Factors

Oxidation-reduction (redox) based regulation of gene expression appears to be a fundamental regulatory mechanism in cell biology. Indeed, it is well established that the level of ROS in the body is rigidly controlled by the antioxidant systems and an excess of free radical production and severe oxidative stress often leads to widespread oxidative damage and cell death. However, a basal level of oxidative stress is essential for cell survival. In fact, a moderate level of oxidative stress, induced by a variety of stressors, can create adaptive responses and improve adaptive ability to stressful challenges [152]. Therefore, a concept of the cellular antioxidant defence has recently been revised with special attention paid to cell signaling. Indeed, in animals redox-signaling pathways use ROS to transfer signals from different sources to the nucleus to regulate a number of various functions including growth, differentiation, proliferation and apoptosis. Various transcription factors are involved in a regulation of the antioxidant defence system [153–156]. These pathways operate in a coordinated manner and several are critically important for animals to cope with oxidative stress insults. They include Keap1/Nrf2, NF-κB, Mapk and AP-1 [157]. In particular great attention has been paid to a basic leucine zipper transcription factor, Nuclear factor-erythroid-2- (NF-E2-) related factor 2 (Nrf2).

5.1. Transcription Factor Nrf2

The recent information on Nrf2 functions can be summarized as follows. Based on the existing evidence, Nrf2 is considered to be the redox-sensitive master regulator of oxidative stress signaling. Growing evidence has demonstrated that the Nrf2 antioxidant response pathway plays an important role in the cellular antioxidant defense by activating a wide variety of genes involved in early defence reactions of higher organisms [158,159]. Indeed, Nrf2 is considered to have a significant role in adaptive responses to oxidative stress, being responsible for the induction of the expression of various antioxidants to combat oxidative and electrophilic stress [160]. In particular, critical components of the cellular antioxidant defense mechanisms include the ROS scavengers, phase II detoxification enzymes, and other detoxification proteins, which contain antioxidant response elements (AREs) in their promoter regions [161–163]. There is considerable experimental evidence suggesting that under normal physiological conditions, Nrf2 is kept in the cytoplasm by forming an inactive complex with the negative regulator, Kelch-like-ECH-associated protein 1 (Keap1), which is anchored to the actin cytoskeleton. In fact, cytosolic protein Keap1 sequesters Nrf2 in the cytoplasm and forwards it to a Cul3-based E3 ligase with the following rapid ubiquitin-proteasome degradation resulting in a short (about 20 min) half-life of Nrf2 under normal conditions (for review see [164]. Therefore, Keap1 acts as a redox sensor and upon exposure to oxidative or electrophilic stress, critical cysteine thiols of Keap1 are modified and Keap1 loses its ability to ubiquitinate Nrf2. Furthermore, phosphorylation of Nrf2 at specific serine and/or tyrosine residues also causes an Nfr2-Keap1 dissociation resulting in Nrf2 release and translocation to the nucleus. In the nucleus, Nrf2 combines with a small musculoaponeurotic fibrosarcoma protein called Maf to form a heterodimer [165]. Indeed, Nrf2 binds to the ARE in the upstream promoter region of genes encoding various antioxidant molecules, leading to the expression of antioxidant proteins, thiol molecules and other protective molecules. This includes enzymes of the first line of the antioxidant defence, namely SOD, GSH-Px and CAT, as well as detoxification enzymes (HO-1, NQO1, GST), GSH-related proteins (γ-GCS), NADPH-producing enzymes and others stress-response proteins contributing to counteracting oxidative and inflammatory damage [166,167]. The aforementioned proteins are vital to the maintenance and responsiveness of a cell antioxidant system. Therefore, an orchestrated change in gene expression via Nrf2 and the ARE is responsible for a synergistic protective effect against oxidative stress [168]. There is considerable experimental evidence suggesting that Nrf2 is controlled through a complex transcriptional/epigenetic and post-translational network that ensures its activity increases in response to redox disturbances, inflammation, growth factor stimulation and nutrient/energy fluxes, orchestrating adaptive responses to diverse forms of stress [169]. Nrf2 can be activated by various mechanisms, including stabilization of Nrf2 via Keap1 cysteine thiol modification and phosphorylation of Nrf2 by upstream kinases [170,171]. Beyond the activation of synthesis of antioxidant molecules, Nrf2 also contributes to adaptation by up-regulating the repair and degradation of damaged macromolecules, and by modulating intermediary metabolism conducting metabolic reprogramming during stress [167]. It is important to note that the expression of Nrf2 has been shown throughout human tissue, with high expression in detoxification organs, including the liver and kidney [160]. It has been suggested that low intensity oxidative stress is mainly sensed by Keap1/Nrf2 system with the following downstream up-regulation of the protective AO genes. Intermediate oxidative stress also increases the activity of antioxidant enzymes, but mainly via NF-κB and AP-1 pathways. At both low and intermediate intensity oxidative stresses, MAP-kinases and other kinases are also involved in signal sensing and cellular response, leading to enhanced antioxidant potential [157]. Emerging evidence also indicates that Nrf2 actively interacts with other transcription factors, including heat shock factor (Hsf1; [172]) providing additional options for AO system regulation.

It has been shown that naturally occurring triterpenoids affect Nrf2/ARE pathway and its downstream targets by triterpenoids were shown to be protective in different diseases [173–175]. Furthermore, Nrf2 activation and subsequent regulation of the transcription of the plethora of protective genes may be increased due to the chemical modifications of the triterpenoid structure [173].

5.1.1. SM and Nrf2 Regulation

An interesting approach to explain health-promoting properties of polyphenols could be to explore the modulation of signaling pathways, e.g., the Nrf2/Keap1/ARE and NF-κB pathways, resulting in increased expression of genes encoding for cytoprotective molecules, including AO enzymes and phase II detoxification enzymes. In addition, decreased expression of NF-κB-regulated genes would reduce production of pro-inflammatory cytokines. Indeed, it has been shown that some phytochemicals (sulforaphane, curcumin, EGCG, diallyl sulfide, resveratrol, *etc.*) at concentrations as low as 10 μM can activate Nrf2 signaling by inducing phosphorylation of Nrf2 via activation of upstream protein kinases and/or through direct interaction with Keap1 cysteine thiols [170]. It is interesting to note that anthocyanins activate the antioxidant transcription factor Nrf2, at a concentration as low as 1 μM [123]. Silymarin's ability to upregulate the master antioxidant coordinator Nrf2 has also been investigated in various *in vitro* and *in vivo* model systems. Initially, the SM-containing supplement Protandim (ashwagandha, bacopa extract, green tea extract, SM, and curcumin) was shown to induce HO-1 in a cell culture primarily through Nrf2 activation [176]. Protandim also induces Nrf2 nuclear localization and antioxidant enzyme expression [177]. However, it is difficult to distinguish the SM effect in the mixture. Furthermore, it was shown that SM (a water soluble formulation Legalon-SIL containing a succinate-modified active compounds) at 50 and 100 μM significantly upregulated Nrf2 protein levels in CON1 cells after 48h treatment [178]. SM (104 μM) protected human cells against the damage from the herbicide paraquat via triggering anti-oxidant related genes, including Nrf2, NQO1 and HO-1 [179]. In an *in vitro* experiment with human hepatic HepG2 cells, SM (100 μM) was shown to be protective against AAPH-induced oxidative stress and apoptosis and the antioxidant potential of SM was correlated with induction of antioxidant genes including HO-1, NQO-1, γ-GCLC and SOD via transcriptional activation of Nrf2 [180].

In a dietary rat model of non-alcoholic steatohepatitis there was an increase in nuclear translocation of Nrf2 protein in SM-fed animals [181] which could enhance the protective effect against oxidative stress. Numerically, there was also a trend in increasing Nrf2 endogenous expression protein in rat liver tissue damaged by CCl4 by pre-treatment with SM (100 mg/kg/BW [182]) as well as in an alcohol-pyrazole-fed rat model (50 mg/kg/BW [183]). It seems likely that the effect of SM on the Nrf2 activation in various oxidative-stress related animal models depends on the dose of SM used. For example, at 10 mg/kg/BW SM showed protective effects against dimerumic acid-induced changes in GSH and HO-1 levels, but did not change Nrf2 levels [166]. Indeed, SB-treatment (75 mg/kg/BW) ameliorated arsenic-induced nephrotoxicity by abrogation of oxidative stress, inflammation and apoptosis in rats [72]. It is interesting to note that beyond the preventive action of SB, in rats fed silibinin alone there was a significant increase in Nrf2 mRNA levels in the kidney. The same SB-treated rats were characterised by increased antioxidant defences as indicated by increased renal SOD and GSH-Px activities, as well as enhanced levels of GSH, vitamins E and C. In fact, SB-induced activation of Nrf2 and HO-1 played a significant role in cellular defence against As-induced oxidative cardiac damage. In a recent study, co-treatment with SM (86 mg/kg/BW) significantly enhanced the antioxidant defence systems of the ethanol-consuming mice via the activation of Nrf2 and HO-1 expression in damaged livers [184]. Furthermore, SM (200 mg/kg/BW) significantly up-regulated HO-1 and Nrf-2 in the liver of AAPH-treated mice [180].

Therefore, it is clear from the aforementioned data that SM/silibinin, similar to other polyphenols, acts as an indirect antioxidant through its ability to induce Nrf2 transactivation, which is responsible for the synthesis of an array of antioxidant molecules decreasing oxidative stress and providing effective protection in various stress conditions. This changes an attitude to SM, a compound that has been assumed to act typically as a free radical scavenger and traditional antioxidant. It is still not clear if physiologically-relevant (up to 2 μM) silybin concentrations would upregulate Nrf2, but *in vivo* results obtained with various model systems clearly indicated an activation of Nrf2 by silibinin. In general, most of the experimental studies suggest that dietary SM/silibinin activates antioxidant pathways such as Nrf2/HO1 and downregulates NFκB (see Section 5.2.1. below), MMPs [44,101,185], PPAR [186,187], HIF-1 [188–190] and STAT [191–193] pathways.

5.1.2. Oxidability and Pro-Oxidant Properties of SM

It has been shown that the oxidation of Keap1 cysteine thiols can be mediated by some polyphenols. It is an interesting fact that among flavonoids, the higher their intrinsic potential to generate oxidative stress and redox cycling, the stronger their potency as inducers of ARE-mediated gene expression [194]. It was hypothesized that low concentrations of polyphenols could generate H_2O_2 and activates Nrf2 signaling, inducing cell adaptation to oxidative stress [195]. Thus, polyphenols act as nutritional "medicines", which might have a preventative nature, rather than functioning as therapeutic agents. Therefore, the activation of Nrf2-ARE signaling by antioxidant polyphenols to induce various AO molecules is probably attributable to their prooxidant activity [170]. It seems likely that the same can be applied for silybin. In fact, pure silybin was found to be unstable whilst silybin in SM was stable in buffers from pH 1.0 to 7.8. The metabolism of silybin was more severe in its pure form compared to silybin in SM, as tested in a range of biological fluids including plasma, intestinal fluid and liver homogenates. It would appear that components in SM have a stabilization effect on its main component silybin [196]. Therefore, it could well be that in some *in vitro* experiments with cell cultures, silibinin could be oxidised and its effects would be a reflection of the oxidised form of the flavonoid or ROS generated during silybin oxidation. For example, silybin and 2,3-dehydrosilybin were shown to chelate transition metals, especially Cu^{2+} and pro-oxidant properties of such a complex *in vitro* have been shown [197]. Similarly, silybin can cause pro-oxidant effects via iron-catalyzed oxidation with the subsequent generation of reactive oxygen species [198]. Silibinin has been shown to be a strong pro-oxidative agent, *i.e.*, it was able to oxidize NADH *in vitro* in the presence of peroxidase and H_2O_2 [50,199]. This pro-oxidative action results from the production of free-radical derivatives and subsequent NADH oxidation [200,201]. Treatment with SM (50, 100, or 200 μM) for 24 h affected the cellular redox status and induced a dose-dependent increase in ROS generation in HepG2 cells. There was also a dose-dependent decrease in intracellular GSH level and decreased total AO potential in HepG2 cells [202]. Silibinin induced cell death in human breast cancer cell lines MCF7 and MDA-MB-231 and it was attenuated by antioxidants, suggesting that the effect of silibinin was dependent on ROS generation [203]. Similarly, treatment of HT29 cells with silibinin increased the intracellular ROS level [204]. It has been suggested that SM at 100 mg/kg/BW without stressors might exert pro-oxidant effects in animals. In fact, the levels of NO and MDA increased in CCl_4 and SM (100 mg/kg/BW)-treated groups, while SM at lower dose levels (25 and 50 mg/kg/BW) did not alter the NO and MDA content in the rat hippocampus. The concentration of total thiol molecules increased in the SM50 group and showed a remarkable decrease in CCl_4 and SM100 groups [117]. From the aforementioned results, we can suggest that silybin can affect Nrf2 activity via its oxidation-reduction potential.

5.2. Transcription Factor NF-κB

The nuclear factor-kappa B (NF-κB) is a widely expressed, inducible transcription factor that has been implicated in regulation of many cellular processes, including inflammation. NF-κB, consisting of a group of five related proteins that are capable of binding to DNA, is activated by a wide range of stimuli including oxidative stress. It regulates the transcription of a range of genes, including pro-inflammatory cytokines and leukocyte adhesion molecules, acute phase proteins and anti-microbial peptides [205–207]. Similar to Nrf2, in normal physiological conditions, NF-κB is found in cytoplasm in an inactive state associated with the inhibitory IκB proteins preventing its binding to target sites. Activation of NF-κB is considered to be an effective mechanism of host defense against infection and stress [208]. Indeed, in response to stimuli, including cytokines and other stressors, IκB proteins are rapidly phosphorylated by IκB kinase on specific serine residues, with following ubiquitination, and degradation by the 26S proteasome. The resulting release of NF-κB and subsequent translocation to the nucleus orchestrates the transcription of target genes, responsible for cell survival and involved with inflammation, immunity, apoptosis, cell proliferation and differentiation [209].

There are also other stimuli implicated in NF-κB activation including cell-surface receptors, inhibitory κB kinases, IκB proteins and factors that regulate the posttranslational modification of the Rel proteins, *etc.* [206–209]. Current information indicates that there are complex interplay/crosstalk mechanisms between Nrf2 and NF-κB pathways. On the one hand, NF-κB pathway is inhibited by several Nrf2 activators. On the other, NF-κB may directly antagonize the transcriptional activity of Nrf2 (for review see [206]). In recent years, several compounds, including various polyphenols, have been isolated from plants that have inhibitory activities against multiple components of NF-*κ*B activation pathway and it seems likely that SM could also have inhibitory activity on NF-κB.

5.2.1. SM and NF-κB Regulation

The anti-inflammatory activity of phenolic compounds has been demonstrated in a number of *in vitro* and *in vivo* studies and polyphenols may affect inflammation mainly as modulators of inflammatory redox signaling pathways [210]. The polyphenolic compounds express anti-inflammatory activity by modulating the expression of pro-inflammatory genes such as cyclooxygenase, lipoxygenase, nitric oxide synthases and several important cytokines, mainly acting through nuclear factor-κB and mitogen-activated protein kinase signaling [210,211]. Due to the large number of studies that have demonstrated regulatory effects of SM/silybin on the expression of NF-κB in various *in vitro* and *in vivo* models there is insufficient space in this review to analyze all of them, so we will focus only on recent investigations addressing the issue. It is well established that various plant-derived polyphenols can suppress TNF-α activated, NF-κB-associated inflammatory pathways both *in vitro* and *in vivo*.

These polyphenols include SM as well as curcumin, resveratrol, genistein, epigallocatechin gallate, flavopiridol, emodin, morin isoliquiritigenin, naringenin, ellagic acid, apigenin, kaempferol, catechins, myricetin, xanthohumol, fisetin, vitexin, escin, mangostin and others [210]. SM has long been known to inhibit the activation of inflammatory NF-κB signaling [212]. Recent studies have confirmed that the protective anti-inflammatory effects of SM/silybin could be mediated by their inhibitory potential on the NF-κB, which is a key transcriptional factor for numerous genes involved in regulation of inflammation, immune system, cell differentiation, survival, apoptosis, *etc.* Here, we give some examples of participation of SM/silibinin in NF-κB pathway inhibition.

5.2.1.1. *In Vitro* Studies

Silybin is a potent inhibitor of NF-κB activation. Indeed, Manna *et al.* [213] tested silybin in a number of *in vitro* human cell experimental systems and found it inhibited TNF-mediated NF-κB activation in a dose-dependent manner with maximum inhibition at 50 μM and it regulated NF-κB 100 times more effectively than aspirin. Furthermore, silibinin (80 μM) protected glial cells against peroxide-induced ROS formation, ATP depletion, and cell damage [214]. Interestingly, the inhibition of peroxide-induced ROS by silybin was partially attenuated by inhibitors of NF-κB or protein kinase C (PKC), suggesting an involvement of NF-κB and PKC signaling pathways. Similarly, a significant decrease in NF-κB was observed, when PBMCs from pre-eclamptic patients were cultured with silibinin at 5 μM and 50 μM concentrations and stimulated with LPS [215]. Moreover, silibinin (50–200 μM) inhibited the nuclear translocation of nuclear factor NF-κB through inhibition of the phosphorylation of IκBα and suppressed NF-κB transcriptional activity in stimulated HMC-1 cells [216]. SM is also effective in inhibiting T cell activation and proliferation, by acting on pathways of NF-κB activation/translocation. In particular, in CD4+ splenocytes from C57/Bl6 mice, SM (50 μM) significantly inhibited CD4+ cells proliferation, inhibited IL-2 and IFN-gamma production and blocked nuclear translocation of transcription factor NF-κB. Moreover, SM inhibited p65/NF-κB phosphorylation in CD4+ T cell [217]. The significant upregulation of oxidative stress biomarkers including MDA, TNF-like, IFN-γ and IL-1β genes was observed as well as NF-κB, COX-2 and iNOS proteins expression occurred upon heat stress in chicken hepatocytes. Furthermore, AO enzyme (SOD, CAT, GR) activities decreased. SM (259 μM) was able to normalize the expression of all of these biomarkers in heat-induced chicken hepatocytes [218]. However, there is a need for more studies with pure silybin at physiologically-relevant concentrations (0.2–2 μM) in order to confirm its modulating effect on the NF-κB pathway.

5.2.1.2. *In Vivo* Studies

It has been shown that silibinin (50 mg/kg/BW) had a protective effect against D-galactose-induced senescence due to promotion of cellular oxidoreductase activities and inhibition of NF-κB activation and ROS production [219]. Histopathological and immunohistochemical studies in the kidney of rats also shows that silibinin (75 mg/kg/BW) markedly reduced the toxicity of As and preserved the normal histological architecture of the renal tissue, inhibited the caspase-3 mediated tubular cell apoptosis and decreased the NADPH oxidase, iNOS and NF-κB overexpression by As [72]. In a model of experimental ischemic stroke in male rats, silibinin (100 mg/kg) significantly downregulated NF-κB in ischemic brain tissue after stroke [189]. In an experimental nonalcoholic steatohepatitis model, mice fed a methionine-choline deficient diet consistently had increased NF-κB p65 and p50 binding activity, while silibinin administration (20 mg/kg/BW/i.p.) significantly decreased the activity of both subunits [220]. It was shown that CCl4 treatment of mice enhanced the NF-κB expression in the liver and induced hepatic fibrosis, while SM (200 mg/kg/BW) reduced the CCl_4-induced hepatic NF-κB and improved CCl_4-induced liver fibrosis [88]. Protective effect of silibinin (150 mg/kg/BW/intragastric) in the rat model of cerebral ischemia was associated with decreased expression of the Bax, NF-κB protein and Bax, NF-κB mRNA in the brain [221]. SM is also shown to have NF-κB inhibiting effects. For example, mRNA and protein expressions of NF-κB were significantly up-regulated in D-galactosamine induced hepatotoxic rats and treatment with SM (25 mg/kg/BW) significantly down-regulated the expressions of these genes [222]. Furthermore, it has been shown that the serum levels of TGF-β1, NF-κB, and IL-6 were significantly elevated in the samples from cirrhosis rats while SM (50 g/kg) significantly prevented the aforementioned changes [223] and SM (250 mg/kg/BW) suppressed NF-κB signaling cascade in alcoholic liver fibrosis in guinea pigs [224,225]. Overall, SM/silibinin targets multiple signaling pathways, including NF-κB, towards inhibiting the secretion of pro-inflammatory cytokines. Together these findings suggest that the inhibitory effect of SM/silibinin on NF-κB signaling in a variety of cells, as well as in *in vivo* studies, could be an important mechanism of its anti-inflammation efficacy, and that targeting this signaling pathway by silibinin may have potential applications in the clinic and veterinary practice.

6. Effect of SM on Vitagene Expression

The term "vitagene" was introduced in 1998 by Rattan [226] who wrote "Our survival and the physical quality of life depends upon an efficient functioning of various maintenance and repair processes. This complex network of the so-called longevity assurance processes is composed of several genes, which may be called *vitagenes*". Later, the vitagene concept has been further developed by Calabrese and colleagues [53,227–237] and major pro-survival mechanisms controlled by homodynamic vitagene network are shown in Table 1. In accordance with Calabrese *et al.* [227–238] the term vitagenes refers to a group of genes that are strictly involved in preserving cellular homeostasis during stress conditions and the vitagene family includes heat shock proteins (Hsps), heme oxygenase-1 (Hsp32, HO-1), Hsp60 and Hsp70, the thioredoxins (Trx)/thioredoxins reductase (TR) system and sirtuins. The list of potential candidates to the vitagene family can be extended. In particular, it seems likely that SOD, a major inducible enzyme of the first level of antioxidant defence, can meet selecting criteria to be included in the vitagene family. The products of the aforementioned genes actively operate in detecting and controlling diverse forms of stress and cell injuries. The cooperative mechanisms of the vitagene network are reviewed in recently published comprehensive reviews [234,237,238] with a major conclusion indicating an essential regulatory role of the vitagene network in cell and whole organism adaptation to various stresses.

Emerging findings suggest a large number of potential mechanisms of action of SM (polyphenols) in preventing disease, which may be beyond their direct conventional antioxidant activities. It seems likely that SM, similar to other flavonoids, can affect the vitagene network. The recent relevant findings are reviewed below. First of all, SM/silybin affects HO-1 activity in different model systems. For example, As-intoxicated rats showed a significant up-regulation of myocardial NADPH (NOX) oxidase sub-units such as NOX2 and NOX4 as well as Keap1 and down-regulation of Nrf2 and vitagene HO-1 protein expressions. Pre-administration of silibinin (75 mg/kg/BW) recovered all these altered parameters to near normalcy in As-induced cardiotoxic rat [71]. Similarly, in a model of liver injury caused by alcohol plus pyrazole, SM administration (50 mg/kg/BW) had a protective effect with a trend in restoring the decreased activity of HO-1 and Nrf2 [183]. SM (250 mg/kg/BW) possesses substantial protective effect against B(a)P-induced damages by increasing (restoring) HO-1 (vitagene) activity [115]. Similarly, *in vitro* SM (500 μM) reduced tBH-induced hepatocyte toxicity by activating HO-1 gene expression [239]. Indeed, the enzyme HO-1 is an important regulatory molecule present in most mammalian cells. In fact, the main function of HO-1 is to break down the pro-oxidant molecule heme into three products; carbon monoxide (CO), biliverdin and free iron and actively participate in the antioxidant defence in the human/animal body [240]. Indeed, HO-1 is a stress-inducible protein and can be induced by various oxidative and inflammatory signals. Therefore, HO-1 expression is regarded as an important adaptive cellular mechanism of AO defence in stress conditions [241,242].

Table 1. Major components of the vitagene network (adapted from [226] and [228]).

Molecular Level	Cellular Level
AO defense	Cell proliferation
DNA-repair systems	Cell differentiation
Transfer of genetic information	Stability of cell membrane
Stress protein synthesis	Stability of intracellular milieu
Proteosomal function	Macromolecular turnover
Tissue and Organ Level	**Physiological and Redox Control Level**
Neutralization and removing toxic chemicals	Neuronal response and synaptic plasticity
Tissue regeneration and wound healing	Stress response
Tumor suppression	Hormonal response
Cell death and cell replacement	Immune response
	Thermoregulation
	HO-1/CO; BVR/BR; UCP; Hsp70; Hsp27; TRXr/TRX

From the data presented above it is clear that SM/silibinin can upregulate HO-1 and improve antioxidant defences. Secondly, SM/silibinin can affect other Hsps including Hsp70. Indeed, in an *in vitro* system based on CHO-K1 cells treated with As, SM (5 μM) significantly decreased Hsp70 expression previously elevated by As [243]. In another *in vitro* system based on heat-induced chicken hepatocytes, SM (259 μM) affected Hsp70 expression significantly, preventing its alleviation by heat stress [218]. A similar protective effect of SM (100 mg/kg/BW) on Hsp70 was seen in rats given SM for 7 days prior to mesenteric ischemia-reperfusion (I-R) compared to I-R group [244]. It is interesting to note that silybin was identified as a novel Hsp90 inhibitor [245]. Therefore, silibinin can decrease Hsp70 expression in stressed cells indicating improved AO defences and decrease stress by other means (e.g., Nrf2-related increased AO synthesis).

Thirdly, there are several recent publications showing a regulatory role of SM/silibinin on the sirtuin functions. In fact, a protective effect of silibinin against chemotherapeutic reagent mitomycin C-induced cell death in A375-S2 cells was studied. It was shown that mitomycin C caused cell apoptosis and over-expression of p53, elevated translocation of p53 into the nucleus, and decreased SIRT1 expression;

and these effects were alleviated by silibinin (100–150 μM) to a large extent [246]. Silibinin protected cardiac myocytes against isoproterenol-induced injury through resuming mitochondrial function and regulating the expression of SIRT1 and Bcl-2 family members [247]. Systemic administration of silibinin (50 mg/kg/BW) reversed streptozotocin-induced downregulation of SIRT1 expression in mice [248]. It is important to note that SIRT1 was activated by SM (500 μM) in UV-irradiated human malignant melanoma, A375-S2 cells [249]. However, silibinin could have completely the opposite action in regulating SIRT1 expression in normal cardiac myocytes and breast cancer MCF-7 cells, where it downragulates SIRT1 and promotes apoptosis [250]. Similarly silybin treatment downregulated SIRT1 in human lung adenocarcinoma cells [251].

Recent findings indicate that sirtuins-mediated signaling pathways are involved in metabolic reprogramming and upregulation of antioxidant defense systems [252]. Sirtuins are a class of proteins possessing deacetylase activity and playing critical roles in cell survival in response to oxidative stress. Seven human sirtuins, SIRT1–SIRT7, have been identified and their localization differ considerably: SIRT1 and SIRT2- nucleus and cytoplasm; SIRT3, SIRT4 and SIRT5- mitochondria; SIRT3- nucleus and cytoplasm and SIRT6 and SIRT7- nucleus [253]. Sirtuins control DNA repair and recombination, chromosomal stability and gene transcription, as well as regulate antioxidant defences. On the one hand, sirtuins decrease ROS levels, inhibit apoptosis and decrease inflammation. On the other, oxidative stress has been shown to decrease SIRT1 expression in the rat hippocampus and cortex, while SIRT1 overexpression prevents oxidative stress-induced apoptosis and increases resistance to oxidative stress (for review see [254]). Sirtuins are responsible for deacetylation of important transcription factors such as p53, FoxO, NF-κB, or PAPARGCα, which control the transcription of pro- and antioxidant enzymes [255]. Therefore, SM-related activation or prevention of inhibition of sirtuins in stress conditions could be considered as an important adaptive mechanism responsible for maintaining redox-regulated homeostasis in the cell and in the whole body. Finally, there is only a single publication devoted to the protective effect of SM on another vitagene called thioredoxins system. Silybin-phospholipid complex containing vitamin E (Realsil®, Indena, IBI-Lorenzini spa, Italy) was administered daily by gavage (15 mg vitamin E and 47 mg silybin complexed with phospholipids) to rats fed a choline-deprived or a high fat diet for 30 days and 60 days, respectively. Realsil was shown to be able to maintain Trx concentration in the rat liver fed on a high fat diet at day 60, while in the control group it dramatically decreased [59]. It should be mentioned that Trx, a small, ubiquitous thiol protein, is one of the most important regulators of redox balance and redox-sensitive cell functions. In fact, the thioredoxin system consisting of Trx, thioredoxin reductase (TrxR), and NADPH, is a key element determining cell redox status and regulating protein dithiol/disulfide balance, signal transduction and gene expression [256]. Indeed, the Trx system together with the glutathione-glutaredoxin (Grx) system (NADPH, glutathione reductase, GSH, and Grx) controls the cellular redox environment and determines a variety of cellular functions including DNA metabolism and repair, transcription, intracellular signaling, cell-cell communication, cell growth, apoptosis inhibition and ultimately cell survival in various stress conditions [256,257]. It is well accepted that the ability of SM to maintain Trx system in stress conditions is an important element of improving the adaptive ability of animals/humans via the vitagene network. Therefore, the vitagene network represents the major cellular pathway involved in the so called "programmed cell life" as an opposite to apoptosis, providing an effective protection against oxidative stress and toxic products of ROS metabolism [230–232]. In particular, nutritional antioxidants can be neuroprotective through the activation of hormetic pathways under the control of the Vitagene protein network [234]. The aforementioned results clearly indicate that, similar to other flavonoids SM, or its active ingredient silibinin, affect vitagene network decreasing detrimental consequences of stresses. Further research is needed to fully understand the molecular mechanisms involved in the vitagene network activation by SM/silybin.

7. Protective Effect of SM in the Gut

It seems likely that the gut is the major place of antioxidant action of polyphenols [258,259]. Indeed, reduction of oxidative damage, modulation of colonic flora and variation in gene expression are involved in the modulation of intestinal function by polyphenols. For example, to study the molecular effects of wine polyphenols at the gene level, the microarray technology was used: rats were treated with 50 mg/kg/BW wine polyphenols for 14 days. Global expression analysis of 5707 genes revealed an extensive down-regulation of genes involved in a wide range of physiological functions, such as metabolism, transport, signal transduction and intercellular signaling [260]. It was shown that two major regulatory pathways were down-regulated in the colon mucosa of polyphenols treated rats: inflammatory response and steroid metabolism. Since flavonoids are consumed in concentrations usually much higher than other antioxidant compounds, their protective effect during digestion is of great importance. For example, it has been shown that flavonoids not only prevented an accumulation of peroxidized lipids but can also switch prooxidant properties of heme-proteins to antioxidant ones [261]. Dietary polyphenols can also modulate *in vivo* oxidative damage in the gastrointestinal tract of rodents [262] supporting the hypothesis that dietary polyphenols might have both a protective and a therapeutic potential in oxidative damage-related pathologies. Indeed, the antioxidant-prooxidant balance (redox status) in various parts of the intestine would ultimately depend on the level of antioxidants and pro-oxidants provided by the diet and released by cells themselves, as well as on the level of absorption of both antioxidants and pro-oxidants [258,259]. In a model system mimicking stomach conditions it was shown that both lipid peroxidation and co-oxidation of vitamin E and beta-carotene were inhibited at pH 3.0 by red wine polyphenols [263]. In the gut, an interaction between different antioxidants could bring about their maximum synergistic effects. In particular, it was found that the mixture of the green tea polyphenol, vitamin E and vitamin C could act synergistically to protect lipid peroxidation [264]. It has been found that consumption of partially oxidized food could increase lipid peroxidation in the stomach and the absorption of cytotoxic lipid peroxidation products into the body. The addition of red wine polyphenols to the meal may alter these outcomes [265], therefore red wine polyphenols exert a beneficial effect by the novel function, absorption inhibition of the MDA [266]. These findings explain the potentially harmful effects of oxidized fats found in foods/feeds [259] and the important benefit of dietary polyphenols in the diet. Indeed, locating the biological site of action of polyphenols in the gastrointestinal tract may lead to a revision in our understanding of how flavonoids, including silybin, work *in vivo* and may help to elucidate the mechanisms of the health-promoting action of silybin and other polyphenols [267].

Redox signaling in gut inflammation is complex and poorly understood. However, it is generally accepted that homeostatic control of the intestinal epithelial redox environment is central for nutrient digestion and absorption, stem cell proliferation, apical enterocyte apoptosis, and immune response [268]. Indeed, polyphenols may play a role in intestinal mucosa integrity, inflammation and permeability [258,269]. For example, wine phenolics were able to prevent or delay the progression of intestinal diseases characterized by oxidative stress and inflammation, acting as both free radical scavengers and modulators of specific inflammation-related genes involved in cellular redox signaling [270]. They exert their effects by modulating cell signaling pathways, mainly activated in response to oxidative and inflammatory stimuli, and Nrf2 and NF-κB are the principal downstream effectors [271]. For example, pigs administered a grape seed and grape marc extract had a lower transactivation of NF-κB and Nrf2 and a lower expression of various target genes of these transcription factors in the duodenal mucosa (but not in the liver and plasma) than control pigs [272]. Furthermore, the ratio of villus height: crypt depth and the gain:feed ratio was higher in the animals fed a gape extract than in control pigs. In mice, the dietary grape seed extract decreased proliferation and enhanced differentiation of epithelial cell and the changes in gut epithelium were associated with the suppression of NF-κB signaling [273]. Polyphenolic compounds could also affect gut integrity. For example, grape seed extract (0.1%) supplied to rats with drinking water for 21 days significantly increased the expression of gut junction protein occludin in the proximal colon and reduced fecal levels of the neutrophil protein calprotectin, compared with control [274]. Recently, it has been reported that a naturally occurring flavonoid procyanidin B2 could modulate cellular redox status and the antioxidant enzyme defence system in colonic cells protecting against oxidative stress and xenobiotics [275,276].

It is possible to suggest that there is a biological reason for some antioxidants not to be absorbed completely and so provide antioxidant protection in lower parts of the intestine [258,259]. Comparatively low bioavailability and antioxidant potential of various flavonoids could be beneficial for humans/animals providing antioxidant protection in various parts of the digestive tract, including the large intestine where levels of other antioxidants would be quite low. The same is true for SM. For example, silibinin concentration in the gut could reach 800 μM [277], a concentration exceeding the requirement for its direct antioxidant activity. Indeed, it is still unclear whether SM has any direct antioxidant effects *in vivo*, although it might be capable of exerting such effects within the gastrointestinal tract, where SM may come into direct contact with cells without having undergone absorption and metabolism [258]. It has been suggested that gastroprotective effects of SM might be related to the protection of gastric mucosal NO and non-protein sulfhydryl groups and the modulation of capsaicin-sensitive gastric sensory afferents [278]. The beneficial effects of SM, on a murine model of colitis was explained [279]. SM effectively reduced colonic damage, lipid peroxidation, neutrophil infiltration and the content of inflammatory cytokines alongside increasing total antioxidant capacity of colonic tissue. In trinitrobenzene sulfonic acid-induced colitis model, silibinin treatment significantly reduced several components of inflammatory colitis such as NF-κB activity, levels of IL-1β, TNFα, TBARS, protein carbonyl, myeloperoxidase activity, and an improvement in antioxidant capability of the colon tissue [280]. Furthermore, silibinin (50 mg/kg/BW) modulates gut microbial enzymes, colonic oxidative stress and Wnt/β-catenin signaling resulting in antiproliferative effect against 1,2-di-methylhydrazine-induced colon carcinogenesis in experimental rats [64].

It is also well known that polyphenols are providing antioxidant protection in the lower intestine and can alter colonic flora [281]. Polyphenols, including silybin, are extensively metabolized by gut bacteria into a complex series of end-products that affect the functional ecology of symbiotic partners that can alter the host physiology [281]. It has been suggested that the consumption of a diet rich in plant foods with high dietary polyphenol content may enhance the gastrointestinal health of the host through microbiota modulation. Indeed, grape antioxidant dietary fiber intake stimulates proliferation of *Lactobacillus* and slightly affects the composition of *Bifidobacterium* species in the cecum of rats [282]. It appears that dietary polyphenols may have the ability to modify the gut microbial balance, but this effect is indirect, *i.e.*, it is mediated by biotransformation products, rather than the original plant compounds [283]. It should be mentioned that polyphenol–microbiota interactions are complex and subject to large inter-individual variability, leading to different polyphenol-metabolizing phenotypes or "metabotypes" [284].

Unfortunately, there is no data available on the effect of SM on gut microbiota, but we can suggest the same changes would occur as described above for other flavonoids. Therefore, it seems likely that reduction of oxidative damage, decreased inflammation, modulation of colonic flora and variation in gene expression are involved in the modulation of intestinal function by polyphenols, including SM.

8. Conclusions

Recent achievements in biochemistry and molecular biology, together with epidemiological data have changed our thinking about food. It has become increasingly clear that our diet plays a pivotal role in maintenance of our health and a misbalanced diet can cause serious health-related problems. It seems likely that antioxidants are among the major regulators of many physiological processes and, therefore, a redox balance between antioxidants and prooxidants in the diet, gastro-intestinal tract, plasma and tissues is an important determinant of the state of our health. Plants consumed by humans amd animals contain thousands of phenolic compounds. Among them, the effects of dietary polyphenols including SM are of great current interest. Indeed, various phytochemicals, including flavonoids, are an essential part of our diet, which are responsible for turning on and maintaining an optimal status of our antioxidant defences. Since flavonoids are not well absorbed in the gut, their active concentration in the plasma and target tissues are comparatively low, but probably sufficient for Nrf2 activation and NF-κB suppression as well as vitagene activation. Indeed, it seems very likely that activation of the Keap1/Nrf2/ARE pathway and inhibition of NF-κB pathway, rather than direct free radical scavenging activity, may be the main mechanisms of the health benefits of phytochemicals [233], including SM. Therefore, consumption of phytochemicals, including SM, could have a pre-conditioning effect on the antioxidant system of the body. This could explain the beneficial health-promoting effects of a diet rich in fruits and vegetables as important sources of the aforementioned chemicals (polyphenols and other phytochemicals) maintaining the body's ability to be highly adaptive to various stresses. SM and its main component silibinin are part of the dietary phytochemical mixture responsible for regulation of the antioxidant defences in the gut and in the whole body. It could well be that some dietary constituents which are not well absorbed could have health-promoting properties by maintaining redox balance in the large intestine, where concentration of other antioxidants (vitamin E, carotenoids, ascorbate) could be low, but prooxidants (iron, oxidized PUFAs, *etc.*) and substrates of oxidation are still present [258,259]. This protective effect in the large intestine could be responsible, for example, for bowl cancer prevention. Therefore, there could be a biological reason for some nutrients not being absorbed, but still being involved in antioxidant protection in the lower gut. Taking into account high concentrations of phytochemicals in the gut, it could well be that they play an essential part in maintaining an optimal antioxidant-prooxidant balance in the digestive tract responsible for additional health effects of phytochemicals including SM. In animal nutrition and disease prevention strategy SM alone, or in combination with other hepatho-active compomds (carnitine, betaine, vitamin B12, *etc.*), could have similar hepatoprotective effects as described in humans with similar mechanisms of protective action.

In conclusion, there are many possible mechanisms by which SM can improve the antioxidant defence mechanisms in the body. They include direct and indirect SM actions. First of all, a direct scavenging free radicals and chelating free Fe and Cu are mainly effective in the gut. Secondly, preventing free radical formation by inhibiting specific ROS-producing enzymes, or improving the integrity of electron-transport chain of mitochondria in stress conditions as a result of SM consumption, is of great importance. Thirdly, maintaining an optimal redox balance in the cell by activating a range of antioxidant enzymes and non-enzymatic antioxidants, mainly via Nrf2 activation, is probably the main driving force of AO action of SM. Fourthly, decreasing inflammatory responses in the gut and other tissues by inhibiting NF-kB pathways is an emerging mechanism of SM protective effect in liver toxicity and diseases. Fifthly, activating vitagenes, responsible for synthesis of protective molecules, including HSP, Trx, sirtuins, *etc.*, and providing additional protection in stress conditions deserves more attention in future research. Finally, affects on the microenvironment of the gut, including SM-bacteria interactions, await future investigation.

Conflicts of Interest

The author declares no conflict of interest.

References

1. Comelli, M.C.; Mengs, U.; Prosdocimi, M.; Schneider, C. Toward the definition of the mechanism of action of silymarin: Activities related to cellular protection from toxic damage induced by chemotherapy. *Integr. Cancer Ther.* **2007**, *6*, 120–129.
2. Biedermann, D.; Vavříková, E.; Cvak, L.; Křen, V. Chemistry of silybin. *Nat. Prod. Rep.* **2014**, *31*, 1138–1157.
3. Křen, V.; Marhol, P.; Purchartová, K.; Gabrielová, E.; Modrianský, M. Biotransformation of silybin and its congeners. *Curr. Drug MeTable* **2013**, *14*, 1009–1021.
4. Hackett, E.S.; Twedt, D.C.; Gustafson, D.L. Milk thistle and its derivative compounds: A review of opportunities for treatment of liver disease. *J. Vet. Intern. Med.* **2013**, *27*, 10–16.
5. Post-White, J.; Ladas, E.J.; Kelly, K.M. Advances in the use of milk thistle (*Silybum marianum*). *Integr. Cancer Ther.* **2007**, *6*, 104–109.
6. Loguercio, C.; Federico, A.; Trappoliere, M.; Tuccillo, C.; de Sio, I.; di Leva, A.; Niosi, M.; D'Auria, M.V.; Capasso, R.; del Vecchio Blanco, C.; *et al.* The effect of a silybin-vitamin E-phospholipid complex on nonalcoholic fatty liver disease: A pilot study. *Dig. Dis. Sci.* **2007**, *52*, 2387–2395.
7. Federico, A.; Niosi, M.; del Vecchio Blanco, C.; Loguercio, C. Emerging drugs for non-alcoholic fatty liver disease. *Exp. Opin. Emerg. Drugs* **2008**, *13*, 145–158.
8. Trappoliere, M.; Caligiuri, A.; Schmid, M.; Bertolani, C.; Failli, P.; Vizzutti, F.; Novo, E.; di Manzano, C.; Marra, F.; Loguercio, C.; *et al.* Silybin, a component of sylimarin, exerts

anti-inflammatory and anti-fibrogenic effects on human hepatic stellate cells. *J. Hepatol.* **2009**, *50*, 1102–1111.

9. Loguercio, C.; Festi, D. Silybin and the liver: From basic research to clinical practice. *World J. Gastroenterol.* **2011**, *17*, 2288–2301.
10. Loguercio, C.; Andreone, P.; Brisc, C.; Brisc, M.C.; Bugianesi, E.; Chiaramonte, M.; Cursaro, C.; Danila, M.; de Sio, I.; Floreani, A.; *et al.* Silybin combined with phosphatidylcholine and vitamin E in patients with nonalcoholic fatty liver disease: A randomized controlled trial. *Free Radic. Biol. Med.* **2012**, *52*, 1658–1665.
11. Stiuso, P.; Scognamiglio, I.; Murolo, M.; Ferranti, P.; de Simone, C.; Rizzo, M.R.; Tuccillo, C.; Caraglia, M.; Loguercio, C.; Federico, A. Serum oxidative stress markers and lipidomic profile to detect NASH patients responsive to an antioxidant treatment: A pilot study. *Oxid. Med. Cell Longev.* **2014**, *2014*, 169216.
12. Testino, G.; Leone, S.; Ansaldi, F.; Borro, P. Silymarin and *S*-adenosyl-L-methionine (SAMe): Two promising pharmacological agents in case of chronic alcoholic hepathopathy. A review and a point of view. *Minerva Gastroenterol. Dietol.* **2013**, *59*, 341–356.
13. Milić, N.; Milosević, N.; Suvajdzić, L.; Zarkov, M.; Abenavoli, L. New therapeutic potentials of milk thistle (*Silybum marianum*). *Nat. Prod. Commun.* **2013**, *8*,1801–1810.
14. Zholobenko, A.; Modriansky, M. Silymarin and its constituents in cardiac preconditioning. *Fitoterapia* **2014**, *97*, 122–132.
15. Madrigal-Santillán, E.; Madrigal-Bujaidar, E.; Álvarez-González, I.; Sumaya-Martínez, M.T.; Gutiérrez-Salinas, J.; Bautista, M.; Morales-González, Á.; García-Luna, Y.; González-Rubio, M.; Aguilar-Faisal, J.L.; *et al.* Review of natural products with hepatoprotective effects. *World J. Gastroenterol.* **2014**, *20*, 14787–14804.
16. Vargas-Mendoza, N.; Madrigal-Santillán, E.; Morales-González, A.; Esquivel-Soto, J.; Esquivel-Chirino, C.; García-Luna, Y.; González-Rubio, M.; Gayosso-de-Lucio, J.A.; Morales-González, J.A. Hepatoprotective effect of silymarin. *World J. Hepatol.* **2014**, *6*, 144–149.
17. Mohd Fozi, N.F.; Mazlan, M.; Shuid, A.N.; Isa Naina, M. Milk thistle: A future potential anti-osteoporotic and fracture healing agent. *Curr. Drug Targets* **2013**, *14*, 1659–1666.
18. Surai, P.F. Polyphenol compounds in the chicken/animal diet: From the past to the future. *J. Anim. Physiol. Anim. Nutr. (Berl.)* **2014**, *98*, 19–31.
19. Hollman, P.C. Unravelling of the health effects of polyphenols is a complex puzzle complicated by metabolism. *Arch. Biochem. Biophys.* **2014**, *559*, 100–105.
20. Li, L.; Zeng, J.; Gao, Y.; He, D. Targeting silibinin in the antiproliferative pathway. *Expert Opin. Investig. Drugs* **2010**, *19*, 243–255.
21. Wen, Z.; Dumas, T.E.; Schrieber, S.J.; Hawke, R.L.; Fried, M.W.; Smith, P.C. Pharmacokinetics and metabolic profile of free, conjugated, and total silymarin flavonolignans in human plasma after oral administration of milk thistle extract. *Drug Metab. Dispos.* **2008**, *36*, 65–72.
22. Zhu, H.J.; Brinda, B.J.; Chavin, K.D.; Bernstein, H.J.; Patrick, K.S.; Markowitz, J.S. An assessment of pharmacokinetics and antioxidant activity of free silymarin flavonolignans in healthy volunteers: A dose escalation study. *Drug Metab. Dispos.* **2013**, *41*, 1679–1685.

23. Wu, J.W.; Lin, L.C.; Hung, S.C.; Chi, C.W.; Tsai, T.H. Analysis of silibinin in rat plasma and bile for hepatobiliary excretion and oral bioavailability application. *J. Pharm. Biomed. Anal.* **2007**, *45*, 635–641.
24. Lorenz, D.; Lucker, P.W.; Mennicke, W.H.; Wetzelsberger, N. Pharmacokinetic studies with silymarin in human serum and bile. *Methods Find. Exp. Clin. Pharmacol.* **1984**, *6*, 655–661.
25. Barzaghi, N.; Crema, F.; Gatti, G.; Pifferi, G.; Perucca, E. Pharmokinetic studies in IdB 1016, a silybin-phosphatidylcholine complex, in healthy human subjects. *Eur. J. Drug. Metab. Pharmacokinet.* **1990**, *15*, 333–338.
26. Flory, P.J.; Krug, G.; Lorenz, D.; Mennicke, W.H. Studies on elimination of silymarin in cholecystectomized patients. I. Biliary and renal elimination after a single oral dose. *Plant. Med.* **1980**, *38*, 227–237.
27. Zhao, J.; Agarwal, R. Tissue distribution of silibinin, the major active constituent of silymarin, in mice and its association with enhancement of phase II enzymes: Implications in cancer chemoprevention. *Carcinogenesis* **1999**, *20*, 2101–2108.
28. Han, Y.H.; Lou, H.X.; Ren, D.M.; Sun, L.R.; Ma, B.; Ji, M. Stereoselective metabolism of silybin diastereoisomers in the glucuronidation process. *J. Pharm. Biomed. Anal.* **2004**, *34*, 1071–1078.
29. D'Andrea, V.; Perez, L.M.; Sanchez Pozzi, E.J. Inhibition of rat liver UDP glucuronosyltransferase by silymarin and the metabolite silibinin-glucuronide. *Life Sci.* **2005**, *77*, 683–692.
30. Weyhenmeyer, R.; Mascher, H.; Birkmayer, J. Study on dose-linearity of the pharmacokinetics of silibinin diastereomers using a new stereospecific assay. *Int. J. Clin. Pharmacol. Ther. Toxicol.* **1992**, *30*, 134–138.
31. Calani, L.; Brighenti, F.; Bruni, R.; del Rio, D. Absorption and metabolism of milk thistle flavanolignans in humans. *Phytomedicine* **2012**, *20*, 40–46.
32. Rickling, B.; Hans, B.; Kramarczyk, R.; Krumbiegel, G.; Weyhenmeyer, R. Two high-performance liquid chromatographic assays for the determination of free and total silibinin diastereomers in plasma using column switching with electrochemical detection and reversed-phase chromatography with ultraviolet detection. *J. Chromatogr. B Biomed. Appl.* **1995**, *670*, 267–277.
33. Surai, P.F.; Fisinin, V.I. Antioxidant Systems of the Body: From Vitamin E to Polyphenols and Beyond. In Proceedings of the 35th Western Nutrition Conference, Edmonton, Canada, 24–25 September 2014; pp. 265–277.
34. Surai, P.F. *Natural Antioxidants in Avian Nutrition and Reproduction*; Nottingham University Press: Nottingham, UK, 2002.
35. Surai, P.F. *Selenium in Nutrition and Health*; Nottingham University Press: Nottingham, UK, 2006.
36. Surai, P.F.; Fisinin, V.I. Ill Health effects of food lipids: Consequences of inadequate food processing, storage and cooking. In *Modern Dietary Fat Intakes in Disease Promotion*; de Meester, F., Zibadi, S., Watson, R.R., Eds.; Springer: New York, NY, USA, 2010; pp. 251–274.
37. Fotina, A.A.; Fisinin, V.I.; Surai, P.F. Recent developments in usage of natural antioxidants to improve chicken meat production and quality. *Bulg. J. Agric. Sci.* **2013**, *19*, 889–896.
38. Dehmlow, C.; Erhard, J.; de Groot, H. Inhibition of Kupffer cell functions as an explanation for the hepatoprotective properties of silibinin. *Hepatology* **1996**, *23*, 749–754.

39. Dehmlow, C.; Murawski, N.; de Groot, H. Scavenging of reactive oxygen species and inhibition of arachidonic acid metabolism by silibinin in human cells. *Life Sci.* **1996**, *58*, 1591–1600.
40. Varga, Z.; Seres, I.; Nagy, E.; Ujhelyi, L.; Balla, G.; Balla, J.; Antus, S. Structure prerequisite for antioxidant activity of silybin in different biochemical systems *in vitro*. *Phytomedicine* **2006**, *13*, 85–93.
41. Fu, H.; Lin, M.; Muroya, Y.; Hata, K.; Katsumura, Y.; Yokoya, A.; Shikazono, N.; Hatano, Y. Free radical scavenging reactions and antioxidant activities of silybin: Mechanistic aspects and pulse radiolytic studies. *Free Radic. Res.* **2009**, *43*, 887–897.
42. Yin, F.; Liu, J.; Ji, X.; Wang, Y.; Zidichouski, J.; Zhang, J. Silibinin: A novel inhibitor of Aβ aggregation. *Neurochem. Int.* **2011**, *58*, 399–403.
43. Cristofalo, R.; Bannwart-Castro, C.F.; Magalhães, C.G.; Borges, V.T.; Peraçoli, J.C.; Witkin, S.S.; Peraçoli, M.T. Silibinin attenuates oxidative metabolism and cytokine production by monocytes from preeclamptic women. *Free Radic. Res.* **2013**, *47*, 268–275.
44. Domitrović, R.; Jakovac, H.; Marchesi, V.V.; Blažeković, B. Resolution of liver fibrosis by isoquinoline alkaloid berberine in CCl_4-intoxicated mice is mediated by suppression of oxidative stress and upregulation of MMP-2 expression. *J. Med. Food* **2013**, *16*, 518–528.
45. Köksal, E.; Gülçin, I.; Beyza, S.; Sarikaya, O.; Bursal, E. *In vitro* antioxidant activity of silymarin. *J. Enzyme Inhib. Med. Chem.* **2009**, *24*, 395–405.
46. Asghar, Z.; Masood, Z. Evaluation of antioxidant properties of silymarin and its potential to inhibit peroxyl radicals *in vitro*. *Pak. J. Pharm. Sci.* **2008**, *21*, 249–254.
47. Dvorak, Z.; Kosina, P.; Walterova, D.; Simanek, V.; Bachleda, P.; Ulrichova, J. Primary cultures of human hepatocytes as a tool in cytotoxicity studies: Cell protection against model toxins by flavonolignans obtained from *Silybum marianum*. *Toxicol. Lett.* **2003**, *137*, 201–212.
48. Crocenzi, F.A.; Mottino, A.D.; Roma, M.G. Regulation of synthesis and trafficking of canalicular transporters and its alteration in acquired hepatocellular cholestasis. Experimental therapeutic strategies for its prevention. *Curr. Med. Chem.* **2004**, *11*, 501–524.
49. Blaising, J.; Lévy, P.L.; Gondeau, C.; Phelip, C.; Varbanov, M.; Teissier, E.; Ruggiero, F.; Polyak, S.J.; Oberlies, N.H.; Ivanovic, T.; *et al.* Silibinin inhibits hepatitis C virus entry into hepatocytes by hindering clathrin-dependent trafficking. *Cell Microbiol.* **2013**, *15*, 1866–1882.
50. Colturato, C.P.; Constantin, R.P.; Maeda, A.S., Jr.; Constantin, R.P.; Yamamoto, N.S.; Bracht, A.; Ishii-Iwamoto, E.L.; Constantin, J. Metabolic effects of silibinin in the rat liver. *Chem. Biol. Interact.* **2012**, *195*, 119–132.
51. Lin, M.T.; Beal, M.F. Mitochondrial dysfunction and oxidative stress in neurodegenerative diseases. *Nature* **2006**, *443*, 787–795.
52. Sekine, S.; Ichijo, H. Mitochondrial proteolysis: Its emerging roles in stress responses. *Biochim. Biophys. Acta* **2015**, *1850*, 274–280.
53. Calabrese, V.; Cornelius, C.; Stella, A.M.; Calabrese, E.J. Cellular stress responses, mitostress and carnitine insufficiencies as critical determinants in aging and neurodegenerative disorders: Role of hormesis and vitagenes. *Neurochem. Res.* **2010**, *35*, 1880–1915.
54. Rolo, A.P.; Oliveira, P.J.; Moreno, A.J.; Palmeira, C.M. Protection against post-ischemic mitochondrial injury in rat liver by silymarin or TUDC. *Hepatol. Res.* **2003**, *26*, 217–224.

55. Zhou, B.; Wu, L.J.; Tashiro, S.; Onodera, S.; Uchiumi, F.; Ikejima, T. Silibinin protects rat cardiac myocyte from isoproterenol-induced DNA damage independent on regulation of cell cycle. *Biol. Pharm. Bull.* **2006**, *29*, 1900–1905.
56. Detaille, D.; Sanchez, C.; Sanz, N.; Lopez-Novoa, J.M.; Leverve, X.; El-Mir, M.Y. Interrelation between the inhibition of glycolytic flux by silibinin and the lowering of mitochondrial ROS production in perifused rat hepatocytes. *Life Sci.* **2008**, *82*, 1070–1076.
57. Ligeret, H.; Brault, A.; Vallerand, D.; Haddad, Y.; Haddad, P.S. Antioxidant and mitochondrial protective effects of silibinin in cold preservation-warm reperfusion liver injury. *J. Ethnopharmacol.* **2008**, *115*, 507–514.
58. Gabrielová, E.; Jabůrek, M.; Gažák, R.; Vostálová, J.; Ježek, J.; Křen, V.; Modrianský, M. Dehydrosilybin attenuates the production of ROS in rat cardiomyocyte mitochondria with an uncoupler-like mechanism. *J. Bioenerg. Biomembr.* **2010**, *42*, 499–509.
59. Grattagliano, I.; Diogo, C.V.; Mastrodonato, M.; de Bari, O.; Persichella, M.; Wang, D.Q.; Liquori, A.; Ferri, D.; Carratù, M.R.; Oliveira, P.J.; *et al.* A silybin-phospholipids complex counteracts rat fatty liver degeneration and mitochondrial oxidative changes. *World J. Gastroenterol.* **2013**, *19*, 3007–3017.
60. Serviddio, G.; Bellanti, F.; Giudetti, A.M.; Gnoni, G.V.; Petrella, A.; Tamborra, R.; Romano, A.D.; Rollo, T.; Vendemiale, G.; Altomare, E. A silybin-phospholipid complex prevents mitochondrial dysfunction in a rodent model of nonalcoholic steatohepatitis. *J. Pharmacol. Exp. Ther.* **2010**, *332*, 922–932.
61. Zhu, S.Y.; Dong, Y.; Tu, J.; Zhou, Y.; Zhou, X.H.; Xu, B. Silybum marianum oil attenuates oxidative stress and ameliorates mitochondrial dysfunction in mice treated with D-galactose. *Pharmacogn. Mag.* **2014**, *10*, S92–S99.
62. Geed, M.; Garabadu, D.; Ahmad, A.; Krishnamurthy, S. Silibinin pretreatment attenuates biochemical and behavioral changes induced by intrastriatal MPP+ injection in rats. *Pharmacol. Biochem. Behav.* **2014**, *117*, 92–103.
63. Mazzio, E.A.; Harris, N.; Soliman, K.F. Food constituents attenuate monoamine oxidase activity and peroxide levels in C6 astrocyte cells. *Plant. Med.* **1998**, *64*, 603–606.
64. Sangeetha, N.; Viswanathan, P.; Balasubramanian, T.; Nalini, N. Colon cancer chemopreventive efficacy of silibinin through perturbation of xenobiotic metabolizing enzymes in experimental rats. *Eur. J. Pharmacol.* **2012**, *674*, 430–438.
65. Tuorkey, M.J.; El-Desouki, N.I.; Kamel, R.A. Cytoprotective Effect of Silymarin against Diabetes-Induced Cardiomyocyte Apoptosis in Diabetic Rats. *Biomed. Environ. Sci.* **2015**, *28*, 36–43.
66. Fan, L.; Ma, Y.; Liu, Y.; Zheng, D.; Huang, G. Silymarin induces cell cycle arrest and apoptosis in ovarian cancer cells. *Eur. J. Pharmacol.* **2014**,*743*, 79–88.
67. Doehner, W.; Landmesser, U. Xanthine oxidase and uric acid in cardiovascular disease: Clinical impact and therapeutic options. *Semin. Nephrol.* **2011**, *31*, 433–440.
68. Varga, Z.; Ujhelyi, L.; Kiss, A.; Balla, J.; Czompa, A.; Antus, S. Effect of silybin on phorbol myristate actetate-induced protein kinase C translocation, NADPH oxidase activity and apoptosis in human neutrophils. *Phytomedicine* **2004**, *11*, 206–212.

69. Pauff, J.M.; Hille, R. Inhibition studies of bovine xanthine oxidase by luteolin, silibinin, quercetin, and curcumin. *J. Nat. Prod.* **2009**, *72*, 725–731.
70. Bedard, K.; Krause, K.H. The NOX family of ROS-generating NADPH oxidases: Physiology and pathophysiology. *Physiol. Rev.* **2007**, *87*, 245–313.
71. Muthumani, M.; Prabu, S.M. Silibinin potentially attenuates arsenic-induced oxidative stress mediated cardiotoxicity and dyslipidemia in rats. *Cardiovasc. Toxicol.* **2014**, *14*, 83–97.
72. Prabu, S.M.; Muthumani, M. Silibinin ameliorates arsenic induced nephrotoxicity by abrogation of oxidative stress, inflammation and apoptosis in rats. *Mol. Biol. Rep.* **2012**, *39*, 11201–11216.
73. Khazim, K.; Gorin, Y.; Cavaglieri, R.C.; Abboud, H.E.; Fanti, P. The antioxidant silybin prevents high glucose-induced oxidative stress and podocyte injury *in vitro* and *in vivo*. *Am. J. Physiol. Renal. Physiol.* **2013**, *305*, F691–F700.
74. Ingawale, D.K.; Mandlik, S.K.; Naik, S.R. Models of hepatotoxicity and the underlying cellular, biochemical and immunological mechanism(s): A critical discussion. *Environ. Toxicol. Pharmacol.* **2014**, *37*, 118–133.
75. Rashid, K.; Sinha, K.; Sil, P.C. An update on oxidative stress-mediated organ pathophysiology. *Food Chem. Toxicol.* **2013**, *62*, 584–600.
76. Muthumani, M.; Prabu, S.M. Silibinin potentially protects arsenic-induced oxidative hepatic dysfunction in rats. *Toxicol. Mech. Methods* **2012**, *22*, 277–288.
77. Soria, E.A.; Eynard, A.R.; Bongiovanni, G.A. Modulation of early stress-related biomarkers in cytoplasm by the antioxidants silymarin and quercetin using a cellular model of acute arsenic poisoning. *Basic Clin. Pharmacol. Toxicol.* **2010**, *107*, 982–987.
78. Jain, A.; Yadav, A.; Bozhkov, A.I.; Padalko, V.I.; Flora, S.J. Therapeutic efficacy of silymarin and naringenin in reducing arsenic-induced hepatic damage in young rats. *Ecotoxicol. Environ. Saf.* **2011**, *74*, 607–614.
79. Manibusan, M.K.; Odin, M.; Eastmond, D.A. Postulated carbon tetrachloride mode of action: A review. *J. Environ. Sci. Health C Environ. Carcinog. Ecotoxicol. Rev.* **2007**, *25*, 185–209.
80. Raj, S.; Gothandam, K.M. Hepatoprotective effect of polyphenols rich methanolic extract of Amorphophallus commutatus var. wayanadensis against CCl4 induced hepatic injury in swiss albino mice. *Food Chem. Toxicol.* **2014**, *67*, 105–112.
81. Alkreathy, H.M.; Khan, R.A.; Khan, M.R.; Sahreen, S. CCl4 induced genotoxicity and DNA oxidative damages in rats: Hepatoprotective effect of Sonchus arvensis. *BMC Complement. Altern. Med.* **2014**, *14*, 452.
82. Krishnappa, P.; Venkatarangaiah, K.; Shivamogga, V.; Rajanna, S.K.; Prakash, K.; Gupta, R. Antioxidant and prophylactic effects of *Delonix elata* L., stem bark extracts, and flavonoid isolated quercetin against carbon tetrachloride-induced hepatotoxicity in rats. *Biomed. Res. Int.* **2014**, *2014*, doi:10.1155/2014/507851.
83. Shaker, E.; Mahmoud, H.; Mnaa, S. Silymarin, the antioxidant component and Silybum marianum extracts prevent liver damage. *Food Chem. Toxicol.* **2010**, *48*, 803–806.
84. Shine, V.J.; Latha, P.G.; Suja, S.N.; Anuja, G.I.; Raj, G.; Rajasekharan, S.N. Ameliorative effect of alkaloid extract of *Cyclea peltata* (Poir.) Hook. f. & Thoms. Roots (ACP) on APAP/CCl4 induced liver toxicity in Wistar rats and *in vitro* free radical scavenging property. *Asian Pac. J. Trop. Biomed.* **2014**, *4*, 143–151.

85. Al-Harbi, N.O.; Imam, F.; Nadeem, A.; Al-Harbi, M.M.; Iqbal, M.; Ahmad, S.F. Carbon tetrachloride-induced hepatotoxicity in rat is reversed by treatment with riboflavin. *Int. Immunopharmacol.* **2014**, *21*, 383–388.
86. Chang, H.C.; Chiu, Y.W.; Lin, Y.M.; Chen, R.J.; Lin, J.A.; Tsai, F.J.; Tsai, C.H.; Kuo, Y.C.; Liu, J.Y.; Huang, C.Y. Herbal supplement attenuation of cardiac fibrosis in rats with CCl_4-induced liver cirrhosis. *Chin. J. Physiol.* **2014**, *57*, 41–47.
87. Abbas, A.T.; El-Shitany, N.A.; Shaala, L.A.; Ali, S.S.; Azhar, E.I.; Abdel-Dayem, U.A.; Youssef, D.T. Red Sea Suberea mollis Sponge Extract Protects against CCl_4-Induced Acute Liver Injury in Rats via an Antioxidant Mechanism. *Evid. Based Complement. Alternat. Med.* **2014**, *2014*, 745606.
88. Li, C.C.; Hsiang, C.Y.; Wu, S.L.; Ho, T.Y. Identification of novel mechanisms of silymarin on the carbon, tetrachloride-induced liver fibrosis in mice by nuclear factor-κB bioluminescent imaging-guided transcriptomic analysis. *Food Chem. Toxicol.* **2012**, *50*, 1568–1575.
89. Jia, R.; Cao, L.; Du, J.; Xu, P.; Jeney, G.; Yin, G. The protective effect of silymarin on the carbon tetrachloride (CCl_4)-induced liver injury in common carp (*Cyprinus carpio*). *Vitro Cell Dev. Biol. Anim.* **2013**, *49*, 155–161.
90. Surai, P.F.; Dvorska, Y.E. Effect of mycotoxins on antioxidant status and immunity. In *The Mycotoxin Blue Book*; Diaz D., Ed.; Nottingham University Press: Nottingham, UK, 2005; pp. 93–137.
91. Sorrenti, V.; di Giacomo, C.; Acquaviva, R.; Barbagallo, I.; Bognanno, M.; Galvano, F. Toxicity of ochratoxin A and its modulation by antioxidants: A review. *Toxins (Basel)* **2013**, *5*, 1742–1766.
92. Wu, Q.H.; Wang, X.; Yang, W.; Nüssler, A.K.; Xiong, L.Y.; Kuča, K.; Dohnal, V.; Zhang, X.J.; Yuan, Z.H. Oxidative stress-mediated cytotoxicity and metabolism of T-2 toxin and deoxynivalenol in animals and humans: An update. *Arch. Toxicol.* **2014**, *88*, 1309–1326.
93. Essid, E.; Dernawi, Y.; Petzinger, E. Apoptosis induction by OTA and TNF-α in cultured primary rat hepatocytes and prevention by silibinin. *Toxins (Basel)* **2012**, *4*, 1139–1156.
94. Al-Anati, L.; Essid, E.; Reinehr, R.; Petzinger, E. Silibinin protects OTA-mediated TNF-alpha release from perfused rat livers and isolated rat Kupffer cells. *Mol. Nutr. Food Res.* **2009**, *53*, 460–466.
95. Shyamal, S.; Latha, P.G.; Suja, S.R.; Shine, V.J.; Anuja, G.I.; Sini, S.; Pradeep, S.; Shikha, P.; Rajasekharan, S. Hepatoprotective effect of three herbal extracts on aflatoxin B1-intoxicated rat liver. *Singapore Med. J.* **2010**, *51*, 326–331.
96. Sozmen, M.; Devrim, A.K.; Tunca, R.; Bayezit, M.; Dag, S.; Essiz, D. Protective effects of silymarin on fumonisin B_1-induced hepatotoxicity in mice. *J. Vet. Sci.* **2014**, *15*, 51–60.
97. Mustafa, H.N.; El Awdan, S.A.; Hegazy, G.A. Protective role of antioxidants on thioacetamide-induced acute hepatic encephalopathy: Biochemical and ultrastructural study. *Tissue Cell* **2013**, *45*, 350–362.
98. Fazal, Y.; Fatima, S.N.; Shahid, S.M.; Mahboob, T. Effects of curcumin on angiotensin-converting enzyme gene expression, oxidative stress and anti-oxidant status in thioacetamide-induced hepatotoxicity. *J. Renin Angiotensin Aldosterone Syst.* **2014**, doi:10.1177/1470320314545777.

99. Kadir, F.A.; Kassim, N.M.; Abdulla, M.A.; Kamalidehghan, B.; Ahmadipour, F.; Yehye, W.A. PASS-predicted hepatoprotective activity of *Caesalpinia sappan* in thioacetamide-induced liver fibrosis in rats. *Sci. World J.* **2014**, *2014*, 301879.
100. Singh, K.; Singh, N.; Chandy, A.; Manigauha, A. *In vivo* antioxidant and hepatoprotective activity of methanolic extracts of *Daucus carota* seeds in experimental animals. *Asian Pac. J. Trop. Biomed.* **2012**, *2*, 385–388.
101. Ali, S.O.; Darwish, H.A.; Ismail, NA. Modulatory effects of curcumin, silybin-phytosome and alpha-R-lipoic acid against thioacetamide-induced liver cirrhosis in rats. *Chem. Biol. Interact.* **2014**, *216*, 26–33.
102. Dos Santos, N.A.; Carvalho Rodrigues, M.A.; Martins, N.M.; dos Santos, A.C. Cisplatin-induced nephrotoxicity and targets of nephroprotection: An update. *Arch. Toxicol.* **2012**, *86*, 1233–1250.
103. Mansour, H.H.; Hafez, H.F.; Fahmy, N.M. Silymarin modulates Cisplatin-induced oxidative stress and hepatotoxicity in rats. *J. Biochem. Mol. Biol.* **2006**, *39*, 656–661.
104. El-Awady, el-S.E.; Moustafa, Y.M.; Abo-Elmatty, D.M.; Radwan, A. Cisplatin-induced cardiotoxicity: Mechanisms and cardioprotective strategies. *Eur. J. Pharmacol.* **2011**, *650*, 335–341.
105. Ninsontia, C.; Pongjit, K.; Chaotham, C.; Chanvorachote, P. Silymarin selectively protects human renal cells from cisplatin-induced cell death. *Pharm. Biol.* **2011**, *49*, 1082–1090.
106. Cho, S.I.; Lee, J.E.; Do, N.Y. Protective effect of silymarin against cisplatin-induced ototoxicity. *Int. J. Pediatr. Otorhinolaryngol.* **2014**, *78*, 474–478.
107. Karki, P.; Lee, E.; Aschner, M. Manganese neurotoxicity: A focus on glutamate transporters. *Ann. Occup. Environ. Med.* **2013**, *25*, 4.
108. Chtourou, Y.; Fetoui, H.; Sefi, M.; Trabelsi, K.; Barkallah, M.; Boudawara, T.; Kallel, H.; Zeghal, N. Silymarin, a natural antioxidant, protects cerebral cortex against manganese-induced neurotoxicity in adult rats. *Biometals* **2010**, *23*, 985–996.
109. Chtourou, Y.; Garoui, E.; Boudawara, T.; Zeghal, N. Therapeutic efficacy of silymarin from milk thistle in reducing manganese-induced hepatic damage and apoptosis in rats. *Hum. Exp. Toxicol.* **2013**, *32*, 70–81.
110. Chtourou, Y.; Fetoui, H.; Garoui, M.; Boudawara, T.; Zeghal, N. Improvement of cerebellum redox states and cholinergic functions contribute to the beneficial effects of silymarin against manganese-induced neurotoxicity. *Neurochem. Res.* **2012**, *37*, 469–479.
111. Chtourou, Y.; Garoui, M.; Boudawara, T.; Zeghal, N. Protective role of silymarin against manganese-induced nephrotoxicity and oxidative stress in rat. *Environ. Toxicol.* **2014**, *29*, 1147–1154.
112. Perumal Vijayaraman, K.; Muruganantham, S.; Subramanian, M.; Shunmugiah, K.P.; Kasi, P.D. Silymarin attenuates benzo(a)pyrene induced toxicity by mitigating ROS production, DNA damage and calcium mediated apoptosis in peripheral blood mononuclear cells (PBMC). *Ecotoxicol. Environ. Saf.* **2012**, *86*, 79–85.
113. Kiruthiga, P.V.; Shafreen, R.B.; Pandian, S.K.; Devi, K.P. Silymarin protection against major reactive oxygen species released by environmental toxins: Exogenous H_2O_2 exposure in erythrocytes. *Basic Clin. Pharmacol. Toxicol.* **2007**, *100*, 414–419.

114. Kiruthiga, P.V.; Pandian, S.K.; Devi, K.P. Silymarin protects PBMC against B(a)P induced toxicity by replenishing redox status and modulating glutathione metabolizing enzymes—An *in vitro* study. *Toxicol. Appl. Pharmacol.* **2010**, *247*, 116–128.
115. Kiruthiga, P.V.; Karthikeyan, K.; Archunan, G.; Karutha Pandian, S.; Pandima Devi, K. Silymarin prevents benzo(a)pyrene-induced toxicity in Wistar rats by modulating xenobiotic-metabolizing enzymes. *Toxicol. Ind. Health.* **2013**, doi:10.1177/0748233713475524.
116. Dirks-Naylor, A.J. The role of autophagy in doxorubicin-induced cardiotoxicity. *Life Sci.* **2013**, *93*, 913–916.
117. Malekinejad, H.; Janbaz-Acyabar, H.; Razi, M.; Varasteh, S. Preventive and protective effects of silymarin on doxorubicin-induced testicular damages correlate with changes in c-myc gene expression. *Phytomedicine* **2012**, *19*, 1077–1084.
118. El-Shitany, N.A.; El-Haggar, S.; El-Desoky, K. Silymarin prevents adriamycin-induced cardiotoxicity and nephrotoxicity in rats. *Food Chem. Toxicol.* **2008**, *46*, 2422–2428.
119. Patel, N.; Joseph, C.; Corcoran, G.B.; Ray, S.D. Silymarin modulates doxorubicin-induced oxidative stress, Bcl-xL and p53 expression while preventing apoptotic and necrotic cell death in the liver. *Toxicol. Appl. Pharmacol.* **2010**, *245*, 143–152.
120. Rašković, A.; Stilinović, N.; Kolarović, J.; Vasović, V.; Vukmirović, S.; Mikov, M. The protective effects of silymarin against doxorubicin-induced cardiotoxicity and hepatotoxicity in rats. *Molecules* **2011**, *16*, 8601–8613.
121. Galicia-Moreno, M.; Gutiérrez-Reyes, G. The role of oxidative stress in the development of alcoholic liver disease. *Rev. Gastroenterol. Mex.* **2014**, *79*, 135–144.
122. Song, Z.; Deaciuc, I.; Song, M.; Lee, D.Y.; Liu, Y.; Ji, X.; McClain, C. Silymarin protects against acute ethanol-induced hepatotoxicity in mice. *Alcohol. Clin. Exp. Res.* **2006**, *30*, 407–413.
123. Zhang, B.; Buya, M.; Qin, W.; Sun, C.; Cai, H.; Xie, Q.; Xu, B.; Wu, Y. Anthocyanins from Chinese bayberry extract activate transcription factor Nrf2 in β cells and negatively regulate oxidative stress-induced autophagy. *J. Agric. Food Chem.* **2013**, *61*, 8765–8772
124. Das, S.K.; Mukherjee, S. Biochemical and immunological basis of silymarin effect, a milk thistle (*Silybum marianum*) against ethanol-induced oxidative damage. *Toxicol. Mech. Methods* **2012**, *22*, 409–413
125. Gupta, Y.K.; Sharma, M.; Chaudhary, G. Pyrogallolinduced hepatotoxicity in rats: A model to evaluate antioxidant hepatoprotective agents. *Methods Find. Exp. Clin. Pharmacol.* **2002**, *24*, 497–500.
126. Upadhyay, G.; Kumar, A.; Singh, M.P. Effect of silymarin on pyrogallol- and rifampicin-induced hepatotoxicity in mouse. *Eur. J. Pharmacol.* **2007**, *565*, 190–201.
127. Upadhyay, G.; Tiwari, M.N.; Prakash, O.; Jyoti, A.; Shanker, R.; Singh, M.P. Involvement of multiple molecular events in pyrogallol-induced hepatotoxicity and silymarin-mediated protection: Evidence from gene expression profiles. *Food Chem. Toxicol.* **2010**, *48*, 1660–1670.
128. Razavi-Azarkhiavi, K.; Ali-Omrani, M.; Solgi, R.; Bagheri, P.; Haji-Noormohammadi, M.; Amani, N.; Sepand, M.R. Silymarin alleviates bleomycin-induced pulmonary toxicity and lipid peroxidation in mice. *Pharm. Biol.* **2014**, *52*, 1267–1271.

129. Taghiabadi, E.; Imenshahidi, M.; Abnous, K.; Mosafa, F.; Sankian, M.; Memar, B.; Karimi, G. Protective Effect of Silymarin against Acrolein-Induced Cardiotoxicity in Mice. *Evid. Based Complement. Alternat. Med.* **2012**, *2012*, 352091.
130. Pradeep, K.; Mohan, C.V.; Gobianand, K.; Karthikeyan, S. Silymarin modulates the oxidant-antioxidant imbalance during diethylnitrosamine induced oxidative stress in rats. *Eur. J. Pharmacol.* **2007**, *560*, 110–116.
131. Nabavi, S.M.; Nabavi, S.F.; Moghaddam, A.H.; Setzer, W.N.; Mirzaei, M. Effect of silymarin on sodium fluoride-induced toxicity and oxidative stress in rat cardiac tissues. *An. Acad. Bras. Cienc.* **2012**, *84*, 1121–1126.
132. Sherif, I.O.; Al-Gayyar, M.M. Antioxidant, anti-inflammatory and hepatoprotective effects of silymarin on hepatic dysfunction induced by sodium nitrite. *Eur. Cytokine Netw.* **2013**, *24*, 114–121.
133. Mateen, S.; Raina, K.; Agarwal, R. Chemopreventive and anti-cancer efficacy of silibinin against growth and progression of lung cancer. *Nutr. Cancer.* **2013**, *65*, 3–11.
134. Deep, G.; Agarwal, R. Targeting tumor microenvironment with silibinin: Promise and potential for a translational cancer chemopreventive strategy. *Curr. Cancer Drug Targets* **2013**, *13*, 486–499.
135. Ting, H.; Deep, G.; Agarwal, R. Molecular mechanisms of silibinin-mediated cancer chemoprevention with major emphasis on prostate cancer. *AAPS J.* **2013**, *15*, 707–716.
136. Kostek, H.; Szponar, J.; Tchórz, M.; Majewska, M.; Lewandowska-Stanek, H. Silibinin and its hepatoprotective action from the perspective of a toxicologist. *Przegl. Lek.* **2012**, *69*, 541–543.
137. Féher, J.; Lengyel, G. Silymarin in the prevention and treatment of liver diseases and primary liver cancer. *Curr. Pharm. Biotechnol.* **2012**, *13*, 210–217.
138. Vaid, M.; Katiyar, S.K. Molecular mechanisms of inhibition of photocarcinogenesis by silymarin, a phytochemical from milk thistle (*Silybum marianum* L. Gaertn.) (Review). *Int. J. Oncol.* **2010**, *36*, 1053–1060.
139. Kim, N.C.; Graf, T.N.; Sparacino, C.M.; Wani, M.C.; Wall, M.E. Complete isolation and characterization of silybins and isosilybins from milk thistle (*Silybum. marianum*). *Org. Biomol. Chem.* **2003**, *1*, 1684–1689.
140. Hoh, C.; Boocock, D.; Marczylo, T.; Singh, R.; Berry, D.P.; Dennison, A.R.; Hemingway, D.; Miller, A.; West, K.; Euden, S.; *et al.* Pilot study of oral silibinin, a putative chemopreventive agent, in colorectal cancer patients: Silibinin levels in plasma, colorectum, and liver and their pharmacodynamic consequences. *Clin. Cancer Res.* **2006**, *12*, 2944– 2950.
141. Halliwell, B. Are polyphenols antioxidants or pro-oxidants? What do we learn from cell culture and *in vivo* studies? *Arch. Biochem. Biophys.* **2008**, *476*, 107–112
142. Bell, J.R.; Donovan, J.L.; Wong, R.; Waterhouse, A.L.; German, J.B.; Walzem, R.L.; Kasim-Karakas, S.E. (+)-Catechin in human plasma after ingestion of a single serving of reconstituted red wine. *Am. J. Clin. Nutr.* **2000**, *71*, 103–108.
143. Lapidot, T.; Harel, S.; Granit, R.; Kanner, J. Bioavailability of red wine anthocyanins as detected in human urine. *J. Agric. Food Chem.* **1998**, *46*, 4297–4302
144. Hu, M.-L. Dietary polyphenols as antioxidants and anticancer agents: More questions than answers. *Chang. Gung Med. J.* **2011**, *34*, 449–460.

145. Howard, A.N.; Constamble, B.J. The metabolism of adrenocorticotrophic hormone and ascorbic acid in the chick. *Biochem. J.* **1958**, *69*, 501–505.
146. McKee, J.S.; Harrison, P.C.; Riskowski, G.L. Effects of supplemental ascorbic acid on the energy conversion of broiler chicks during heat stress and feed withdrawal. *Poultry Sci.* **1997**, *76*, 1278–1286.
147. Brenes, A.; Viveros, A.; Goni, I.; Centeno, C.; Sayago-Ayerdi, S.G.; Arija, I.; Saura-Calixto, F. Effect of grape pomace concentrate and vitamin E on digestibility of polyphenols and antioxidant activity in chickens. *Poultry Sci.* **2008**, *87*, 307–316.
148. Willemsen, H.; Swennen, Q.; Everaert, N.; Geraert, P.A.; Mercier, Y.; Stinckens, A.; Decuypere, E.; Buyse, J. Effects of dietary supplementation of methionine and its hydroxy analog DL-2-hydroxy-4-methylthio butanoic acid on growth performance, plasma hormone levels, and the redox status of broiler chickens exposed to high temperatures. *Poultry Sci.* **2011**, *90*, 2311–2320.
149. Benzie, I.F.; Szeto, Y.T.; Strain, J.J.; Tomlinson, B. Consumption of green tea causes rapid increase in plasma antioxidant power in humans. *Nutr. Cancer* **1999**, *34*, 83–87.
150. Halliwell, B.; Rafter, J.; Jenner, A. Health promotion by flavonoids, tocopherols, tocotrienols, and other phenols: Direct or indirect effects? Antioxidant or not? *Am. J. Clin. Nutr.* **2005**, *81*, 268S–276S.
151. Lim, R.; Morwood, C.J.; Barker, G.; Lappas, M. Effect of silibinin in reducing inflammatory pathways in *in vitro* and *in vivo* models of infection-induced preterm birth. *PLoS One* **2014**, *9*, e92505.
152. Yan, L.J. Positive oxidative stress in aging and aging-related disease tolerance. *Redox Biol.* **2014**, *2*, 165–169.
153. Ma, Q.; He, X. Molecular basis of electrophilic and oxidative defense: Promises and perils of Nrf2. *Pharmacol. Rev.* **2012**, *64*, 1055–1081.
154. Majzunova, M.; Dovinova, I.; Barancik, M.; Chan. J.Y. Redox signaling in pathophysiology of hypertension. *J. Biomed. Sci.* **2013**, *18*, 69.
155. Song, P.; Zou, M.H. Redox regulation of endothelial cell fate. *Cell Mol. Life Sci.* **2014**, *71*, 3219–3239.
156. Kweider, N.; Huppertz, B.; Kadyrov, M.; Rath, W.; Pufe, T.; Wruck, C.J. A possible protective role of Nrf2 in preeclampsia. *Ann. Anat.* **2014**, *196*, 268–277.
157. Lushchak, V.I. Adaptive response to oxidative stress: Bacteria, fungi, plants and animals. *Comp. Biochem. Physiol. C Toxicol. Pharmacol.* **2011**, *153*, 175–190.
158. Ma, Q. Role of nrf2 in oxidative stress and toxicity. *Annu. Rev. Pharmacol. Toxicol.* **2013**, *53*, 401–426.
159. Van der Wijst, M.G.; Brown, R.; Rots, M.G. Nrf2, the master redox switch: The Achilles' heel of ovarian cancer? *Biochim. Biophys. Acta* **2014**, *1846*, 494–509.
160. Tang, W.; Jiang, Y.F.; Ponnusamy, M.; Diallo, M. Role of Nrf2 in chronic liver disease. *World J. Gastroenterol.* **2014**, *20*, 13079–13087.
161. Howden, R. Nrf2 and cardiovascular defense. *Oxid. Med. Cell Longev.* **2013**, 104308.
162. Vriend, J.; Reiter R.J. The Keap1-Nrf2-antioxidant response element pathway: A review of its regulation by melatonin and the proteasome. *Mol. Cell. Endocrinol.* **2015**, *401*, 213–220.

163. Keum, Y.S.; Choi, B.Y. Molecular and chemical regulation of the Keap1-Nrf2 signaling pathway. *Molecules* **2014**, *19*, 10074–10089.
164. Choi, B.H.; Kang, K.S.; Kwak, M.K. Effect of redox modulating NRF2 activators on chronic kidney disease. *Molecules* **2014**, *19*, 12727–12759.
165. Bhakkiyalakshmi, E.; Sireesh, D.; Rajaguru, P.; Paulmurugan, R.; Ramkumar, K.M. The emerging role of redox-sensitive Nrf2-Keap1 pathway in diabetes. *Pharmacol. Res.* **2015**, *91*, 104–114.
166. Lee, B.H.; Hsu, W.H.; Hsu, Y.W.; Pan, T.M. Dimerumic acid attenuates receptor for advanced glycation endproducts signal to inhibit inflammation and diabetes mediated by Nrf2 activation and promotes methylglyoxal metabolism into D-lactic acid. *Free Radic. Biol. Med.* **2013**, *60*, 7–16.
167. Zhou, S.; Sun, W.; Zhang, Z.; Zheng, Y. The role of Nrf2-mediated pathway in cardiac remodeling and heart failure. *Oxid. Med. Cell Longev.* **2014**, *2014*, 260429.
168. Lee, J.M.; Calkins, M.J.; Chan, K.; Kan, Y.W.; Johnson, J.A. Identification of the NF-E2-related factor-2-dependent genes conferring protection against oxidative stress in primary cortical astrocytes using oligonucleotide microarray analysis. *J. Biol. Chem.* **2003**, *278*, 12029–12038.
169. Hayes, J.D.; Dinkova-Kostova, A.T. The Nrf2 regulatory network provides an interface between redox and intermediary metabolism. *Trends Biochem. Sci.* **2014**, *39*, 199–218.
170. Surh, Y.J. NF-κB and Nrf2 as potential chemopreventive targets of some anti-inflammatory and antioxidative phytonutrients with anti-inflammatory and antioxidative activities. *Asia Pac. J. Clin. Nutr.* **2008**, *17*, 269–272.
171. Surh, Y-J.; Kundu, J.K.; Na, H.-K. Nrf2 as a master redox switch in turning on the cellular signaling involved in the induction of cytoprotective genes by some chemopreventive phytochemicals. *Plant. Med.* **2008**, *74*, 1526–1539.
172. Dayalan Naidu, S.; Kostov, R.V.; Dinkova-Kostova, A.T. Transcription factors Hsf1 and Nrf2 engage in crosstalk for cytoprotection. *Trends Pharmacol. Sci.* **2015**, *36*, 6–14.
173. Loboda, A.; Rojczyk-Golebiewska, E.; Bednarczyk-Cwynar, B.; Lucjusz, Z.; Jozkowicz, A.; Dulak, J. Targeting Nrf2-mediated gene transcription by triterpenoids and their derivatives. *Biomol. Ther.* **2012**, *20*, 499–505.
174. Owusu-Ansah, A.; Choi, S.H.; Petrosiute, A.; Letterio, J.J.; Huang, A.Y. Triterpenoid inducers of Nrf2 signaling as potential therapeutic agents in sickle cell disease: A review. *Front. Med.* **2015**, *9*, 46–56.
175. Copple, I.M.; Shelton, L.M.; Walsh, J.; Kratschmar, D.V.; Lister, A.; Odermatt, A.; Goldring, C.E.; Dinkova-Kostova, A.T.; Honda, T.; Park, B.K. Chemical tuning enhances both potency toward Nrf2 and *in vitro* therapeutic index of triterpenoids. *Toxicol. Sci.* **2014**, *140*, 462–469.
176. Velmurugan, K.; Alam, J.; McCord, J.M.; Pugazhenthi, S. Synergistic induction of heme oxygenase-1 by the components of the antioxidant supplement Protandim. *Free Radic. Biol. Med.* **2009**, *46*, 430–440.
177. Donovan, E.L.; McCord, J.M.; Reuland, D.J.; Miller, B.F.; Hamilton, K.L. Phytochemical activation of Nrf2 protects human coronary artery endothelial cells against an oxidative challenge. *Oxid. Med. Cell Longev.* **2012**, *2012*, 132931.
178. Mehrab-Mohseni, M.; Sendi, H.; Steuerwald, N.; Ghosh, S.; Schrum, L.W.; Bonkovsky, H.L. Legalon-SIL downregulates HCV core and NS5A in human hepatocytes expressing full-length HCV. *World J. Gastroenterol.* **2011**, *17*, 1694–1700.

179. Podder, B.; Kim, Y.S.; Zerin, T.; Song, H.Y. Antioxidant effect of silymarin on paraquat-induced human lung adenocarcinoma A549 cell line. *Food Chem. Toxicol.* **2012**, *50*, 3206–3214.
180. Gokila Vani, M.; Kumar, K.J.; Liao, J.W.; Chien, S.C.; Mau, J.L.; Chiang, S.S.; Lin, C.C.; Kuo, Y.H.; Wang, S.Y. Antcin C from Antrodia cinnamomea Protects Liver Cells Against Free Radical-Induced Oxidative Stress and Apoptosis *In Vitro* and *In Vivo* through Nrf2-Dependent Mechanism. *Evid. Based Complement. Alternat. Med.* **2013**, *2013*, 296082.
181. Kim, M.; Yang, S.G.; Kim, J.M.; Lee, J.W.; Kim, Y.S.; Lee, J.I. Silymarin suppresses hepatic stellate cell activation in a dietary rat model of non-alcoholic steatohepatitis: Analysis of isolated hepatic stellate cells. *Int. J. Mol. Med.* **2012**, *30*, 473–479.
182. Wu, S.; Yue, Y.; Tian, H.; Li, Z.; Li, X.; He, W.; Ding, H. Carthamus red from *Carthamus tinctorius* L. exerts antioxidant and hepatoprotective effect against CCl_4-induced liver damage in rats via the Nrf2 pathway. *J. Ethnopharmacol.* **2013**, *148*, 570–578.
183. Choi, M.K.; Han, J.M.; Kim, H.G.; Lee, J.S.; Lee, J.S.; Wang, J.H.; Son, S.W.; Park, H.J.; Son, C.G. Aqueous extract of Artemisia capillaris exerts hepatoprotective action in alcohol-pyrazole-fed rat model. *J. Ethnopharmacol.* **2013**, *147*, 662–670.
184. Cao, Y.W.; Jiang, Y.; Zhang, D.Y.; Wang, M.; Chen, W.S.; Su, H.; Wang, Y.T.; Wan, J.B. Protective effects of *Penthorum* chinense Pursh against chronic ethanol-induced liver injury in mice. *J. Ethnopharmacol.* **2015**, *161*, 92–98.
185. Yousefi, M.; Ghaffari, S.H.; Soltani, B.M.; Nafissi, S.; Momeny, M.; Zekri, A.; Behmanesh, M.; Alimoghaddam, K.; Ghavamzadeh, A. Therapeutic efficacy of silibinin on human neuroblastoma cells: Akt and NF-κB expressions may play an important role in silibinin-induced response. *Neurochem. Res.* **2012**, *37*, 2053–2063.
186. Pferschy-Wenzig, E.M.; Atanasov, A.G.; Malainer, C.; Noha, S.M.; Kunert, O.; Schuster, D.; Heiss, E.H.; Oberlies, N.H.; Wagner, H.; Bauer, R.; *et al.* Identification of isosilybin a from milk thistle seeds as an agonist of peroxisome proliferator-activated receptor gamma. *J. Nat. Prod.* **2014**, *77*, 842–847.
187. Prakash, P.; Singh, V.; Jainm M.; Rana, M.; Khanna, V.; Barthwal, M.K.; Dikshit, M. Silymarin ameliorates fructose induced insulin resistance syndrome by reducing de novo hepatic lipogenesis in the rat. *Eur. J. Pharmacol.* **2014**, *727*, 15–28.
188. Lin, C.H.; Li, C.H.; Liao, P.L.; Tse, L.S.; Huang, W.K.; Cheng, H.W.; Cheng, Y.W. Silibinin inhibits VEGF secretion and age-related macular degeneration in a hypoxia-dependent manner through the PI-3 kinase/Akt/mTOR pathway. *Br. J. Pharmacol.* **2013**, *168*, 920–931.
189. Wang, C.; Wang, Z.; Zhang, X.; Zhang, X.; Dong, L.; Xing, Y.; Lim Y.; Liu, Z.; Chen, L.; Qiao, H.; *et al.* Protection by silibinin against experimental ischemic stroke: Up-regulated pAkt, pmTOR, HIF-1α and Bcl-2, down-regulated Bax, NF-κB expression. *Neurosci. Lett.* **2012**, *529*, 45–50.
190. Jung, H.J.; Park, J.W.; Lee, J.S.; Lee, S.R.; Jang, B.C.; Suh, S.I.; Suh, M.H.; Baek, W.K. Silibinin inhibits expression of HIF-1alpha through suppression of protein translation in prostate cancer cells. *Biochem. Biophys. Res. Commun.* **2009**, *390*, 71–76.
191. Tyagi, A.; Agarwal, C.; Dwyer-Nield, L.D.; Singh, R.P.; Malkinson, A.M.; Agarwal, R. Silibinin modulates TNF-α and IFN-γ mediated signaling to regulate COX2 and iNOS expression in tumorigenic mouse lung epithelial LM2 cells. *Mol. Carcinog.* **2012**, *51*, 832–842.

192. Hou, Y.C.; Liou, K.T.; Chern, C.M.; Wang, Y.H.; Liao, J.F.; Chang, S.; Chou, Y.H.; Shen, Y.C. Preventive effect of silymarin in cerebral ischemia-reperfusion-induced brain injury in rats possibly through impairing NF-κB and STAT-1 activation. *Phytomedicine* **2010**, *17*, 963–973.
193. Agarwal, C.; Tyagi, A.; Kaur, M.; Agarwal, R. Silibinin inhibits constitutive activation of Stat3, and causes caspase activation and apoptotic death of human prostate carcinoma DU145 cells. *Carcinogenesis* **2007**, *28*, 1463–1470.
194. Lee-Hilz, Y.Y.; Boerboom, A.M.; Westphal, A.H.; Berkel, W.J.; Aarts, J.M.; Rietjens, I.M. Pro-oxidant activity of flavonoids induces EpRE-mediated gene expression. *Chem. Res. Toxicol.* **2006**, *19*, 1499–1505.
195. Erlank, H.; Elmann, A.; Kohen, R.; Kanner, J. Polyphenols activate Nrf2 in astrocytes via H_2O_2, semiquinones, and quinones. *Free Radic. Biol. Med.* **2011**, *51*, 2319–2327.
196. Tang, N.; Wu, D.; Lu, Y.; Chen, J.; Zhang, B.; Wu, W. A comparative study on the stability of silybin and that in silymarin in buffers and biological fluids. *Drug Metab. Lett.* **2009**, *3*, 115–119.
197. Zatloukalová, M.; Křen, V.; Gažák, R.; Kubala, M.; Trouillas, P.; Ulrichová, J.; Vacek, J. Electrochemical investigation of flavonolignans and study of their interactions with DNA in the presence of Cu(II). *Bioelectrochemistry* **2011**, *82*, 117–124.
198. Gharagozloo, M.; Khoshdel, Z.; Amirghofran, Z. The effect of an iron (III) chelator, silybin, on the proliferation and cell cycle of Jurkat cells: A comparison with desferrioxamine. *Eur. J. Pharmacol.* **2008**, *589*, 1–7.
199. Fan, S.; Qi, M.; Yu, Y.; Li, L.; Yao, G.; Tashiro, S.; Onodera, S.; Ikejima, T. P53 activation plays a crucial role in silibinin induced ROS generation via PUMA and JNK. *Free Radic. Res.* **2012**, *46*, 310–319.
200. Awad, H.M.; Boersma, M.G.; Boeren, S.; van Bladeren, P.J.; Vervoort, J.; Rietjens, I.M. Structure-activity study on the quinone/quinone methide chemistry of flavonoids. *Chem. Res. Toxicol.* **2001**, *14*, 398–408.
201. Awad, H.M.; Boersma, M.G.; Boeren, S.; van Bladeren, P.J.; Vervoort, J.; Rietjens, I.M. The regioselectivity of glutathione adduct formation with flavonoid quinone/quinone methides is pH-dependent. *Chem. Res. Toxicol.* **2002**, *15*, 343–351.
202. Zhang, S.; Yang, Y.; Liang, Z.; Duan, W.; Yang, J.; Yan, J.; Wang, N.; Feng, W.; Ding, M.; Nie, Y.; *et al.* Silybin-mediated inhibition of notch signaling exerts antitumor activity in human hepatocellular carcinoma cells. *PLoS One* **2013**, *8*, e83699.
203. Kim, T.H.; Woo, J.S.; Kim, Y.K.; Kim, K.H. Silibinin induces cell death through reactive oxygen species-dependent downregulation of notch-1/ERK/Akt signaling in human breast cancer cells. *J. Pharmacol. Exp. Ther.* **2014**, *349*, 268–278.
204. Woo, S.M.; Min, K.J.; Kim, S.; Park, J.W.; Kim, D.E.; Chun, K.S.; Kim, Y.H.; Lee, T.J.; Kim, S.H.; Choi, Y.H.; *et al.* Silibinin induces apoptosis of HT29 colon carcinoma cells through early growth response-1 (EGR-1)-mediated non-steroidal anti-inflammatory drug-activated gene-1 (NAG-1) up-regulation. *Chem. Biol. Interact.* **2014**, *211*, 36–43.
205. Tkach, K.E.; Oyler, J.E.; Altan-Bonnet, G. Cracking the NF-κB code. *Sci Signal.* **2014**, *7*, pe5.
206. Buelna-Chontal, M.; Zazueta, C. Redox activation of Nrf2 & NF-κB: A double end sword? *Cell Signal.* **2013**, *25*, 2548–2557.

207. Pedruzzi, L.M.; Stockler-Pinto, M.B.; Leite, M., Jr.; Mafra, D. Nrf2-keap1 system *versus* NF-κB: The good and the evil in chronic kidney disease? *Biochimie* **2012**, *94*, 2461–2466.

208. Pal, S.; Bhattacharjee, A.; Ali, A.; Mandal, N.C.; Mandal, S.C.; Pal, M. Chronic inflammation and cancer: Potential chemoprevention through nuclear factor kappa B and p53 mutual antagonism. *J. Inflamm. (Lond.)*. **2014**, *11*, 23.

209. Hayden, M.S.; Ghosh, S. Regulation of NF-κB by TNF family cytokines. *Semin. Immunol.* **2014**, *26*, 253–266.

210. Gupta, S.C.; Tyagi, A.K.; Deshmukh-Taskar, P.; Hinojosa, M.; Prasad, S.; Aggarwal, B.B. Downregulation of tumor necrosis factor and other proinflammatory biomarkers by polyphenols. *Arch. Biochem. Biophys.* **2014**, *559*, 91–99.

211. Stevenson, D.E.; Hurst, R.D. Polyphenolic phytochemicals—Just antioxidants or much more? *Cell Mol. Life. Sci.* **2007**, *64*, 2900–2916.

212. Ramasamy, K.; Agarwal, R. Multitargeted therapy of cancer by silymarin. *Cancer Lett.* **2008**, *269*, 352–362.

213. Manna, S.K.; Mukhopadhyay, A.; Van, N.T.; Aggarwal, B.B. Silymarin suppresses TNF-induced activation of NF-κB, c-Jun *N*-terminal kinase, and apoptosis. *J. Immunol.* **1999**, *163*, 6800–6809.

214. Tsai, M.J.; Liao, J.F.; Lin, D.Y.; Huang, M.C.; Liou, D.Y.; Yang, H.C.; Lee, H.J.; Chen, Y.T.; Chi, C.W.; Huang, W.C.; *et al.* Silymarin protects spinal cord and cortical cells against oxidative stress and lipopolysaccharide stimulation. *Neurochem. Int.* **2010**, *57*, 867–875.

215. Giorgi, V.S.; Peracoli, M.T.; Peracoli, J.C.; Witkin, S.S.; Bannwart-Castro, C.F. Silibinin modulates the NF-κB pathway and pro-inflammatory cytokine production by mononuclear cells from preeclamptic women. *J. Reprod. Immunol.* **2012**, *95*, 67–72.

216. Kim, B.R.; Seo, H.S.; Ku, J.M.; Kim, G.J.; Jeon, C.Y.; Park, J.H.; Jang, B.H.; Park, S.J.; Shin, Y.C.; Ko, S.G. Silibinin inhibits the production of pro-inflammatory cytokines through inhibition of NF-κB signaling pathway in HMC-1 human mast cells. *Inflamm. Res.* **2013**, *62*, 941–950.

217. Gharagozloo, M.; Velardi, E.; Bruscoli, S.; Agostini, M.; di Sante, M.; Donato, V.; Amirghofran, Z.; Riccardi, C. Silymarin suppress CD4+ T cell activation and proliferation: Effects on NF-κB activity and IL-2 production. *Pharmacol. Res.* **2010**, *61*, 405–409.

218. Oskoueian, E.; Abdullah, N.; Idrus, Z.; Ebrahimi, M.; Goh, Y.M.; Shakeri, M.; Oskoueian, A. Palm kernel cake extract exerts hepatoprotective activity in heat-induced oxidative stress in chicken hepatocytes. BMC Complement *Altern. Med.* **2014**, *14*, 368.

219. Wang, Q.; Zou, L.; Liu, W.; Hao, W.; Tashiro, S.; Onodera, S.; Ikejima, T. Inhibiting NF-κB activation and ROS production are involved in the mechanism of silibinin's protection against D-galactose-induced senescence. *Pharmacol. Biochem. Behav.* **2011**, *98*, 140–149.

220. Salamone, F.; Galvano, F.; Marino Gammazza, A.; Paternostro, C.; Tibullo, D.; Bucchieri, F.; Mangiameli, A.; Parola, M.; Bugianesi, E.; Li Volti, G. Silibinin improves hepatic and myocardial injury in mice with nonalcoholic steatohepatitis. *Dig. Liver Dis.* **2012**, *44*, 334–342.

221. Liu, B.N.; Hanm B.X.; Liu, F. Neuroprotective effect of pAkt and HIF-1 α on ischemia rats. *Asian Pac. J. Trop. Med.* **2014**, *7*, 221–225.

222. Aristatile, B.; Al-Assaf, A.H.; Pugalendi, K.V. Carvacrol suppresses the expression of inflammatory marker genes in D-galactosamine-hepatotoxic rats. *Asian Pac. J. Trop. Med.* **2013**, *6*, 205–211.
223. Salama, S.M.; Abdulla, M.A.; Alrashdi, A.S.; Hadi, A.H. Mechanism of Hepatoprotective Effect of *Boesenbergia rotunda* in Thioacetamide-Induced Liver Damage in Rats. *Evid. Based Complement. Alternat. Med.* **2013**, *2013*, 157456.
224. Abhilash, P.A.; Harikrishnan, R.; Indira, M. Ascorbic acid is superior to silymarin in the recovery of ethanol-induced inflammatory reactions in hepatocytes of guinea pigs. *J. Physiol. Biochem.* **2013**, *69*, 785–798.
225. Abhilash, P.A.; Harikrishnan, R.; Indira, M. Ascorbic acid suppresses endotoxemia and NF-κB signaling cascade in alcoholic liver fibrosis in guinea pigs: A mechanistic approach. *Toxicol. Appl. Pharmacol.* **2014**, *274*, 215–224.
226. Rattan, S.I. The nature of gerontogenes and vitagenes. Antiaging effects of repeated heat shock on human fibroblasts. *Ann. N. Y. Acad. Sci.* **1998**, *854*, 54–60.
227. Calabrese, V.; Boyd-Kimball, D.; Scapagnini, G.; Butterfield, D.A. Nitric oxide and cellular stress response in brain aging and neurodegenerative disorders: The role of vitagenes. *Vivo* **2004**, *18*, 245–267.
228. Calabrese, V.; Guagliano, E.; Sapienza, M.; Panebianco, M.; Calafato, S.; Puleo, E.; Pennisi, G.; Mancuso, C.; Butterfield, D.A.; Stella, A.G. Redox regulation of cellular stress response in aging and neurodegenerative disorders: Role of vitagenes. *Neurochem. Res.* **2007**, *32*, 757–773.
229. Calabrese, V.; Calafato, S.; Puleo, E.; Cornelius, C.; Sapienza, M.; Morganti, P.; Mancuso, C. Redox regulation of cellular stress response by ferulic acid ethyl ester in human dermal fibroblasts: Role of vitagenes. *Clin. Dermatol.* **2008**, *26*, 358–363.
230. Calabrese, V.; Cornelius, C.; Mancuso, C.; Barone, E.; Calafato, S.; Bates, T.; Rizzarelli, E.; Kostova, A.T. Vitagenes, dietary antioxidants and neuroprotection in neurodegenerative diseases. *Front. Biosci.* **2009**, *14*, 376–397.
231. Calabrese, V.; Cornelius, C.; Dinkova-Kostova, A.T.; Calabrese, E.J. Vitagenes, cellular stress response, and acetylcarnitine: Relevance to hormesis. *Biofactors* **2009**, *35*, 146–160.
232. Calabrese, V.; Cornelius, C.; Dinkova-Kostova, A.T.; Calabrese, E.J.; Mattson, M.P. Cellular stress responses, the hormesis paradigm, and vitagenes: Novel targets for therapeutic intervention in neurodegenerative disorders. *Antioxid. Redox Signal.* **2010**, *13*, 1763–1811.
233. Calabrese, V.; Cornelius, C.; Dinkova-Kostova, A.T.; Iavicoli, I.; di Paola, R.; Koverech, A.; Cuzzocrea, S.; Rizzarelli, E.; Calabrese, E.J. Cellular stress responses, hermetic phytochemicals and vitagenes in aging and longevity. Biochim. *Biophys. Acta* **2012**, *1822*, 753–783.
234. Calabrese, V.; Scapagnini, G.; Davinelli, S.; Koverech, G.; Koverech, A.; de Pasquale, C.; Salinaro, A.T.; Scuto, M.; Calabrese, E.J.; Genazzani, A.R. Sex hormonal regulation and hormesis in aging and longevity: Role of vitagenes. *J. Cell Commun. Signal.* **2014**, *8*, 369–384.
235. Cornelius, C.; Perrotta, R.; Graziano, A.; Calabrese, E.J.; Calabrese, V. Stress responses, vitagenes and hormesis as critical determinants in aging and longevity: Mitochondria as a “chi”. *Immun. Ageing* **2013**, *10*, 15.

236. Cornelius, C.; Graziano, A.; Calabrese, E.J.; Calabrese, V. Hormesis and vitagenes in aging and longevity: Mitochondrial control and hormonal regulation. *Horm. Mol. Biol. Clin. Investig.* **2013**, *16*, 73–89.
237. Cornelius, C.; Koverech, G.; Crupi, R.; di Paola, R.; Koverech, A.; Lodato, F.; Scuto, M.; Salinaro, A.T.; Cuzzocrea, S.; Calabrese, E.J.; *et al.* Osteoporosis and Alzheimer pathology: Role of cellular stress response and hormetic redox signaling in aging and bone remodeling. *Front. Pharmacol.* **2014**, *5*, 120.
238. Trovato Salinaro, A.; Cornelius, C.; Koverech, G.; Koverech, A.; Scuto, M.; Lodato, F.; Fronte, V.; Muccilli, V.; Reibaldi, M.; Longo, A.; *et al.* Cellular stress response, redox status, and vitagenes in glaucoma: A systemic oxidant disorder linked to Alzheimer's disease. *Front. Pharmacol.* **2014**, *5*, 129.
239. Cerný, D.; Canová, N.K.; Martínek, J.; Horínek, A.; Kmonícková, E.; Zídek, Z.; Farghali, H. Effects of resveratrol pretreatment on tert-butylhydroperoxide induced hepatocyte toxicity in immobilized perifused hepatocytes: Involvement of inducible nitric oxide synthase and hemoxygenase-1. *Nitric Oxide* **2009**, *20*, 1–8.
240. Venditti, C.C.; Smith, G.N. Involvement of the heme oxygenase system in the development of preeclampsia and as a possible therapeutic target. *Womens. Health (Lond. Engl.)* **2014**, *10*, 623–643.
241. Naito, Y.; Takagi, T.; Higashimura, Y. Heme oxygenase-1 and anti-inflammatory M2 macrophages. *Arch. Biochem. Biophys.* **2014**, *564*, 83–88.
242. Ambegaokar, S.S.; Kolson, D.L. Heme oxygenase-1 dysregulation in the brain: Implications for HIV-associated neurocognitive disorders. *Curr. HIV Res.* **2014**, *12*, 174–188.
243. Bongiovanni, G.A.; Soria, E.A.; Eynard, A.R. Effects of the plant flavonoids silymarin and quercetin on arsenite-induced oxidative stress in CHO-K1 cells. *Food Chem. Toxicol.* **2007**, *45*, 971–976.
244. Demir, M.; Amanvermez, R.; Kamalı Polat, A.; Karabıçak, I.; Cınar, H.; Kesicioğlu, T.; Polat, C. The effect of silymarin on mesenteric ischemia-reperfusion injury. *Med. Princ. Pract.* **2014**, *23*, 140–144.
245. Zhao, H.; Brandt, G.E.; Galam, L.; Matts, R.L.; Blagg, B.S. Identification and initial SAR of silybin: An Hsp90 inhibitor. *Bioorg. Med. Chem. Lett.* **2011**, *21*, 2659–2664.
246. Jiang, Y.Y.; Wang, H.J.; Wang, J.; Tashiro, S.; Onodera, S.; Ikejima, T. The protective effect of silibinin against mitomycin C-induced intrinsic apoptosis in human melanoma A375-S2 cells. *J. Pharmacol. Sci.* **2009**, *111*, 137–146.
247. Zhou, B.; Wu, L.J.; Li, L.H.; Tashiro, S.; Onodera, S.; Uchiumi, F.; Ikejima, T. Silibinin protects against isoproterenol-induced rat cardiac myocyte injury through mitochondrial pathway after up-regulation of SIRT1. *J. Pharmacol.Sci.* **2006**, *102*, 387–395.
248. Wang, Q.; Liu, M.; Liu, W.W.; Hao, W.B.; Tashiro, S.; Onodera, S.; Ikejima, T. *In vivo* recovery effect of silibinin treatment on streptozotocin-induced diabetic mice is associated with the modulations of Sirt-1 expression and autophagy in pancreatic β-cell. *J. Asian Nat. Prod. Res.* **2012**, *14*, 413–423.

249. Li, L.H.; Wu, L.J.; Tashiro, S.I.; Onodera, S.; Uchiumi, F.; Ikejima, T. Activation of the SIRT1 pathway and modulation of the cell cycle were involved in silymarin's protection against UV-induced A375-S2 cell apoptosis. *J. Asian Nat. Prod. Res.* **2007**, *9*, 245–252.
250. Wang, H.J.; Tashiro, S.; Onodera, S.; Ikejima, T. Inhibition of insulin-like growth factor 1 receptor signaling enhanced silibinin-induced activation of death receptor and mitochondrial apoptotic pathways in human breast cancer MCF-7 cells. *J. Pharmacol. Sci.* **2008**, *107*, 260–269.
251. Wu, Y.T.; Wu, S.B.; Wei, Y.H. Roles of sirtuins in the regulation of antioxidant defense and bioenergetic function of mitochondria under oxidative stress. *Free Radic. Res.* **2014**, *48*, 1070–1084.
252. Wu, Y.T.; Wu, S.B.; Wei, Y.H. Metabolic reprogramming of human cells in response to oxidative stress: Implications in the pathophysiology and therapy of mitochondrial diseases. *Curr. Pharm. Des.* **2014**, *20*, 5510–5526.
253. Horio, Y.; Hayashi, T.; Kuno, A.; Kunimoto, R. Cellular and molecular effects of sirtuins in health and disease. *Clin. Sci. (Lond.)* **2011**, *121*, 191–203.
254. Jayasena, T.; Poljak, A.; Smythe, G.; Braidy, N.; Münch, G.; Sachdev, P. The role of polyphenols in the modulation of sirtuins and other pathways involved in Alzheimer's disease. *Ageing Res. Rev.* **2013**, *12*, 867–883.
255. Radak, Z.; Koltai, E.; Taylor, A.W.; Higuchi, M.; Kumagai, S.; Ohno, H.; Goto, S.; Boldogh, I. Redox-regulating sirtuins in aging, caloric restriction, and exercise. *Free Radic. Biol. Med.* **2013**, *58*, 87–97.
256. Lu, J.; Holmgren, A. The thioredoxin antioxidant system. *Free Radic. Biol. Med.* **2014**, *66*, 75–87.
257. Penney, R.B.; Roy, D. Thioredoxin-mediated redox regulation of resistance to endocrine therapy in breast cancer. *Biochim. Biophys. Acta* **2013**, *1836*, 60–79.
258. Surai, K.P.; Surai, P.F.; Speake, B.K.; Sparks, N.H.C. Antioxidant-prooxidant balance in the intestine: Food for thought. 2. Antioxidants. *Curr. Top. Nutraceutical. Res.* **2004**, *2*, 27–46.
259. Surai, K.P.; Surai, P.F.; Speake, B.K.; Sparks, N.H.C. Antioxidant-prooxidant balance in the intestine: Food for thought. 1. Prooxidants. *Nutr. Genomics Funct. Foods* **2003**, *1*, 51–70.
260. Dolara, P.; Luceri, C.; de Filippo, C.; Femia, A.P.; Giovannelli, L.; Caderni, G.; Cecchini, C.; Silvi, S.; Orpianesi. C.; Cresci, A. Red wine polyphenols influence carcinogenesis, intestinal microflora, oxidative damage and gene expression profiles of colonic mucosa in F344 rats. *Mutat. Res.* **2005**, *591*, 237–246.
261. Kanner, J.; Lapidot, T. The stomach as a bioreactor: Dietary lipid peroxidation in the gastric fluid and the effects of plant-derived antioxidants. *Free Radic. Biol. Med.* **2001**, *31*, 1388–1395.
262. Giovannelli, L.; Testa, G.; de Filippo, C.; Cheynier, V.; Clifford, M.N.; Dolara, P. Effect of complex polyphenols and tannins from red wine on DNA oxidative damage of rat colon mucosa *in vivo*. *Eur. J. Nutr.* **2000**, *39*, 207–212.
263. Gorelik, S.; Lapidot, T.; Shaham, I.; Granit, R.; Ligumsky, M.; Kohen, R.; Kanner, J. Lipid peroxidation and coupled vitamin oxidation in simulated and human gastric fluid inhibited by dietary polyphenols: Health implications. *J. Agric. Food Chem.* **2005**, *53*, 3397–3402.
264. Dai, F.; Chen, W.F.; Zhou, B. Antioxidant synergism of green tea polyphenols with alpha-tocopherol and L-ascorbic acid in SDS micelles. *Biochimie* **2008**, *90*, 1499–1505.

265. Gorelik, S.; Ligumsky, M.; Kohen, R.; Kanner, J. The stomach as a "bioreactor": When red meat meets red wine. *J. Agric. Food Chem.* **2008**, *56*, 5002–5007.

266. Gorelik, S.; Ligumsky, M.; Kohen, R.; Kanner, J. A novel function of red wine polyphenols in humans: Prevention of absorption of cytotoxic lipid peroxidation products. *FASEB J.* **2008**, *22*, 41–46.

267. Kanner, J.; Gorelik, S.; Roman, S.; Kohen, R. Protection by polyphenols of postprandial human plasma and low-density lipoprotein modification: The stomach as a bioreactor. *J. Agric. Food Chem.* **2012**, *60*, 8790–8796.

268. Circu, M.L.; Aw, T.Y. Intestinal redox biology and oxidative stress. *Semin. Cell Dev. Biol.* **2012**, *23*, 729–737.

269. Martínez, J.A.; Etxeberría, U.; Galar, A.; Milagro, F.I. Role of dietary polyphenols and inflammatory processes on disease progression mediated by the gut microbiota. *Rejuvenation Res.* **2013**, *16*, 435–437.

270. Biasi, F.; Deiana, M.; Guina, T.; Gamba, P.; Leonarduzzi, G.; Poli, G. Wine consumption and intestinal redox homeostasis. *Redox Biol.* **2014**, *2*, 795–802.

271. Biasi, F.; Astegiano, M.; Maina, M.; Leonarduzzi, G.; Poli, G. Polyphenol supplementation as a complementary medicinal approach to treating inflammatory bowel disease. *Curr. Med. Chem.* **2011**, *18*, 4851–4865.

272. Gessner, D.K.; Fiesel, A.; Most, E.; Dinges, J.; Wen, G.; Ringseis, R.; Eder, K. Supplementation of a grape seed and grape marc meal extract decreases activities of the oxidative stress-responsive transcription factors NF-κB and Nrf2 in the duodenal mucosa of pigs. *Acta Vet. Scand.* **2013**, *55*, 18.

273. Yang, G.; Wangm H.; Kang, Y.; Zhu, M.J. Grape seed extract improves epithelial structure and suppresses inflammation in ileum of IL-10-deficient mice. *Food Funct.* **2014**, *5*, 2558–2563.

274. Goodrich, K.M.; Fundaro, G.; Griffin, L.E.; Grant, A.; Hulver, M.W.; Ponder, M.A.; Neilson, A.P. Chronic administration of dietary grape seed extract increases colonic expression of gut tight junction protein occludin and reduces fecal calprotectin: A secondary analysis of healthy Wistar Furth rats. *Nutr. Res.* **2012**, *32*, 787–794.

275. Rodríguez-Ramiro, I.; Ramos, S.; Bravo, L.; Goya, L.; Martín, M.Á. Procyanidin B2 and a cocoa polyphenolic extract inhibit acrylamide-induced apoptosis in human Caco-2 cells by preventing oxidative stress and activation of JNK pathway. *J. Nutr. Biochem.* **2011**, *22*, 1186–1194.

276. Rodríguez-Ramiro, I.; Martín, M.A.; Ramos, S.; Bravo, L.; Goya, L. Comparative effects of dietary flavanols on antioxidant defences and their response to oxidant-induced stress on Caco2 cells. *Eur. J. Nutr.* **2011**, *50*, 313–322.

277. Köck, K.; Xie, Y.; Hawke, R.L.; Oberlies, N.H.; Brouwer, K.L. Interaction of silymarin flavonolignans with organic anion-transporting polypeptides. *Drug Metab. Dispos.* **2013**, *41*, 958–965.

278. Shin, J.H.; Lee, C.W.; Oh, S.J.; Yun, J.; Lee, K.; Park, S.K.; Kim, H.M.; Han, S.B.; Kim, Y.; Kim, H.C.; *et al.* Protective effect of silymarin against ethanol-induced gastritis in rats: Role of sulfhydryls, nitric oxide and gastric sensory afferents. *Food Chem. Toxicol.* **2013**, *55*, 353–357.

279. Esmaily, H.; Hosseini-Tabatabaei, A.; Rahimian, R.; Khorasani, R.; Baeeri, M.; Barazesh-Morgani, A.; Yasa, N.; Khademi, Y.; Abdollahi, M. On the benefits of silymarin in

murine colitis by improving balance of destructive cytokines and reduction of toxic stress in the bowel cells. *Central Eur. Kournal. Biol.* **2009**, *4*, 204–213.

280. Esmaily, H.; Vaziri-Bami, A.; Miroliaee, A.E.; Baeeri, M.; Abdollahi, M. The correlation between NF-κB inhibition and disease activity by coadministration of silibinin and ursodeoxycholic acid in experimental colitis. *Fundam. Clin. Pharmacol.* **2011**, *25*, 723–733.
281. Moco, S.; Martin, F.P.; Rezzi, S. Metabolomics view on gut microbiome modulation by polyphenol-rich foods. *J. Proteome. Res.* **2012**, *11*, 4781–4790.
282. Pozuelo, M.J.; Agis-Torres, A.; Hervert-Hernández, D.; Elvira López-Oliva, M.; Muñoz-Martínez, E.; Rotger, R.; Goñi, I. Grape antioxidant dietary fiber stimulates *Lactobacillus* growth in rat cecum. *J. Food Sci.* **2012**, *77*, H59–H62.
283. Parkar, S.G.; Trower, T.M.; Stevenson, D.E. Fecal microbial metabolism of polyphenols and its effects on human gut microbiota. *Anaerobe* **2013**, *23*, 12–19.
284. Bolca, S.; van de Wiele, T.; Possemiers, S. Gut metabotypes govern health effects of dietary polyphenols. *Curr. Opin. Biotechnol.* **2013**, *24*, 220–225.

14

Mechanisms of Photoaging and Cutaneous Photocarcinogenesis, and Photoprotective Strategies with Phytochemicals

Ricardo Bosch [1,2], Neena Philips [3,*], Jorge A. Suárez-Pérez [1,2], Angeles Juarranz [4], Avani Devmurari [3], Jovinna Chalensouk-Khaosaat [3] and Salvador González [5,6]

1 Department of Dermatology, Virgen de la Victoria University Hospital, Málaga 29010, Spain; E-Mails: ricardobosch@aedv.es (R.B.); jasuape@hotmail.com (J.A.S.-P.)

2 Dermatology and Medicine Department, University of Málaga, Málag 29071, Spain

3 School of Natural Sciences, Fairleigh Dickinson University, 1000 River Road, Teaneck, NJ 07666, USA; E-Mails: avaniben@student.fdu.edu (A.D.); jovinna.ck323@yahoo.com (J.C.-K.)

4 Biology Department, Universidad Autónoma de Madrid, Madrid 28903, Spain; E-Mail: angeles.juarranz@uam.es

5 Dermatology Service, Memorial Sloan-Kettering Cancer Center, New York, NY 10022, USA; E-Mail: gonzals6@mskcc.org

6 Ramon y Cajal Hospital, Alcala University, Madrid 28034, Spain

* Author to whom correspondence should be addressed; E-Mail: nphilips@fdu.edu or neenaphilips@optonline.net

---

Abstract: Photoaging and photocarcinogenesis are primarily due to solar ultraviolet (UV) radiation, which alters DNA, cellular antioxidant balance, signal transduction pathways, immunology, and the extracellular matrix (ECM). The DNA alterations include UV radiation induced thymine-thymine dimers and loss of tumor suppressor gene p53. UV radiation reduces cellular antioxidant status by generating reactive oxygen species (ROS), and the resultant oxidative stress alters signal transduction pathways such as the mitogen-activated protein kinase (MAPK), the nuclear factor-kappa beta (NF-κB)/p65, the janus kinase (JAK), signal transduction and activation of transcription (STAT) and the nuclear factor erythroid 2-related factor 2 (Nrf2). UV radiation induces pro-inflammatory genes and causes immunosuppression by depleting the number and activity of the epidermal Langerhans cells. Further, UV radiation remodels the ECM by increasing matrixmetalloproteinases (MMP) and reducing structural collagen and elastin. The photoprotective strategies to prevent/treat photoaging and photocarcinogenesis include oral or topical agents that act as sunscreens or counteract the effects of UV radiation on DNA, cellular antioxidant balance, signal transduction pathways, immunology and the ECM. Many of these agents are phytochemical derivatives and include polyphenols and non-polyphenols. The flavonoids are polyphenols and include catechins, isoflavones, proanthocyanidins, and anthocyanins, whereas the non-flavonoids comprise mono phenolic acids and stilbenes. The natural sources of polyphenols include tea, cocoa, grape/wine, soy, pomegranate, and *Polypodium leucotomos*. The non-phenolic phytochemicals include carotenoids, caffeine and sulphoraphance (SFN). In addition, there are other phytochemical derivatives or whole extracts such as baicalin, flavangenol, raspberry extract, and *Photomorphe umbellata* with photoprotective activity against UVB radiation, and thereby carcinogenesis.

Keywords: skin aging; skin cancer; ultraviolet radiation; signal transduction pathways; extracellular matrix; polyphenols

---

1. Introduction

Premature skin aging and development of malignant cutaneous tumors, melanoma and non-melanoma, are interrelated issues that are increasingly important problems in the field of dermatology. Skin aging is important aesthetically, whereas skin cancer is a direct threat to the health of the patient. Hence, research aimed at providing knowledge in these areas has increased exponentially to develop preventive and therapeutic approaches.

We review the mechanisms of photoaging and photocarcinogenesis, the photoprotective strategies, and the phytochemicals that can provide photoprotection. The photoaging and photocarcinogenic mechanisms are predominantly the effect of solar ultraviolet (UV) radiation that induces reactive oxygen species (ROS) and alters DNA/cellular homeostasis, which together alter signal transduction pathways and inflammatory cascade and induce immunosuppression and extracellular matrix (ECM) remodeling. The photoprotective strategies include the blockade of UV photon incidence, DNA repair, removal of ROS (antioxidant), anti-inflammation, and immunomodulation. The photochemical derivatives that are effective for these photoprotective strategies are polyphenols, flavonoids and non-flavonoids, non-phenolic derivatives, and whole plant extracts.

2. Mechanisms of Photoaging and Photocarcinogenesis

Aging is a natural process leading to the progressive deterioration of the organs and its resultant clinical and histological changes. A primary cause is the imbalance between ROS production and their neutralization by natural antioxidant systems, which generates oxidative stress. ROS promote peroxidation of the lipid components of the cell membrane, alter the structure and function of several enzymatic systems, and promote carbohydrate oxidation. Incident UV radiation is the predominant cause of the oxidative stress in the skin and the histological differences between UV radiation exposed and non-exposed skin areas. The whole array of changes caused by UV radiation in exposed skin is termed photo aging, whereas changes from other factors that contribute to aging, such as metabolic or hormonal, are termed "chronologic" or "intrinsic" aging [1–4].

Cancer is a disease caused by the alteration of certain genes, resulting in uncontrolled cellular proliferation and loss of normal control mechanisms to inhibit such growth. Its development begins at the cross roads between genetics and environment, the latter being more important in certain types of cancer, e.g., skin tumors. The chief promoter of skin cancer, which underlies the use of the term photo carcinogenesis, is sun radiation; although other factors may contribute, such as viruses or chemicals.

The photoaging and photocarcinogenic mechanisms through UV radiation induced ROS and DNA damage, and the resultant cellular damage, inflammation, immunosuppression and ECM remodeling/angiogenesis are illustrated in Figure 1.

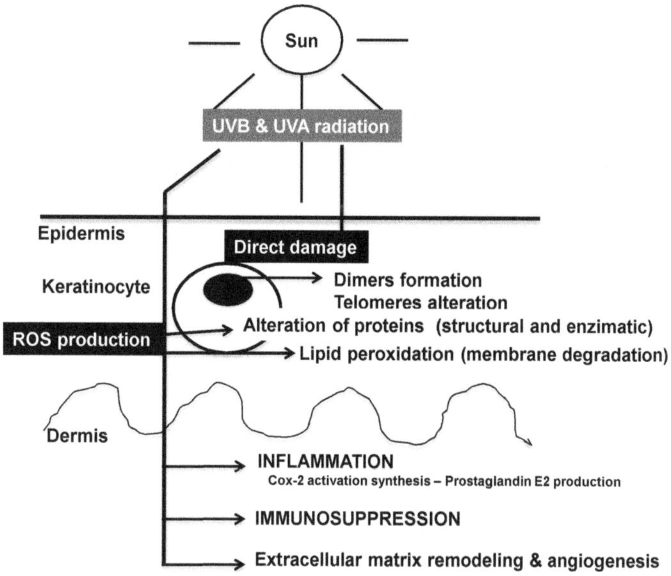

Figure 1. Summary of the major deleterious effects of sun-generated ultraviolet (UV) radiation in skin.

2.1. Effects of Solar Ultraviolet Radiation

Photon energy carried in UV (particularly UVB at 280–315 nm, and UVA at 315–400 nm) induces alterations that accumulate and promote the majority of the typical manifestations of skin aging and cancer. UVB makes up only 5% of the UV radiation that reaches the surface of earth and has little penetrance, but it displays great biological activity. UVA makes up the remaining 95% of incident light and is more penetrating, promoting photo aging. However, UVA carries less energy and therefore promote carcinogenesis to a lower extent than UVB [5].

The main effects of acute and chronic exposure to UV radiation are DNA damage, inflammation and immunosuppression. These effects are direct as well as indirect due to ROS production. The ROS are particularly harmful in that they destabilize other molecules and promote chain reactions that damage biomolecules rapidly, such as telomere shortening and deterioration, mitochondrial damage, membrane degradation and oxidation of structural and enzymatic proteins [4].

UV radiation directly and through ROS participates in the three stages of the carcinogenic process [6–8]. During initiation, it produces genetic damage through direct effect on the DNA or by activating other factors [9,10]. In the promotion stage, it favors the proliferation of malignant cells by inhibiting the mechanisms of immune controls and promoting genomic instability [11]. Finally, it also boosts progression and dissemination of tumors by promoting protease release and angiogenesis [12,13].

2.2. DNA and Cellular Homeostasis

UV radiation, particularly UVB, alters DNA by promoting the formation of thymine-thymine dimers and pyrimidine-pyrimidone dimers, and generates ROS [14–16]. The thymine-thymine dimers are particularly crucial when they affect the tumor suppressor gene p53. Mutated p53 appears in skin displaying chronic sun damage, actinic keratosis and skin cancer [17]. The p53 mutation make cells resistant to apoptosis and the cells enter mitosis without having undergone DNA repair [18]. Further, the ROS participate in p53-independent apoptotic pathways [19].

2.3. Signal Transduction Pathways

The predominant pathways regulated by photooxidative stress include the mitogen-activated protein kinase (MAPK), the nuclear factor-kappa beta (NF-κB)/p65, the JAK/STAT (Signal Transduction and Activation of Transcription) and the nuclear factor erythroid 2-related factor 2 (Nrf2) [20,21]. The activation of MAP kinase pathway, through the receptor tyrosine kinase, results in the activation of transcription factor activator protein-1 (AP-1) that activates expression of MMPs and is comprised of the extracellular signal-regulated kinase 1/2 (ERK1/2), c-Jun-N-terminal- kinase (JNK) and p38 proteins [21]. The JNK and p38 pathways play a major role in the UV radiation mediated increase in AP-1 and cyclooxygenase-2 (COX-2) expression, and are targets for chemoprevention of skin cancer [5]. The transcription factor Nrf2 regulates the expression of phase II key protective enzymes through the antioxidant-response element. Nrf2 and several of its target genes are significantly down-regulated, causing oxidative stress in human prostate cancer [22].

The NF-κB pathway is also activated by oxidative stress through the activation of cytoplasmic I-κB kinase. Active I-κB kinase phosphorylates and degrades I-κB, the inhibitor of NF-κB transcription factor [20]. The NF-κB activation is associated with UVA and UVB radiation mediated oxidative modification of cellular membrane components [3]. The release of NF-κB from its inhibitor (I-κB), results in the translocation of active NF-κB to the nucleus to activate the inflammatory cytokines and prostaglandins [20]. In general, inhibition of NF-κB by use of antioxidants, proteasome inhibitors, prevention of Ikb phosphorylation or expression of overactivated, mutant (Ikb) mitigate UV-induced damage (Reviewed in [21]).

2.4. Role of Mitochondria and Cellular Bioenergetics

Mitochondria are important players in the oxidative response. Mitochondrial DNA accumulates UV-induced mutations that track exposure to UV light (reviewed in [22]). UV alters mitochondrial function, including decreased O2 consumption and ATP production, which effects cellular processes such as cell migration and division (Reviewed in [23]). The mechanism of UV-induced mitochondrial dysfunction and toxicity includes interrelated steps of caspase activation, membrane depolarization and cytochrome C release (Reviewed in [24]). Overall mitochondrial dysfunction, in addition to the depletion of cellular energy required activities such as motility and DNA repair, increases the levels of oxidative stress from increased ROS production at mitochondrial complexes (reviewed in [25]). UV radiation targets Nrf2. Nrf2 is a master regulator of the antioxidant response as it controls the expression of several endogenous antioxidant systems, e.g., the enzymes involved in NADPH synthesis such as glucose 6-phosphate dehydrogenase, synthesis of thioredoxin such as thioredoxin reductase, and GSH, such as glutathione *S*-transferases and peroxidase (reviewed in [26]). Nrf2 also controls the bioavailability of mitochondrial respiratory (reviewed in [27]), underlining the role of bioenergetics in Nrf2-mediated antioxidant protection.

2.5. *Inflammation Cascade*

UV radiation induces pro-inflammatory genes. Inflammation is an important mediator of photoaging and photocarcinogenesis [28–31]. The inflammatory mediators are released from keratinocytes, fibroblasts, tumor cells, leukocytes, and the endothelial lining of blood vessels. The mediators include the plasma mediators (bradykinin, plasmin, fibrin), lipid mediators (prostaglandins, leukotrienes, and platelet activating factor), and the inflammatory cytokines [interleukin-1 (IL-1), IL-6, and tumor necrosis factor (TNF)-α]. The lipid mediators, COX-2 (cyclooxigenase-2) and prostaglandin E2 (PGE2) are also activated by ROS [32–34]. UV radiation also participates in the activation of the enzyme ornithine decarboxylase, which decreases the activity of different polyamines that regulate cell proliferation [35]. The inflammatory process triggers ROS and RNS (reactive nitrogen species), which generates peroxynitrite that triggers DNA deletion and rearrangement [36,37]. The processes of DNA repair, cell cycle and apoptosis are altered to favor tumor progression. Further, UV radiation alters the expression of transforming growth factor-b (TGF-β), which is the predominant regulator of the matrix metalloproteinases that remodel the extracellular matrix for skin photoaging and tumor dissemination [38].

2.6. *Immunosuppression*

Immunosuppressed patients are more prone to develop tumors, likely due to decreased local immunosurveillance. UV radiation mainly decreases the cellular response, but the humoral response is also affected (reviewed in [39]). UV radiation depletes epidermal Langerhans cells (LC), which are crucial mediators of the cellular immune response due to their role in antigen presentation [40] (Figure 2). Importantly, UV radiation not only decreases the number of LC but also impairs their functions. e.g., migration and antigen presentation in lymph nodes. Whereas the former is not well understood mechanistically, impaired antigenic presentation is due to a loss of co-stimulatory molecules. e.g., B7. The main mechanism involved in this process is the isomerization of *trans*-urocanic acid to the *cis* form, which directly reduces LC migration and activity (reviewed in [41–43]). In addition, UV radiation promotes secretion of the immunosuppressive cytokines (IL-10). Il-10 is secreted by keratinocytes in response to *cis*-UCA or CPD. This cytokine is of particular interest because of its crucial role in immunosuppression in skin, not only induced by UV radiation but in other skin pathologies, e.g., melanoma, in which high IL-10 correlates with bad prognosis (reviewed in [44]). Also, depletion of LC and the proinflammatory microenvironment induced by UV causes influx of macrophages that activate regulatory T cells (TREG) and polarizes the Th1/Th2 response towards Th2 [37,38] (Figure 2). The role of the Th2 response in immunosuppression in response to UV light is likely related with the expression of IL12 by LC because IL-12 depletion skews T cell activation towards Th2 while promoting TREG activation (reviewed in [45–47]).

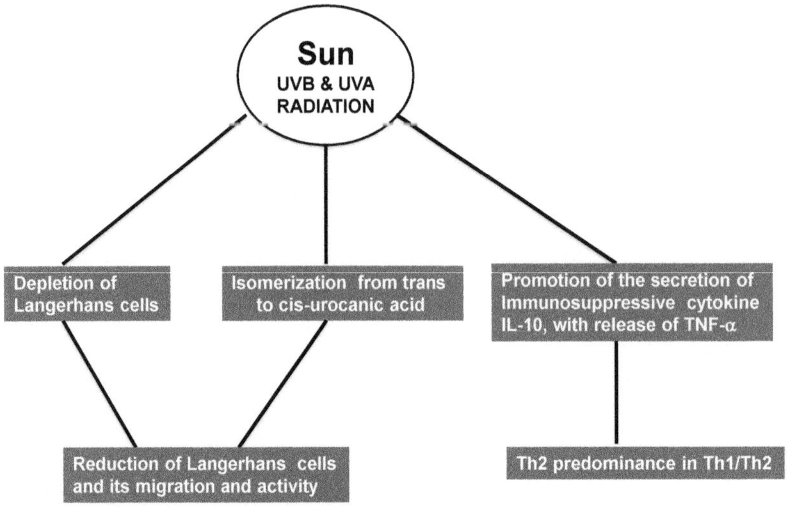

Figure 2. Immunosuppressive effects of UV radiation.

2.7. Extracellular Remodeling: Collagen, Elastin and Matrix Metalloproteinases Network

Collagen and elastin are the structural proteins of the ECM. The deterioration/remodeling of the collagen and elastin fibers facilitates angiogenesis and metastasis, and the damaged collagen and elastin proteins serve as additional sensitizers of photooxidative stress [48]. The specific properties of the different collagens are based on the lengths of the triple helical segments, interruptions to the triple helix, and amino acid modifications [28]. The elastin fibers that provide stretch-recoil properties to skin are composed predominantly of an elastin core (90%) surrounded by fibrillin microfibrils. The loss of proper elastin fibers occurs with the exposure of skin to UV radiation [4]. UV radiation also depletes the microfibrillar network in the epidermal-dermal layer and the dermis, which contributes to the aberrant elastic fibers [49].

The ECM proteolytic enzymes (MMPs/elastases) are produced by epidermal keratinocytes, fibroblasts, in the mediation of ECM remodeling and skin cancer [50–52]. Their basal levels increase with aging, and are further increased by environmental pollutants and UV radiation, resulting in the fragmentation of collagen and elastin fiber proteins for carcinogenesis. MMPs are categorized on the basis of the presence of AP-1 or TATA nucleotide sequences in the promoters into group I MMPs (MMP-1, 3, 7, 9, 10, 12, 13, 19, and 26) that contain TATA box and activator protein-1 (AP-1 site), group II MMPs (MMP-8, 11, 21) without the AP-1 site, and group III (MMP-2, 14, 28) without the TATA box and AP-1 site [53,54].

The transcription factor AP-1, stimulated largely by the MAPK pathway, stimulates the transcription of several MMPs that collectively degrade the ECM, such as MMP-1, MMP-2/9, and MMP-3 [3]. Further, AP-1 inhibits the transcription of type I collagen gene [28]. Hence, the damage to the ECM and tissue integrity is from the enhanced degradation of ECM by MMPs as well as the reduced expression of the structural ECM proteins. The pro- and active forms of MMPs are inhibited by the tissue inhibitors of MMPs or tissue inhibitor of matrixmetalloproteinases (TIMPs) [53,54]. The remodeling of collagen and elastin, for angiogenesis, metastasis, and tissue destruction, is largely from the increased expression or activation of MMPs and reduced expression of TIMPs [53,54].

3. Photoprotective Strategies

The photoprotective strategies to prevent and/or repair the deleterious effect of UV radiation leading to photoaging and photocarcinogenesis from direct blockade of UV photons, to counteracting the direct or indirect effects of UV radiation through DNA repair systems and antioxidants, are illustrated in Figure 3.

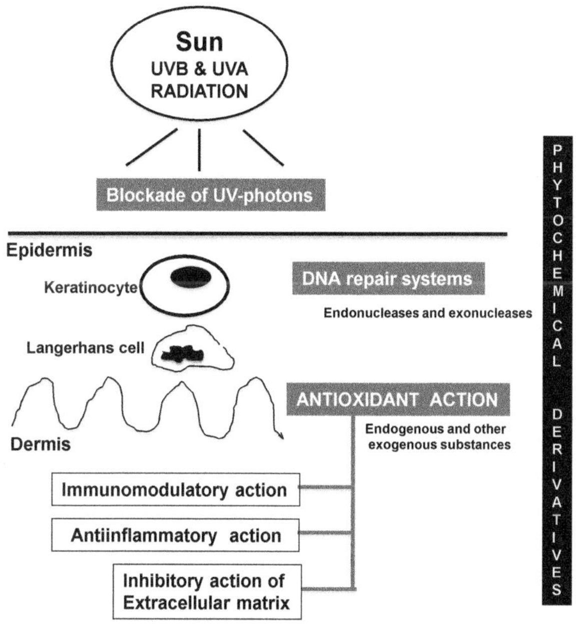

Figure 3. Major beneficial effects of phytochemical derivatives against UV-induced photodamage.

3.1. Blockade of UV Photon Incidence

The most obvious strategy to prevent the deleterious effects of UV radiation is to prevent its incidence on the skin; thus, physical blockers and screens are the most widely accepted and used countermeasures. Topical sunscreens can be divided into reflective and absorbing substances. Their use requires application of the correct amount, and frequent replenishment upon changing environment conditions, e.g., increased perspiration, water immersion, *etc.* Additional difficulties to their use include displeasing sensitivity, e.g., stickiness, aesthetic issues due to whitening, *etc.* Finally, complete blockade in cases of extreme photoprotection may lead to Vitamin D deficiency, which can promote carcinogenesis [55].

Sunscreens are the most important preventive measures against photoaging and photocarcinogenesis. In fact, proper use of sunscreens mitigates the chances of developing UV-induced skin cancer (reviewed in [56–58]). Some phytochemical derivatives fulfill this function, administered either topically or systemically, which increases adherence to the application regime as well as its uniformity.

3.2. DNA Repair Systems

Oxidative DNA damage is repaired by multiple, overlapping DNA repair pathways. Two major mechanisms exist to repair oxidatively induced DNA lesions: base-excision repair (BER) and nucleotide-excision repair (NER). In BER-mediated repair, DNA glycosylase usually detects the damaged base and mediates base removal prior to nuclease, polymerase, and ligase proteins bridging the gap and completing the repair process. On the other hand, NER-mediated repair recognizes base lesions that distort the helical structure. The damaged base is excised as a 22–30 base oligonucleotide resulting in single-stranded DNA that is repaired by proteins such as DNA polymerase before proceeding to ligation. An endogenous enzymatic system corrects and repairs the formation of cyclobutane pyrimidine dimers. The absence of the DNA repair systems produces diseases such as xeroderma pigmentosum, and underlies a large percentage of skin cancers in the general population.

Different preparations containing DNA repair enzymes have been assayed and shown to decrease DNA damage caused by UV radiation [59,60]. Among these, photoliase promotes a process termed DNA photo-reactivation that repairs T-T, cyclobutane pyrimidine dimers and 6-4 photoproducts [61]. Another preparation containing T4 endonuclease N5 (Dimericine) promotes DNA repair in xeroderma pigmentosum patients [62].

3.3. Antioxidant Activity

Oxidative stress is a key process underlying photo aging and photo carcinogenesis [63]. Endogenous systems to prevent its deleterious effects include enzymes, e.g., superoxide dismutase, catalase, ceruloplasmin, transferrin, *etc.*, and other substances obtained from the diet, e.g., Vitamin E (α-tocopherol), Vitamin C (ascorbic acid) and carotenoids (β-carotene). Non-melanoma skin cancers display a significant decrease of the enzymatic antioxidant systems [64]. However, whether exogenous supplementation of non-enzymatic antioxidants, e.g., Vitamin E, C or carotenes, is useful to prevent photocarcinogenesis is currently controversial [65–68]. There is disparity in the parameters evaluated, including erythema, immunosuppression, photo aging or tumor formation, and the lack of homogeneity may account for the controversy [69,70]. A largely beneficial substance to prevent skin cancer is Vitamin A (retinol, derived from β-carotene) and its derivatives (isotretinoin and acitretin) [71–73].

3.4. Anti-Inflammatory Action

Aspirin (acetyl salicylic acid) and other NSAIDs (non-steroidal anti-inflammatory drugs), e.g., indomethacin, piroxicam, sulindac, diclofenac, are useful to decrease the incidence of skin tumors and to treat actinic keratosis [74,75]. These molecules decrease prostaglandin production by inhibiting COX-1 and COX-2 [76]. Colecoxib is another inhibitor specific of COX-2 [77]. Several studies support the anti-inflammatory effect of different phytochemicals [78,79].

3.5. Immunomodulatory Action

Another strategy against photo carcinogenesis is to revert immunosuppression induced by UVB. For example, supplementation with Vitamin B3 (niacin) reduced the immunosuppressive effect of

UVB [80]. Other phytochemicals decrease immunosuppression, e.g., by preventing Langerhans cell depletion [81].

3.6. Inhibitory Activity of ECM Remodeling

An emerging strategy against photo aging is to prevent the changes induced by UV, particularly UVA, in the dermis. These changes include alterations of the extracellular matrix proteins that form fibrillar structures, e.g., collagen and elastin. Metalloproteinase and elastase expression and/or activation underlie these alterations, which also promote angiogenesis and tumor progression. Several phytochemicals orally and/or topically administered provide protection against these changes [82].

4. Photoprotective Activity of Phytochemical Derivatives

A myriad of extracts or isolated/purified substances from different parts of plants, including roots, leaves, flowers, seeds, *etc.*, have been traditionally used to prevent and treat skin cancer [2]. Although some of these substances have been used topically, its route of administration is mainly oral, as food supplements, concentrates and purified extracts. Many of these substances contain active principles of the polyphenol group (antioxidants), or other antioxidants with diverse chemical structures [83]. The active phytochemicals or the extracts of their sources have become major photoprotective strategies. Although they mainly function as antioxidants, they also display anti-inflammatory and immunomodulatory activity and also control dermal extracellular matrix remodeling.

4.1. Polyphenols

Polyphenols are chemicals characterized by the presence of more than one phenolic group (a hydroxyl group bound to an aromatic ring) per molecule. Their intrinsic antioxidant function resides in the hydroxyl (−OH) group that, bound to the aromatic ring, act as a hydrogen or an electron donor, giving it to a free radical or other reactive species. This underlies the inhibition of ROS and ROS mediated damage on DNA, proteins and lipids; production of inflammatory cytokines; and the activation of the signal transduction pathways such as mitogen activated protein kinase and nuclear factor kappa-B (NF-κB)/p65 that regulate transcriptional activity [84].

The typical classification of these molecules takes into account the number and type of phenolics, which determine their biological properties. According to this, polyphenols are either flavonoids (the most numerous) or non-flavonoids, appearing in numerous plants (Table 1). In addition to their antioxidant capability, some of them display metal (Cu and Fe) chelating properties, thereby preventing the Fenton reaction, which involves formation of free radicals from hydrogen peroxide (H_2O_2). Non-flavonoids comprise mono phenolic acids and alcohols, benzoic and cinnamic acid and stilbenes.

4.1.1. Flavonoids

The flavonoids include catechins, isoflavones, proanthocyanidins, and anthocyanins (Table 1).

The catechins are mainly present in tea leaves; they contain a pyrocatechol group and comprise of the following: catechin, epicatechin, galactocatechin, epicatechingallate and epigallocatechin-3-gallate.

In humans, their topical administration decreases UV-induced changes in epidermis, specifically increased p53 expression and apoptosis [85].

Table 1. Main sources of polyphenols with antioxidant activity.

Polyphenol	Major Sources
Flavonoids	
Catechins: catechin, epicatechin, galactocatechin, epicatechingallate, epigallocatechin-3-gallate	Tea
Isoflavones: Genistein	Soy
Sylimarin	Thistle
Proanthocyanidins (tannins)	Grapeseed
Anthocyanins	Pomegranate
Non-flavonoids	
Phenolic acids	Grape & derivatives
Benzoic acids: Galic acid	Tea
Cinnamic acids	*Polypodium leucotomos*
Stilbene	Grape & derivatives
Resveratrol	Nuts, peanuts

The isoflavones contain a phloroglucinol group. The most well-known isoflavones are genistein, derived from soybean, and silymarin, derived from the milk thistle (*Silybum marianum*). Genistein exerts a photoprotective effect and halts skin photocarcinogenesis in animal models [86]. The major active principle of silymarin is silibinin. It is widely used as a liver protector, but it has been shown to be also photoprotective in animals due to ROS reduction as it decreases infiltration of CD11b+ lymphocytes in UV-irradiated areas [87].

The proanthocyanidins are also known as condensed tannins; this is a group of substances widely represented in grape seeds. An extract of grape seeds prevents tumor induction in response to UV radiation in mice. These effects are likely due to their antioxidant and anti-inflammatory properties [88,89].

The anthocyanins are water-soluble blue dyes and confer this color to leaves and fruits and seeds, e.g., grapes, to protect them against solar radiation. It has an important role for the color/appearance of wine.

4.1.2. Non-Flavonoids

The non-flavonoids include phenolic acids and stilbene (Table 1). The phenolic acids include benzoic, galic, and cinnamic (caffeic, ferulic and p-cumaric) acids. They appear in wine (more in red wine) and in tea. They exhibit antioxidant and anti-neoplastic properties. Caffeic acid, not related to caffeine, belongs to the hidroxycinnamic group and is widely present in all plants. It protects against UVA-induced photo damage [90].

Stilbene represents anti-microbial substances secreted by plants. The most important is glycosylated *trans*-resveratrol (in grapes, grape juice and wine, peanuts and others). It is a good antioxidant with anti-aging and anti-photocarcinogenic properties in animals [91,92].

4.1.3. Natural Sources of Polyphenols

The polyphenols are part of a normal diet, occurring in vegetables, fruits, beans and cereals. They are often used as concentrated dietary supplements obtained from common or uncommon vegetables. The topical use of polyphenols depends on their lipophilic or hydrophilic nature. In general, polyphenols are hydrophilic and hardly penetrate the epidermal barrier (an exception is silymarin). The steps to address this have mainly been the improvement of penetrance by iontophoresis or use of liposomes. The natural sources of polyphenols include tea, cocoa, grape/wine, soy, pomegranate, and *Polypodium leucotomos* (Table 1).

Tea is the second most consumed liquid in the world, after water. It is an infusion of the leave of *Camellia sinensis*. Its name depends on the main features of the plant leaves used to brew the beverage, white (young leaves), green (non-oxidated), yellow or blue (medium oxidation) or black (high degree of oxidation, which increases its theaflavin content). Other denominations are geographical. Green tea is the most consumed tea around the world. It mainly contains catechins, particularly epigallocatechin-3-gallate and simple phenolic acids. It protects against sun damage, and anecdotal epidemiological evidence suggests that it reduces the development of certain tumors [93,94].

Cocoa extracts contain polyphenols, particularly catechins and proanthocyanidin flavanols, as well as smaller amounts of gallocatechin and epicagallo-catechin. In addition, cocoa contains large amounts of theobromine, a methylxanthine with ROS scavenging properties in skin upon UV treatment [95].

Grape, *Vitis vinífera*, contains numerous polyphenols in its seeds and grape peels. These pass to the fermented form, wine. Specifically, they contain antocyanins (which provide color) and other flavonoids, e.g., proanthocyanidins. Some of these include tannins that cause constipation. Grape/wine contains several non-flavonoid polyphenols, e.g., cinnamic acid and resveratrol, which are well-known photoprotectors.

Soy, *Glycine maxi* (soybean plant), is enriched in proteins that are consumed in several forms, e.g., boiled bean pods, soybean cake, milk and sauce. It is high in protein content, which enables its use as meat or fish substitute. It also contains large amounts of genistein, which has photoprotective activity for the prevention/treatment of photoaging, and photocarcinogenesis [96,97].

Pomegranate extract contains anthocyanins, ellagitannins and hydrolyzable tannins. Its oral administration reduces UVB-induced carcinogenesis in mice [98].

Polypodium leucotomos extracts, from the tropical fern, contain a high concentration of antioxidant phenolic acids, e.g., caffeic and ferulic acids [99]. In addition, it contains monosaccharides, e.g., fructose and glucose, and many other components [88]. It is safe when administered orally and can undergo topical absorption. It displays great efficacy against photo aging and photo carcinogenesis [89]. It prevents lipid peroxidation, UV-induced membrane damage, transcriptional activation of proinflammatory AP1 and NF-κB factors, and induction of enzymes that generate nitric oxide [100]. It inhibits UV-mediated actin disarray and loss of cell-extracellular matrix focal adhesion and also prevents keratinocyte apoptosis [101–104]. It is also used as an adjuvant in PUVA (Psoralens + UVA) therapy to prevent the deleterious side effects of irradiation [105]. Its mechanism of action involves antioxidant, anti-inflammatory and immunomodulating activities [81,106–108]. In addition, it modulates metalloproteinase activity by inducing TIMP (tissue inhibitor of metalloproteinase) and induces elastin and collagen to counteract skin aging and photocarcinogenesis [109].

4.2. Non-Polyphenols with Photoprotective Sctivity

The non-phenolic phytochemicals include carotenoids, caffeine, and sulphoraphance (SFN) as well as several other whole extracts or components.

Carotenoids are pigments synthesized by plants and exhibit clear antioxidant action. They are mainly present in yellow or orange vegetables and fruits. They require additional fat for their absorption. B-carotene has been used as a photo protector in patients with pathological photosensitivity (porphyria and some others), but their sustained administration does not display any beneficial effect against non-melanoma cancer incidence compared to a control group [110]. Lycopene is a carotenoid most abundant in tomatoes. It is not a Vitamin A precursor, yet it protects against various skin alterations induced by UV radiation [111].

Caffeine is present in large amounts in coffee. Some epidemiological studies indicate that drinking large amounts of coffee decreases the incidence of skin cancer, particularly BCC (basal cell carcinoma) [112,113]. Caffeine underlies this effect since decaffeinated coffee does not replicate this finding [114]. The mechanism of this effect involves increased apoptosis of cells with defective DNA repair [115].

Sulphoraphane (SFN) is present largely in broccoli, extracts of which are metabolised into isothiocyanates. They main isothiocyanate from broccoli is SFN that has been shown to decrease UV-induced skin erythema in humans [116] and reduce the risk of skin cancer in mice [117]. Glucoraphanin, also known as SFN glucosinolate (SGS), is the precursor of SFN. It seems that SFN induces transcriptional activation of Nrf2. Other protective mechanisms in cells include inhibition of the activation of procarcinogens, disposal of damaged and potentially neoplastic cells by cell cycle arrest and apoptosis, and the suppression of inflammatory responses.

Other phytochemical derivatives or whole extracts with photoprotective activity against UVB radiation are represented in Table 2 [118–122]. In addition, hypericin (Saint John Wort) can lower erythema in photodyanamic therapy, without being phototoxic [123].

Most of these studies have been carried out in animal models using topical or systemic administration.

Table 2. Other phytochemical derivatives with photoprotective activity against UVB radiation in mice (Murine model).

Substance and Origin	Activity	Reference
Topical "Baicalin" Genus *Scutellaria*	Inhibition of Ki67, PCNA and COX-2 expression	[118]
Oral "Flavangenol" French maritime pine bark extract	Reduction of Ki-67, and (8-OHdG)-positive cells and VEGF expression	[119]
Topical black raspberry extract	Reduction of edema, p53 levels and neutrophil activation	[120]
Topical *Photomorphe umbellata* extract	Inhibition of the hyperplastic reaction and p53-positive cells	[121]
Oral and topical Brown algae polyphenols	Inhibition of ciclooxygenase-2 activity and cell proliferation	[122]

5. Conclusions

We have reviewed the mechanisms of photoaging and photocarcinogenesis, the photoprotective strategies, and the phytochemicals that can provide photoprotection. The photoaging and photocarcinogenic mechanisms are primarily through UV radiation induced ROS and DNA damage and the resultant cellular damage, alterations in inflammatory/oxidative stress mediating signal transduction pathways, inflammation, immunosuppression and ECM remodeling/angiogenesis. The photoprotective strategies include the blockade of UV photon incidence, DNA repair through DNA repair enzymes, removal of ROS with antioxidant agents, and anti-inflammation/immunomodulation with anti-inflammatory agents. Many of these photoprotective strategies involve photochemical derivatives, including polyphenols (flavonoids and non-flavonoids), non-phenolic derivatives and whole plant extracts. An increase in the blockage of UV radiation as well as the strengthening of cellular antioxidant balance will reduce the incidence of photoaging and photocarcinogenesis.

Author Contributions

Dr. Bosch originated the manuscript with his colleagues Dr. Suárez-Pérez and Dr. Juarranz. Dr González and Dr. Philips revised and re-revised the manuscript. Ms. Devmurari and Ms Chalensouk-Khaosaat contributed through editing or literature searches. The authors also want to thank Fairleigh Dickinson University students, Andrew Avedano, Tam Ho, Rinat Khundoker, and Christian Gonzalez, for contributing to the formatting of references.

Conflicts of Interest

The authors declare no conflict of interest.

References

1. Uliasz, A.; Spencer, J.M. Chemoprevention of skin cancer and photoaging. *Clin. Dermatol.* **2004**, *22*, 178–182.
2. Millsop, J.W.; Sivamani, R.K.; Fazel, N. Botanical agents for the treatment of nonmelanoma skin cancer. *Dermatol. Res. Pract.* **2013**, *2013*, doi:10.1155/2013/837152.
3. Fisher G.J.; Datta S.C.; Talwar, H.S.; Wang, Z.Q.; Varani, J.; Kang, S; Voorhes J.J. Molecular basis of sun-induced premature skin ageing and retinoid antagonism. **1996**, *379*, 335–339.
4. Yaar, M.; Gilchrest, B.A. Photoageing: Mechanism, prevention and therapy. *Br. J. Dermatol.* **2007**, *157*, 874–887.
5. Bachelor, M.A.; Bowden, G.T. UVA-mediated activation of signaling pathways involved in skin tumor promotion and progression. *Sem. Cancer Biol.* **2004**, *14*, 131–138.
6. Dreher, D.; Junod, A.F. Role of oxygen free radicals in cancer development. *Eur. J. Cancer* **1996**, *32A*, 30–38.
7. Hussain, S.P.; Hofseth, L.J.; Harris, C.C. Radical causes of cancer. *Nat. Rev. Cancer* **2003**, *3*, 276–285.
8. Sander, C.S.; Chang, H.; Hamm, F.; Elsner, P.; Thiele, J.J. Role of oxidative stress and antioxidant network in cutaneous carcinogenesis. *Int. J. Dermatol.* **2004**, *43*, 326–335.
9. Perchellet, J.P.; Perchellet, E.M.; Gali, H.U.; Gao, X.M. Oxidant stress and multistage skin carcinogenesis. In *Skin Cancer: Mechanisms and Human Relevance*; CRC Press: Boca Raton, FL, USA, 1995; pp. 145–180.

10. Hofseth, L.J.; Hussain, S.P.; Wogan, G.N.; Harris, C.C. Nitric oxide in cancer and chemoprevention. *Free Radic. Biol. Med.* **2003**, *34*, 955–968.
11. Kawanishi, S.; Hiraku, Y.; Oikawa, S. Mechanism of guanine-specific DNA damage by oxidative stress and its role in carcinogenesis and aging. *Mutat. Res.* **2001**, *488*, 65–76.
12. Athar, M.; Lloyd, J.R.; Bickers, D.R.; Mukhtar, H. Malignant conversion of UV radiation and chemically induced mouse skin benign rumors by free-radical-generating compounds. *Carcinogenesis* **1989**, *10*, 1841–1845.
13. Lelkes, P.I.; Hahn, K.L.; Sukovich, D.A.; Karmiol, S.; Schmidt, D.H. On the possible role of reactive oxygen species in angiogenesis. *Adv. Exp. Med. Biol.* **1998**, *454*, 295–310.
14. Lippke, J.A.; Gordon, L.K.; Brash, D.E.; Haseltine, W.A. Distribution of UV light-induced damage in a defined sequence of human DNA: Detection of alkaline-sensitive lesions at pyrimidine nucleoside-cytidine sequences. *Proc. Natl. Acad. Sci. USA* **1981**, *78*, 3388–3392.
15. Mitchell, D.L.; Nairn, R.S. The biology of the (6–4) photoproduct. *Photochem. Photobiol.* **1989**, *49*, 805–819.
16. Nishigori, C. Cellular aspects of photocarcinogenesis. *Photochem. Photobiol. Sci.* **2006**, *5*, 208–214.
17. Einspahr, J.G.; Alberts, D.S.; Warneke, J.A.; Bozzo, P.; Basye J.; Grogan, T.M.; Nelson, M.A.; Bowden, G.T. Relationship of p53 mutations to epidermal cell proliferation and apoptosis in human UV-induced skin carcinogenesis. *Neoplasia* **1999**, *1*, 468–475.
18. Cross, S.M.; Sanchez, C.A.; Morgan, C.A.; Schimke, M.K.; Ramel, S.; Idzerda, R.L.; Raskind, W.H.; Reid, B.J. A p53-dependent mouse spindle checkpoint. *Science* **1995**, *267*, 1353–1356.
19. Shindo, Y.; Hashimoto, T. Ultraviolet B-induced cell death in four cutaneous cell lines exhibiting different enzymatic antioxidant defences: Involvement of apoptosis. *J. Dermatol. Sci.* **1998**, *17*, 140–150.
20. Lodish, H.; Berk, A.; Kaiser, C.A.; Krieger, M.; Scott, M.P.; Bretscher, A. *Molecular Cell Biology*, 6th ed.; W.H. Freeman and Company: New York, NY, USA, 2008.
21. Magné, N.; Toillon, R.A.; Bottero, V.; Didelot, C.; Houtte, P.V.; Gérard, J.P.; Peyron, J.F. NF-κB modulation and ionizing radiation: Mechanisms and future directions for cancer treatment. *Cancer Lett.* **2006**, *231*, 158–168.
22. Birch-Machin, M.A.; Swalwell, H. How mitochondria record the effects of UV exposure and oxidative stress using human skin as a model tissue. *Mutagenesis* **2010**, *25*, 101–107.
23. Shimmura, S.; Tsubota K. Ultraviolet B-induced mitochondrial dysfunction is associated with decreased cell detachment of corneal epithelial cells *in vitro*. *Investig. Ophthalmol. Vis. Sci.* **1997**, *38*, 620–626.
24. Denning, M.F.; Wang, Y.; Tibudan, S.; Alkan, S.; Nickoloff, B.J.; Qin, J.Z. Caspaseactivation and disruption of mitochondrial membrane potential during UV radiation induced apoptosis of human keratinocytes requires activation of protein kinase C. *Cell Death Differ.* **2002**, *9*, 40–52.
25. Gniadecki, R.; Thorn, T.; Vicanova, J.; Petersen, A.; Wulf, H.C. Role of mitochondria in ultraviolet-induced oxidative stress. *J Cell Biochem.* **2000**, *80*, 216–222.
26. Gorrini, C.; Harris, I.S.; Mak, T.W. Modulation of oxidative stress as an anticancerstrategy. *Nat. Rev. Drug Discov.* **2013**, *12*, 931–947.

27. Holmström, K.M.; Baird, L.; Zhang, Y.; Hargreaves, I.; Chalasani, A.; Land, J.M.; Stanyer, L.; Yamamoto, M.; Dinkova-Kostova, A.T.; Abramov, A.Y. Nrf2 impacts cellular bioenergetics by controlling substrate availability for mitochondrial respiration. *Biol. Open* **2013**, *2*, 761–770.
28. Callaghan, T.M.; Wilhelm, K.P. A review of ageing and an examination of clinical methods in the assessment of ageing skin. *Int. J. Cosmet. Sci.* **2008**, *30*, 323–332.
29. Tulah, A.S.; Birch-Machin, M.A. Stressed out mitochondria: The role of mitochondria in ageing and cancer focussing on strategies and opportunities in human skin. *Mitochondrion* **2013**, *13*, 444–453.
30. Grossman, D.; Leffell, D.J. The molecular basis of nonmelanoma skin cancer: New understanding. *Arch. Dermatol.* **1997**, *133*, 1263–1270.
31. Hofseth, L.J.; Saito, S.; Hussain, S.P.; Espey, M.G.; Miranda, K.M.; Araki, Y.; Jahppan, C.; Higashimoto, Y.; He, P.; Linke, S.P.; *et al.* Nitric oxide-induced cellular stress and p53 activation in chronic inflammation. *Proc. Natl. Acad. Sci. USA* **2003**, *100*, 143–148.
32. Athar, M.; An, K.P.; Morel, K.D.; Kim, A.L.; Aszterbaum, M.; Longley, J. Ultraviolet B(UVB)-induced cox-2 expression in murine skin: An immunohistochemical study. *Biochem. Biophys. Res. Commun.* **2001**, *280*, 1042–1047.
33. Buckman, S.Y.; Gresham, A.; Hale, P.; Hruza, G.; Anast, J.; Masferrer, J.; Pentland, A.P. COX-2 expression is induced by UVB exposure in human skin: Implications for the development of skin cancer. *Carcinogenesis* **1998**, *19*, 723–729.
34. Wang, D.; Dubois, R.N. Prostaglandins and cancer. *Gut* **2006**, *55*, 115–122.
35. Einspahr, J.G.; Xu, M.J.; Warneke, J.; Saboda, K.; Ranger-Moore, J.; Bozzo, P. Reproducibility and expression of skin bio markers in sun-damaged skin and actinic keratoses. *Cancer Epidemiol. Biomark. Prev.* **2006**, *15*, 1841–1848.
36. Maeda, H.; Akaike, T. Nitric oxide and oxygen radicals in infection, inflammation, and cancer. *Biochemistry* **1998**, *637*, 854–865.
37. Coussens, L.M.; Werb, Z. Inflammatory cells and cancer: Think different! *J. Exp. Med.* **2001**, *193*, F23–F26.
38. Derynck, R.; Akhurst, R.J.; Balmain, A. TGF-β signaling in tumor suppression and cancer progression. *Nat. Genet.* **2001**, *29*, 117–129.
39. Ullrich, S.E. Mechanisms underlying UV-induced immune suppression. *Mutat. Res.* **2005**, *571*, 185–205.
40. Simon, J.C.; Tigelaar, R.E.; Bergstresser, P.R.; Edelbaum, D.; Cruz, P.D., Jr. Ultraviolet B radiation converts Langerhans cells from immunogenic to tolergenic antigen-presenting cells. *J. Immunol.* **1991**, *146*, 485–491.
41. Norval, M.; Simpson, T.J.; Ross, J.A. Urocanic acid and immunosuppression. *Photochem. Photobiol.* **1989**, *50*, 267–275.
42. Noonan, F.P.; de Fabo, E.C. Immunosuppression by ultraviolet B radiation: Initiation by urocanic acid. *Immunol. Today* **1992**, *13*, 250–254.
43. Kurimoto, I.; and Streilein, J.W. Deleterious effects of *cis*-urocanic acid and UVB radiation on Langerhans cells and on induction of contact hypersensitivity are mediated by tumor necrosis factor-alpha. *J. Investig. Dermatol.* **1992**, *99*, 69S–70S.

44. Kruger-Krasagakes, S.; Krasagakis, K.; Garbe, C.; Schmitt, E.; Huls, C.; Blankenstein, T.; Diamantstein, T. Expression of interleukin 10 in human melanoma. *Br. J. Cancer* **1994**, *70*, 1182–1185.
45. Baadsgaard, O.; Fox, D.A.; Cooper, K.D. Human epidermal cells from ultraviolet light-exposed skin preferentially activate autoreactive CD4+2H4+ suppressor- inducer lymphocytes and CD8+ suppressor/cytotoxic lymphocytes. *J. Immunol.* **1988**, *140*, 1738–1744.
46. Simon, J.C.; Cruz, P.D., Jr.; Bergstresser, P.R.; Tigelaar, R.E. Low dose ultraviolet B-irradiated Langerhans cells preferentially activate CD4+ cells of the T helper 2 subset. *J. Immunol.* **1990**, *145*, 2087–2091.
47. Schmitt, D.A.; Owen-Schaub, L.; Ullirch, S.E. Effect of IL-12 on immune suppression and suppressor cell induction by ultraviolet radiation. *J. Immunol.* **1995**, *154*, 5114–5120.
48. Wondrak, G.T.; Roberts, M.J.; Cervantes-Laurean, D.; Jacobson, M.K.; Jacobson, E.L. Proteins of the extracellular matrix are sensitizers of photo-oxidative stress in human skin cells. *J. Investig. Dermatol.* **2003**, *121*, 578–586.
49. Watson, R.E.; Griffiths, C.E.; Craven, N.M.; Shuttleworth, C.A.; Kielty, C.M. Fibrillin-rich microfibrils are reduced in photo aged skin. Distribution at the dermal-epidermal junction. *J. Investig. Dermatol.* **1999**, *112*, 782–787.
50. Philips, N.; Burchill, D.; O'Donoghue D.; Keller, T.; Gonzalez, S. Identification of benzene metabolites in dermal fibroblasts as nonphenolic: Regulation of cell viability, apoptosis, lipid peroxidation and expression of matrix metalloproteinase 1 and elastin by benzene metabolites. *Skin Pharmacol. Physiol.* **2004**, *17*, 147–152.
51. Philips, N.; Hwang, H.; Chauhan, S.; Leonardi, D.; Gonzalez, S. Stimulation of cell proliferation and expression of matrixmetalloproteinase-1 and interleukin-8 genes in dermal, fibroblasts by copper. *Connect. Tissue Res.* **2010**, *51*, 224–229.
52. Philips, N.; Tuason, M.; Chang, T.; Lin, Y.; Tahir, M.; Rodriguez, S.G. Differential effects of ceramide on cell viability and extracellular matrix remodeling in keratinocytes and fibroblasts. *Skin Pharmacol. Physiol.* **2009**, *22*, 151–157.
53. Herouy, Y. Matrix metalloproteinases in skin pathology (Review). *Int. J. Mol. Med.* **2001**, *7*, 3–12.
54. Yan, C.; Boyd, D.D. Regulation of matrix metalloproteinase gene expression. *J. Cell. Physiol.* **2007**, *211*, 19–26.
55. Holick, M.F. Sunlight, UV-radiation, vitamin D and skin cancer: How much sunlight do we need? *Adv. Exp. Med. Biol.* **2008**, *624*, 1–15.
56. Kligman, L.H.; Akin, F.J.; Kligman, A.M. Sunscreens prevent ultraviolet photocarcinogenesis. *J. Am. Acad. Dermatol.* **1980**, *3*, 30–35.
57. Forbes, P.D. Photocarcinogenesis: An overview. *J. Investig. Dermatol.* **1981**, *77*, 139–143.
58. Elmets, C.A.; Anderson, C.Y. Sunscreens and photocarcinogenesis: An objective assessment. *Photochem. Photobiol.* **1996**, *63*, 435–440.
59. Stege, H. Effect of xenogenic repair enzymes on photoimmunology and photocarcinogenesis. *J. Photochem. Photobiol.* **2001**, *65*, 105–108.

60. Emanuele, E.; Altabas, V.; Altabas, K.; Berardesca, E. Topical application of preparations containing DNA repair enzymes prevents ultraviolet-induced telomere shortening and c-FOS proto-oncogene hyperexpression in human skin: An experimental pilot study. *J. Drugs Dermatol.* **2013**, *12*, 1017–1021.
61. Weber, S. Light-driven enzymatic catalysis of DNA repair: A review of recent biophysical studies on photolyase. *Biochimica et Biophysica Acta* **2005**, *1707*, 1–23.
62. Zahid, S.; Brownell, I. Repairing DNA damage in xeroderma pigmentosum: T4N5 lotion and gene therapy. *J. Drugs Dermatol.* **2008**, *7*, 405–408.
63. Polefka, T.G.; Meyer, T.A.; Agin, P.P.; Bianchini, R.J. Cutaneous oxidative stress. *J. Cosmet. Dermatol.* **2012**, *11*, 55–64.
64. Sander, C.S.; Hamm, F.; Elsner, P.; Thiele, J.J. Oxidative stress in malignant melanoma and non-melanoma skin cancer. *Br. J. Dermatol.* **2003**, *148*, 913–922.
65. Mathews-Roth, M.M.; Pathak, M.A.; Parrish, J.; Fitzpatrick, T.B.; Kass, E.H.; Toda, K.; Clemens, W. A clinical trial of the effects of oral β-carotene on the responses of human skin to solar radiation. *J. Investig. Dermatol.* **1972**, *59*, 349–353.
66. Greenberg, E.R.; Baron, J.A.; Stukel, T.A.; Stevens, M.M.; Mandel, J.S.; Spencer, S.K.; Elisa, P.M.; Lowe, N.; Nierenberg, D.W.; Bsyrd, G. A clinical trial of beta carotene to prevent basal-cell and squamous-cell cancers of the skin. *N. Eng. J. Med.* **1990**, *323*, 789–795.
67. Werninghaus, K.; Meydani, M.; Bhawan, J.; Margolis, R.; Blumberg, J.B.; Gilchrest, B.A. Evaluation of the photoprotective effect of oral vitamin E supplementation. *Arch. Dermatol.* **1994**, *130*, 1257–1261.
68. Lin, F.H.; Lin, J.Y.; Gupta, R.D.; Tournas, J.A.; Burch, J.A.; Selim, M.A.; Monterio-Rivire N.A.; Grichnik, J.M.; Zielinski, J.; Pinnell, S.R. Ferulic acid stabilizes a solution of vitamins C and E and doubles its photoprotection of skin. *J. Investig. Dermatol.* **2005**, *125*, 826–832.
69. Fuchs, J. Potentials and limitations of the natural antioxidants RRR-α-tocopherol, L-ascorbic acid and β-carotene in cutaneous photoprotection. *Free Radic. Biol. Med.* **1998**, *25*, 848–873.
70. De Gálvez, M.V. Antioxidantes en fotoprotección, ¿realmente funcionan? *Actas Dermo Sifiliogr.* **2010**, *101*, 197–200.
71. Kraemer, K.H.; DiGiovanna, J.J.; Moshell, A.N.; Tarone, R.E.; Peck, G.L. Prevention of skin cancer in xeroderma pigmentosum with the use of oral isotretinoin. *N. Eng. J. Med.* **1988**, *318*, 1633–1637.
72. De Graaf, Y.G.; Euvrard, S.; Bouwes Bavinck, J.N. Prevention of skin cancer in xeroderma pigmentosum with the use of oral isotretinoin. *Dermatol. Surg.* **2004**, *304*, 656–661.
73. Kalil, C.L.; Fachinello, F.Z.; Lamb, F.M.; Comunello, L.N. Use of oral isotretinoin in photoaging therapy. *Skinmed* **2008**, *7*, 10–14.
74. Marks, F.; Furstenberger, G. Cancer chemoprevention through interruption of multistage carcinogenesis. The lessons learnt by comparing mouse skin carcinogenesis and human large bowel cancer. *Eur. Cancer* **2000**, *36*, 314–329.
75. Peterson, S.R.; Goldberg, L.H. New and emerging treatments for nonmelanomas and actinic keratoses. *J. Drugs Dermatol.* **2003**, *2*, 429–432.

76. Tripp, C.S.; Blomme, E.A.; Chinn, K.S.; Hardy, M.M.; LaCelle, P.; Pentland, A.P. Epidermal COX-2 induction following ultraviolet irradiation: Suggested mechanism for the role of COX-2 inhibition in photoprotection. *J. Investig. Dermatol.* **2003**, *121*, 853–861.
77. Elmets, C.A.; Viner, J.L.; Pentland, A.P.; Cantrell, W.; Lin, H.Y.; Bailey, H. Chemoprevention of nonmelanoma skin cancer with celecoxib: A randomized, double-blind, placebo-controlled trial. *J. Natl. Cancer Inst.* **2010**, *102*, 1835–1844.
78. Katiyar, S.K.; Mukhtar, H. Green tea polyphenol (−)-epigallocatechin-3-gallate treatment to mouse skin prevents UVB-induced infiltration of leukocytes, depletion of antigen-presenting cells, and oxidative stress. *J. Leukoc. Biol.* **2001**, *69*, 719–726.
79. Sharma, S.D.; Katiyar, S.K. Dietary grape seed proanthocyanidins inhibit UVB-induced cyclooxygenase-2 expression and other inflammatory mediators in UVB-exposed skin and skin tumors of SKH-1 hairless mice. *Pharm. Res.* **2010**, *27*, 1092–1102.
80. Gensler, H.L.; Williams, T.; Huang, A.C.; Jacobson, E.L. Oral niacin prevents photocarcinogenesis and photoimmunosuppression in mice. *Nutr. Cancer* **1999**, *34*, 36–41.
81. Mulero, M.; Rodriguez-Yanes, E.; Nogues, M.R.; Giralt, M.; Romeu, M.; Gonzalez, S. *Polypodium leucotomos* extract inhibits glutathione oxidation and prevents Langerhans cell depletion induced by UVB/UVA radiation in a hairless rat model. *Exp. Dermatol.* **2008**, *17*, 653–658.
82. Philips, N.; Conte, J.; Chen, Y.J.; Natrajan, P.; Taw, M.; Keller, T. Beneficial regulation of matrixmetalloproteinases and their inhibitors, fibrillar collagens and transforming growth factor-beta by *Polypodium leucotomos*, directly or in dermal fibroblasts, ultraviolet radiated fibroblasts, and melanoma cells. *Arch. Dermatol. Res.* **2009**, *301*, 487–495.
83. Afaq, F.; Katiyar, S.K. Polyphenols: Skin photoprotection and inhibition of photocarcinogenesis. Mini reviews in medicinal chemistry. *Mini Rev. Med. Chem.* **2011**, *11*, 1200–1215.
84. Nichols, J.A.; Katiyar, S.K. Skin photoprotection by natural polyphenols: Anti-inflammatory, antioxidant and DNA repair mechanisms. *Arch. Dermatol. Res.* **2010**, *302*, 71–83.
85. Mnich, C.D.; Hoek, K.S.; Virkki, L.V.; Farkas, A.; Dudli, C.; Laine, E.; Urosevik, M.; Dummer, R. Green tea extract reduces induction of p53 and apoptosis in UVB-irradiated human skin independent of transcriptional controls. *Exp. Dermatol.* **2009**, *18*, 69–77.
86. Wei, H.; Saladi, R.; Lu, Y.; Wang, Y.; Palep, S.R.; Moore, J.; Phelps, R.; Shyong, E.; Lebwohl, M.G. Isoflavone genistein: Photoprotection and clinical implications in dermatology. *J. Nutr.* **2003**, *133*, 3811S–3819S.
87. Katiyar, S.K.; Meleth, S.; Sharma, S.D. Silymarin, a flavonoid from milk thistle (*Silybum marianum* L.), inhibits UV-induced oxidative stress through targeting infiltrating CD11b+ cells in mouse skin. *Photochem. Photobiol.* **2008**, *84*, 266–271.
88. Mittal, A.; Elmets, C.A.; Katiyar, S.K. Dietary feeding of proanthocyanidins from grape seeds prevents photocarcinogenesis in SKH-1 hairless mice:relationship to decreased fat and lipid peroxidation. *Carcinogenesis* **2003**, *24*, 1379–1388.
89. Li, W.G.; Zhang, X.Y.; Wu, Y.J.; Tian, X. Anti-inflammatory effect and mechanism of proanthocyanidins from grape seeds. *Acta Pharmacol. Sin.* **2001**, *22*, 1117–1120.
90. Pluemsamran, T.; Onkoksoong, T.; Panich, U. Caffeic acid and ferulic acid inhibit UVA-induced matrix metalloproteinase-1 through regulation of antioxidant defense system in keratinocyte HaCaT cells. *Photochem. Photobiol.* **2012**, *88*, 961–968.

91. Aziz, M.H.; Reagan-Shaw, S.; Wu, J.; Longley, B.J.; Ahmad, N. Chemoprevention of skin cancer by grape constituent resveratrol: Relevance to human disease. *FASEB J.* **2005**, *19*, 1193–1195.
92. Caddeo, C.; Teskac, K.; Sinico, C.; Kristl, J. Effect of resveratrol incorporated in liposomes on proliferation and UV-B protection of cells. *Int. J. Pharm.* **2008**, *363*, 183–191.
93. Elmets, C.A.; Singh, D.; Tubesing, K.; Matsui, M.; Katiyar, S.; Mukhtar, H. Cutaneous photoprotection from ultraviolet injury by green tea polyphenols. *Am. Acad. Dermatol.* **2001**, *44*, 425–432.
94. Zheng, W.; Doyle, T.J.; Kushi, L.H.; Sellers, T.A.; Hong, C.P.; Folsom, A.R. Tea consumption and cancer incidence in a prospective cohort study of postmenopausal women. *Am. J. Epidemiol.* **1996**, *144*, 175–182.
95. F'Guyer, S.; Afaq, F.; Mukhtar, H. Photochemoprevention of skin cancer by botanical agents. *Photodermatol. Photoimmunol. Photomed.* **2003**, *19*, 56–72.
96. Moore, J.O.; Wang, Y.; Stebbins, W.G.; Gao, D.; Zhou, X.; Phelps, R. Photoprotective effect of isoflavone genistein on ultraviolet B-induced pyrimidine dimer formation and PCNA expression in human reconstituted skin and its implications in dermatology and prevention of cutaneous carcinogenesis. *Carcinogenesis* **2006**, *27*, 1627–1635.
97. Huang, C.C.; Hsu, B.Y.; Wu, N.L.; Tsui, W.H.; Lin, T.J.; Su, C.C. Anti-photoaging effects of soy isoflavone extract (aglycone and acetylglucoside form) from soybean cake. *Int. J. Mol. Sci.* **2010**, *11*, 4782–4795.
98. Afaq, F.; Khan, N.; Syed, D.N.; Mukhtar, H. Oral feeding of pomegranate fruit extract inhibits early biomarkers of UVB radiation-induced carcinogenesis in SKH-1 hairless mouse epidermis. *Photochem. Photobiol.* **2010**, *86*, 1318–1326.
99. Koshihara, Y.; Neichi, T.; Murota, S.; Lao, A.; Fujimoto, Y.; Tatsuno, T. Caffeic acid is a selective inhibitor for leukotriene biosynthesis. *Biochimica et Biophysica Acta* **1984**, *792*, 92–97.
100. Gonzalez, S.; Gilaberte, Y.; Philips, N. Mechanistic insights in the use of a *Polypodium leucotomos* extract as an oral and topical photoprotective agent. *Photochem. Photobiol. Sci.* **2010**, *9*, 559–563.
101. Gonzalez, S.; Gilaberte, Y.; Philips, N.; Juarranz, A. Fernblock, a nutriceutical with photoprotective properties and potential preventive agent for skin photoaging and photoinduced skin cancers. *Int. J. Mol. Sci.* **2011**, *12*, 8466–8475.
102. Janczyk, A.; Garcia-Lopez, M.A.; Fernandez-Penas, P.; Alonso-Lebrero, J.L.; Benedicto, I.; Lopez-Cabrera, M.A. *Polypodium leucotomos* extract inhibits solar-simulated radiation-induced TNF-α and iNOS expression, transcriptional activation and apoptosis. *Exp. Dermatol.* **2007**, *16*, 823–829.
103. Alonso-Lebrero, J.L.; Domínguez-Jiménez, C.; Tejedor, R.; Brieva, A.; Pivel, J.P. Photoprotective properties of a hydrophilic extract of the fern *Polypodium leucotomos* on human skin cells. *J. Photochem. Photobiol.* **2003**, *70*, 31–37.
104. Mittelbrunn, M.; Tejedor, R.; de la Fuente, H.; Garcia-Lopez, M.A.; Alonso-Lebrero, J.L.; Pivel J.P.; Gonzalez S.; Gonzalez-Amaro R.; Sanchez–Madrid, F. *Polypodium leucotomos* induces pretection of UV-induced apoptosis in human skin cells. The Society of Investigative Dermatology. *J. Investig. Dermatol.* **2005**, *125*, 334–342.

105. Middelkamp-Hup, M.A.; Pathak, M.A.; Parrado, C.; Garcia-Caballero, T.; Rius-Diaz, F.; Fitzpatrick, T.B. Orally administered *Polypodium leucotomos* extract decreases psoralen-UVA-induced phototoxicity, pigmentation, and damage of human skin. *J. Am. Acad. Dermatol.* **2004**, *50*, 41–49.
106. Capote, R.; Alonso-Lebrero, J.L.; Garcia, F.; Brieva, A.; Pivel, J.P.; Gonzalez, S. *Polypodium leucotomos* extract inhibits *trans*-urocanic acid photoisomerization photodecomposition. *J. Photochem. Photobiol.* **2006**, *82*, 173–179.
107. Gonzalez, S.; Pathak, M.A.; Cuevas, J.; Villarrubia, V.G.; Fitzpatrick, T.B. Topical or oral administration with an extract of *Polypodium leucotomos* prevents acute sunburn and psoralen-induced phototoxic reactions as well as depletion of Langerhans cells in human skin. *Photodermatol. Photoimmunol. Photomed.* **1997**, *13*, 50–60.
108. Middel kamp-Hup, M.A.; Pathak, M.A.; Parrado, C.; Goukassian, D.; Rius-Diaz, F.; Mihm, M.C. Oral *Polypodium leucotomos* extract decreases ultraviolet-induced damage of human skin. *J. Am. Acad. Dermatol.* **2004**, *51*, 910–918.
109. Philips, N.; Auler, S.; Hugo, R.; Gonzalez, S. Beneficial regulation of matrix metalloproteinases for skin health. *Enzym. Res.* **2011**, *2011*, doi:10.4061/2011/427285.
110. Frieling, U.M.; Schaumberg, D.A.; Kupper, T.S.; Muntwyler, J.; Hennekens, C.H. A randomized, 12-year primary-prevention trial of β carotene supplementation for nonmelanoma skin cancer in the physician's health study. *Arch. Dermatol.* **2000**, *136*, 179–184.
111. Rizwan, M.; Rodriguez-Blanco, I.; Harbottle, A.; Birch-Machin, M.A.; Watson, R.E.; Rhodes, L.E. Tomato paste rich in lycopene protects against cutaneous photodamage in humans *in vivo*: A randomized controlled trial. *Br. J. Dermatol.* **2011**, *164*, 154–162.
112. Song, F.; Qureshi, A.A.; Han, J. Increased caffeine intake is associated with reduced risk of basal cell carcinoma of the skin. *Cancer Res.* **2012**, *72*, 3282–3289.
113. Lou, Y.; Peng, Q.; Li, T.; Nolan, B.; Bernard, J.J.; Wagner, G.C.; Lin, Y.; Shih, W.J.; Conny, A.H.; Lou, Y.P. Oral caffeine during voluntary exercise markedly inhibits skin carcinogenesis and decreases inflammatory cytokines in UVB-treated mice. *Nutr. Cancer* **2013**, *65*, 1002–1013.
114. Abel, E.L.; Hendrix, S.O.; McNeeley, S.G.; Johnson, K.C.; Rosenberg, C.A.; Mossavar-Rahmani, Y.; Vitolins. M.; Kruger, M. Daily coffee consumption and prevalence of nonmelanoma skin cancer in Caucasian women. *Eur. J. Cancer Prev.* **2007**, *16*, 446–452.
115. Han, W.; Ming, M.; He, Y.Y. Caffeine promotes ultraviolet B-induced apoptosis in human keratinocytes without complete DNA repair. *J. Biol. Chem.* **2011**, *286*, 22825–22832.
116. Talalay, P.; Fahey, J.W.; Healy, Z.R.; Wehage, S.L.; Benedict, A.L.; Min, C. Sulforaphane mobilizes cellular defenses that protect skin against damage by UV radiation. *Proc. Natl. Acad. Sci. USA* **2007**, *104*, 17500–17505.
117. Dickinson, S.E.; Melton, T.F.; Olson, E.R.; Zhang, J.; Saboda, K; Bowden, G.T. Inhibition of activator protein-1 by sulforaphane involves interaction with cysteine in the cFos DNA-binding domain: Implications for chemoprevention of UVB-induced skin cancer. *Cancer Res.* **2009**, *69*, 7103–7110.
118. Zhou, B.R.; Liu, W.L.; Luo, D. Protective effect of baicalin against multiple ultraviolet B exposure-mediated injuries in C57BL/6 mouse skin. *Arch. Pharm. Res.* **2011**, *34*, 261–268.

119. Kimura, Y.; Sumiyoshi, M. French maritime pine bark (*Pinus maritima* Lam.) extract (Flavangenol) prevents chronic UVB radiation-induced skin damage and carcinogenesis in melanin-possessing hairless mice. *Photochem. Photobiol.* **2010**, *86*, 955–963.
120. Duncan, F.J.; Martin, J.R.; Wulff, B.C.; Stoner, G.D.; Tober, K.L.; Oberyszyn, T.M.; Kusevitt, D.F.; van Buskirk, A.M. Topical treatment with black raspberry extract reduces cutaneous UVB-induced carcinogenesis and inflammation. *Cancer Prev. Res.* **2009**, *2*, 665–672.
121. Da Silva, V.V.; Ropke, C.D.; Miranda, D.V.; de Almeida, R.L.; Sawada, T.C.; Rivelli, D.P. Photoprotective effect of Pothomorphe umbellata on UVB radiation-induced biomarkers involved in carcinogenesis of hairless mouse epidermis. *Cutan. Ocul. Toxicol.* **2009**, *28*, 54–60.
122. Hwang, H.; Chen, T.; Nines, R.G.; Shin, H.C.; Stoner, G.D. Photochemoprevention of UVB-induced skin carcinogenesis in SKH-1 mice by brown algae polyphenols. *Int. J. Cancer* **2006**, *119*, 2742–2749.
123. Schempp, C.M.; Winghofer, B.; Müller, K.; Schulte-Mönting, J.; Mannel, M.; Schöpf, E.; Simon, J.C. Effect of oral administration of *Hypericum perforatum* extract (St. John's Wort) on skin erythema and pigmentation induced by UVB, UVA, visible light and solar simulated radiation. *Phytother. Res.* **2003**, *17*, 141–146.

15

Discrimination between Leave of *Apocynum venetum* and Its Adulterant, *A. pictum* Based on Antioxidant Assay and Chemical Profiles Combined with Multivariate Statistical Analysis

Chi-On Chan [1,2], Ching-Ching Lau [2], Yam-Fung Ng [2], Li-Jia Xu [3], Si-Bao Chen [1,2,3,*], Shun-Wan Chan [1,2] and Daniel Kam-Wah Mok [1,2,*]

1 State Key Laboratory of Chinese Medicine and Molecular Pharmacology, Department of Applied Biology and Chemical Technology, The Hong Kong Polytechnic University, Shenzhen 518057, China; E-Mails: on.chan@polyu.edu.hk (C.-O.C.); sw.chan@polyu.edu.hk (S.-W.C.)

2 Department of Applied Biology and Chemical Technology, The Hong Kong Polytechnic University, Hong Kong, China; E-Mails: christy.lau@connect.polyu.hk (C.-C.L.); yam-fung@hotmail.com (Y.-F.N.)

3 The Institute of Medicinal Plant Development, Chinese Academy of Medical Sciences and Peking Union Medical College, Beijing, 100193, China; E-Mail: xulijia@hotmail.com

* Authors to whom correspondence should be addressed;
E-Mails: sibao.chen@ polyu.edu.hk (S.-B.C.); daniel.mok@polyu.edu.hk (D.K.-W.M.)

Academic Editors: Antonio Segura-Carretero and David Arráez-Román

---

Abstract: An integrated approach including chemical and biological assessments was developed to investigate the differences between *Apocynum venetum* L. (AV) and its adulterant, *Apocynum pictum* Schrenk (AP). Ten flavonoids were tentatively identified by ultra-visible and mass spectra data. The chemical component, hyperoside, was identified as a critical parameter for discrimination of two species from the results of principal component analysis (PCA) and quantitative analysis. The anti-oxidative power of the herbal extracts were determined using 2,2-diphenyl-1-(2,4,6-trinitrophenyl) hydrazyl (DPPH) assay and H_2O_2-induced cell damage on LO2 cells. The results of the biological assays suggested that the chemical differences between AV and AP do lead to difference in activity and AV is demonstrated to have higher anti-oxidant activity.

Keywords: *Apocynum venetum* L.; *Apocynum pictum* Schrenk; HPLC fingerprints; free radical scavenging capacity

---

1. Introduction

Traditional Chinese medicines (TCMs) have attracted global attention for their various applications in therapeutic agents or diet supplements; the increasing trend of consuming TCMs has resulted in a serious resource problem. Therefore, related plant species from the same genus are often used as substitute in many folk medicine practices assuming similar species usually contain the same classes of compounds and similar therapeutic activities. Although the species from same genus may contain the same classes of compounds, the detailed chemical components present are very often different. A comparison of their chemical composition and pharmacological activities is certainly needed [1] to support the use of an alternative species. In this study, the chemical composition and biological activities of two related species *Apocynum venetum* L. (AV) and *A. pictum* Schrenk (AP) are studied to compare their similarity and differences in chemical composition and biological activities.

Apocynum venetum L. (AV), (*Luobuma* in Chinese, *Apocynaceae*), is a perennial herbaceous or half-shrub plant that grows ubiquitously in central to northwestern China [2]. Its leave have a long history as a Chinese traditional medicine with uses to calm the liver, soothe the nerves, dissipate heat, and promote dieresis [3]. Previous studies reported that AV extract exhibits a series of pharmacological activities, such as anti-oxidation [4,5], anti-hypertensive [6,7] and anti-depressant effects [8,9], cholesterol-lowering [10] and AGEs-inhibitory activity [5], as well as its main active fractions were found to be phenolic acid, flavones and flavan-3-ols components [11,12]. As a result, AV has been applied and appeared to be a popular daily beverage in China, Japan and the US [13]. Various formulations of this herb, including tea bags, powder, and crimped and curled roasted leaves, have recently been commercialized as a sedative and anti-aging supplement, which has become increasingly popular in North American and East Asian health food markets [14].

A. pictum (AP), an easily confusable herb for AV, is similar to AV in terms of plant morphological characteristics and geographical distribution. Generally, AP is known locally as White Hemp, while AV is known as Red Hemp, probably due to the different colors and shapes of their flowers and leaves. The industrial applications of AP are limited to fibers for spinning and papermaking rather than medicinal purposes [13]. Due to the excessive exploitation, wild AV has dropped in recent years even though some protective measures have been enacted [15]. Because of the scarcity of AV, and since AP has a rather similar appearance to AV, AP is very often used as a substitute in the market. In order to distinguish both species and guarantee the quality of products derivated from this herb, the present study establishes an integrative approach to investigate the differences of chemical constituents and antioxidant activities between AV and AP. High performance liquid chromatography linked with diode array detection and mass spectrometer was applied to identify the flavonoids presented in AV and AP. The anti-oxidative capacity of the extracts was also determined using chemical (DPPH test) and biological (*in vitro* H_2O_2-induced cell model) methods.

2. Experimental Section

2.1. Chemicals and Reagents

Reference standards of hyperoside and isoquercetrin, (purity > 98%) were obtained from the National Institute for the Control of Pharmaceutical and Biological Products (Beijing, China). Analytical grade formic acid and trifluoroacetic acid were purchased from Wako (Japan) and Aldrich (USA), respectively, while HPLC grade acetonitrile was obtained from Tedia (USA). Double deionized water was prepared by a Milli-Q water-purification system (Millipore, MA, USA). Dulbecco's Modified Eagle Medium (DMEM), fetal bovine serum (FBS), Phosphate buffer saline (PBS), antibiotics, antimycotics, and trypsin were purchased from Invitrogen (Carlsbad, CA, USA). 2,2-diphenyl-1-(2,4,6-trinitrophenyl) hydrazyl (DPPH) and ascorbic acid were obtained from Sigma Chemicals (St. Louis, MO, USA). 3-(4,5-dimethylthiazol-2yl)-2,5-diphenyltetrazolium Bromide (MTT) was purchased from Invitrogen (Grand Island, NY, USA).

2.2. Sample Collection and Authentication

All samples were collected from different growing locations and Hong Kong drug stores. A list of sample collection information is shown in Table 1. Samples from different growing locations (AV5-AV10, AP1-AP2) were authenticated based on herbarium specimen (deposited in the Herbarium of State Key Laboratory of Chinese Medicine and Molecular Pharmacology, Shenzhen, China) and their morphological and microscopic characteristics; while those from Hong Kong drug stores were authenticated based only on their morphological and microscopic characteristics. All samples were identified as the leaves of *Apocynum venetum* L. or *A. pictum* Schrenk.

Table 1. A list of *Apocynum venetum* and *Apocynum pictum* sample collected for this study.

Sample Codes	Geographical/Collection Sources	Morphological and Microscopic Characteristics			Species [b]
		Lateral Vein	Texture	Papillae of Epidermal Cell [a]	
AV1	Hong Kong Drug stores	distinct	thin, fragile	–	AV
AV2	Hong Kong Drug stores	distinct	thin, fragile	–	AV
AV3	Hong Kong Drug stores	distinct	thin, fragile	–	AV
AV4	Hong Kong Drug stores	distinct	thin, fragile	–	AV
AV5	Tianjin province, China	distinct	thin, fragile	–	AV
AV6	Hebei province, China	distinct	thin, fragile	–	AV
AV7	Shanxi province, China	distinct	thin, fragile	–	AV
AV8	Shanxi province, China	distinct	thin, fragile	–	AV
AV9	Shanxi province, China	distinct	thin, fragile	–	AV
AV10	Shanxi province, China	distinct	thin, fragile	–	AV
AP1	Xinjiang province, China	indistinct	thick, hard	+	AP
AP2	Xinjiang province, China	indistinct	thick, hard	+	AP
AP3	Hong Kong Drug stores	indistinct	thick, hard	+	AP
AP4	Hong Kong Drug stores	indistinct	thick, hard	+	AP
AP5	Hong Kong Drug stores	indistinct	thick, hard	+	AP
AP6	Hong Kong Drug stores	indistinct	thick, hard	+	AP

[a] +: present, –: absent; [b] AV: *Apocynum venetum* L., AP: *Apocynum pictum* Schrenk.

2.3. Sample Preparation for Antioxidant Analysis

The extracts prepared for antioxidant analyses were prepared by accurately weighing 10 g fine powder of samples (AV and AP) and mixing with 100 mL distilled water. After shaking in a horizontal shaker at 37 °C, 300 rpm for 2 h, the solution was centrifuged at 3000 rpm for 30 min. The supernatant was collected, and the residue was re-extracted two more times with the same volume of distilled water. Finally, the pooled supernatant was lyophilized by freeze-dryer (LABCONCO FreeZone® 6, Houston, TX, USA). The dried extracts were stored at −20 °C before use.

2.4. Fingerprint Analysis by HPLC-DAD-MS

About 0.5 g of accurately weighed sample was sonicated with 20 mL 70% ethanol for 30 min. Then, the mixture was centrifuged at about 3000× *g* for 5 min. Finally, the sample solution was filtered through a 0.45 μm PTFE filter before HPLC-DAD-MS analysis.

Chromatographic analysis was carried out on a Agilent Zorbax column (250 mm × 4.6 mm, 5 μm, Agilent Corp., Wilmington, DE, USA) at 25 °C using an Agilent 1100 liquid chromatography system, equipped with a quaternary solvent deliver system, an auto-sampler and a DAD system. The detection wavelength was set at 360 nm. The gradient elution of the mobile phase consisting of (A) acetonitrile and (B) 0.1% (v/v) formic acid is as follows: 12%–17% of A at 0–10 min; 17% of A isocratic at 10–23 min; from 17% to 20% of A at 23–35 min; from 20% to 30% of A at 35–45 min; from 30% to 70% of A at 45–50 min; 70% of A isocratic at 50–60 min. 15 min re-equilibrium was allowed between injections. The flow rate was 1.0 mL/min and aliquots of 10 μL were injected into the HPLC.

An Agilent MSD Trap VL module mass spectrometer was connected to the Agilent 1100 HPLC instrument via an orthogonal ESI interface. The LC effluent was introduced into the ESI source in a post-column splitting 5:1. Ultra high purity helium (He) was used as the collision gas, while high-purity nitrogen (N_2) was used as the nebulizing gas. The ionization parameters of negative ion mode were set as follows: ion spray voltage 3.5 kV, nebulizing gas of nitrogen at 30.0 psi; drying gas of 9.0 L/min at 350 °C. Data acquisition was performed in the full scan mode form m/z 100 to 1500 for MS and with an accumulation time of 200 ms and 7 microscans was averaged per recorded scan.

2.5. Quantitative Analysis of Hyperoside and Isoquercetrin by HPLC-UV

Quantitative analysis of hyperoside and isoquercetrin in samples were carried out using HPLC-UV method cited in the Hong Kong Chinese Materia Medica Standard [16]. Briefly, 0.2 g sample, accurately weighed, was refluxed with 40 mL 70% methanol for 3 hours. Then, the mixture was centrifuged at about 3000× *g* for 5 min. Afterward, the residue was washed with 8 mL 70% methanol and was centrifuged at about 3000× *g* for 5 min. All extracts was combined and marked to 50 mL volumetric flask. Finally, the sample solution filtered through a 0.45 μm PTFE filter before HPLC-DAD analysis. Chromatographic analysis was carried out on a Agilent Zorbax column (250 mm × 4.6 mm, 5 μm) at 25 °C using an Agilent 1100 liquid chromatography system, equipped with a quaternary solvent deliver system, an auto-sampler and a DAD system. The detection wavelength was 254 nm. The gradient elution of the mobile phase consisting of (A) acetonitrile and (B) 0.05% (v/v) trifluoroacetic acid is as follows: 17%–18% (A) at 0–22 min; from 18% to 35% (A) at 22–30 min;

10 min re-equilibrium was allowed between injections. The flow rate was 1.0 mL/min and aliquots of 10 μL were injected into the HPLC.

2.6. Free Radical Scavenging Capacity

The 2,2-diphenyl-1-(2,4,6-trinitrophenyl) hydrazyl (DPPH) assay was used to compare the antioxidant activities of the AV and AP extracts. The two extracts of different concentrations were prepared with ddH_2O for dilution. Fifty microliters of the aqueous dilutions of the two extracts were mixed with 950 μL DPPH solution (0.024 mg/mL DPPH in absolute methanol) in duplicate. Fifty microliters ddH_2O was mixed with 950 μL DPPH solution in duplicate as the control. All these mixtures were incubated in the dark for 2 h. For the measurement, absolute methanol was used to set blank at 515 nm. The absorbance values of the control and the aqueous dilutions of AV and AP sample extracts were measured at 515 nm and recorded. The percentage free radical scavenging capacity (SR %) was calculated as $(1 - A_{Sample}/A_{Control}) \times 100\%$. The values of the concentration that scavenged half the quantity of DPPH free radical (SC50) of the two extracts were determined using GraphPad Prism version 5.01 for Windows (GraphPad Software, San Diego, CA, USA, www.graphpad.com).

2.7. Cell Culture and Cell Viability Assay

Human embryo liver LO2 cells were cultured in high glucose DMEM medium with 10% FBS (v/v) and 1% Penicillin/Streptomycin (v/v) in a humidified incubator maintained at 37 °C and 5% CO_2. Cells were subcultured with 0.25% trypsin every three days with a subcultivation ratio of 1:4. The protective effects of the two extracts against H_2O_2 induced cell damage were evaluated with MTT cell viability assay. Cells for viability assay were collected by trypsinization. They were seeded in a 96-well plate with 1×10^4 cells in 100 μL complete medium per well and incubated overnight. Then, the culture medium was removed and the seeded cells were treated with or without different concentrations of two extracts dissolved in complete medium and incubated for 2 h in the incubator at 37 °C. After that, the culture medium with AV and AP extracts was changed with complete medium and 300 μM H_2O_2 and further incubated for 24 h. After 24 h treatment, the medium with 300 μM H_2O_2 was replaced with MTT solution (0.5 mg/mL in serum-free high glucose DMEM medium) and cells were incubated for 3 h. At the end, the MTT solution was removed and 100 μL DMSO was added into each well to dissolve formazan in cells. After complete dissolution, the absorbance was measured at 570 nm, with 655 nm as reference. The percentage cell viability was calculated by: (Absorbance of drug treated group/Absorbance of model group) × 100%.

2.8. Statistical Analysis

All biological data are presented as means standard errors of means (S.E.M), and *n* denotes the number of replications for each data point. Statistical analysis was performed by using analysis of variance (ANOVA) to detect significant differences between multiple groups. A value of probability (p) < 0.05 was considered to be statistically significant. All statistical analysis tests were performed using GraphPad Prism 5.02 for Windows (GraphPad Software, San Diego, CA, USA). As for chromatographic fingerprint data, all the calculations and chemometrics data treatments were carried out using MATLAB versions 6.5. (Mathworks, Natick, MA, USA).

3. Results and Discussion

3.1. Chromatographic Fingerprint Studies between AV and AP by HPLC-DAD-MS

Prior to the chemical fingerprints comparison between AV and AP, optimization of extraction procedures and chromatographic conditions have to be established. In order to obtain the optimal extraction conditions, the influences of the solvents, extraction methods and durations were investigated while two markers, hyperoside and isoquercetrin [11,17] served as assessment indicators. Different percentage of ethanol (30%–95%) and methanol (30%–100%) were used for the preliminary studies. As a result, shown in Figure 1, 70% ethanol and 70% methanol were found to be the most effective by comparing the peak areas in HPLC-DAD chromatograms. Due to environmental concerns, 70% ethanol was selected as extraction solvent. Sonication and heat-reflux extraction methods were also examined, and both methods brought out similar extraction effects. Because of the easier operation, sonication was selected. Finally, different durations (10–60 min) were carried out and the results suggested that 30 min was optimal. Thus, sonication with 70% ethanol for 30 min was the optimal extracting condition for flavonoids in sample.

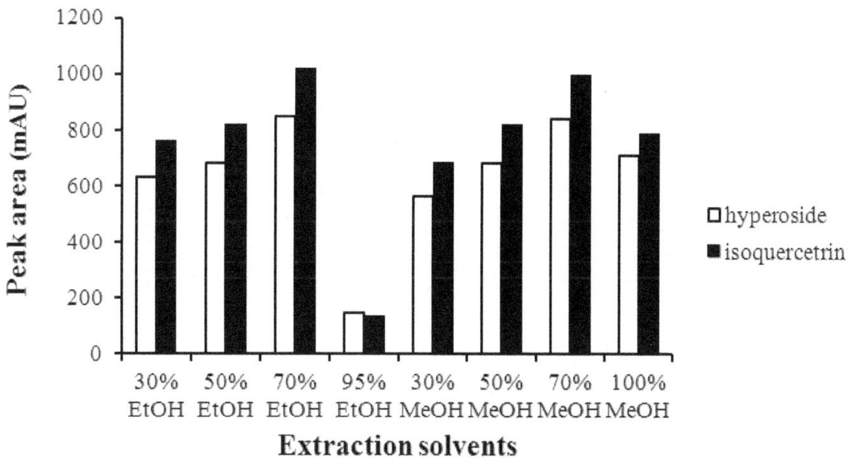

Figure 1. Effect of solvent in extracting hyperoside and isoquercitrin in AV samples.

In this study, acid was added into the mobile phase to minimize the tailing effect and achieved a better separation between hyperoside and isoquercetrin during the separation process. Formic acid was used in our work because it was applicable in the MS detection with good intensity. In addition, different chromatographic columns including Agilent Zorbax SB C18, ACE C18 and Grace Alltima C18 were tested. Agilent Zorbax was finally selected as it provided a shortened duration in separating the two markers. As for the detection wavelength, 360 nm was utilized as most of the compounds detected in the chromatogram possessed strong UV absorbance.

Figure 2 shows typical chromatograms of the standard solution containing the two flavonoids markers (50 ppm of hyperoside and 60 ppm of isoquercetrin) (Figure 2a) and the extracts of AV (Figure 2b) and AP (Figure 2c). The presence of the two flavonoids was confirmed by comparing the retention time and UV spectrum of the corresponding peak with those of the standards. The retention time of hyperoside and isoquercetrin are 20.79 min and 22.00 min, respectively. It was observed that chromatograms of AV and AP were different from each other. Due to the limited chemical information obtained from the UV spectra, LC-MS was adopted to identify the chemical constituents of AV and AP samples. In our work, electrospray ionization (ESI) methods in both negative and positive modes were compared in the detection of flavonoids. Negative mode ESI was found to be more sensitive for flavonoids and further utilized for flavonoids identification. Table 2 summarized the UV and MS data obtained from the chromatograms for AV and AP samples. A total of 10 flavonoids were characterized from the 70% ethanol extracts of the two species, and tentatively identified by comparing their mass spectra with those of reference isolated flavonoids reported in some previous literatures [12,18].

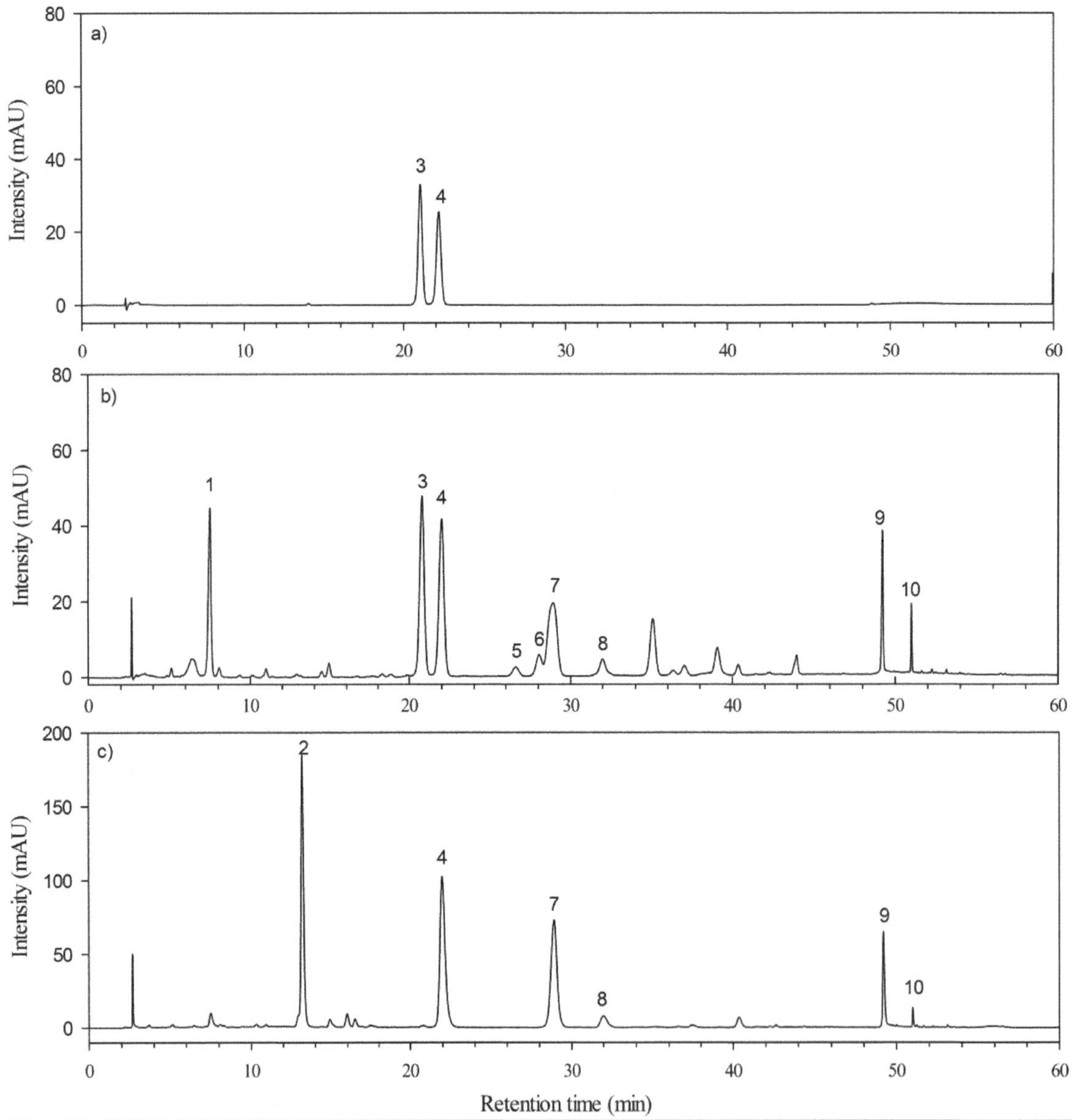

Figure 2. HPLC/UV (366 nm) chromatograms of the studied flavonoids components: (**a**) two flavonoids markers (50 ppm of hyperoside and 60 ppm of isoquercetrin) and (**b**) the extracts of *Apocynum venetum* L.; (**c**) extracts of *Apocynum pictum* : (1) Chlorogenic acid; (2) Rutin; (3) Hyperoside; (4) Isoquercetrin; (5) Acetylated hyperoside; (6) Trifolin; (7) Acetylated isoquercetrin; (8) Astragalin; (9) Quercetin; (10) Kaempferol.

Table 2. HPLC-DAD-MS data of major flavonoids identified in the extract of *Apocynum venetum* and *Apocynum pictum*.

Name of Flavonoids	Peak	Retention Time (min)	UV λ_{max}	MW	MS Fragmentation Data (ve)
Chlorogenic acid	1	7.50	240, 326	354	352.7, 190.7
Rutin	2	18.94	256, 354	610	608.7, 300.8
Hyperoside	3	20.79	256, 354	464	462.6, 300.8
Isoquercetrin	4	22.00	256, 354	464	462.6. 300.7
Acetylated hyperoside	5	26.59	256, 354	506	504.6, 462.8
Trifolin	6	28.04	264, 346	448	446.7, 283.7
Acetylated isoquercetrin	7	28.92	366, 354	506	504.6, 462.6
Astragalin	8	31.99	266, 346	448	446.6, 283.6
Quercetin	9	49.20	256, 370	302	300.8, 150.7
Kaempferol	10	51.00	266, 366	286	284.8, 150.8

3.2. Similarity Index Evaluation and Principal Component Analysis

To examine the chemical composition sameness and difference between AV and AP samples, the entire chromatographic fingerprints were applied and evaluated by similarity index (SI) calculation, recommended by State Food and Drug Administration of China (SFDA) since 2004. Before analysis, it is necessary to correct the retention time shift in all chromatograms obtained because multivariate analysis is highly sensitive to retention time shift among chromatograms. Local least square method [19], which has been successfully applied in chromatographic alignment in quality control of TCM, was utilized in this work.

Correlation coefficient is used as a measure for similarity index analysis. The chromatogram of a standardized extract is recommended to serve as the reference profile, but unfortunately it is rarely available. Therefore, the mean or median chromatographic profile of the data set is usually used. The results of correlation coefficient of each chromatogram with the mean HPLC fingerprint of AV and AP are shown in Table 3. As expected, chromatograms of the samples (shown Figure 3a) from the same species showed much better similarity as indicated by the correlation coefficient higher than 0.9. From Table 3, it is obvious that chromatograms of AV are different from AP and can be distinguish from those of the other species. Since AV samples (as well in AP samples) are similar to each other, sample number AV10 and AP2 are selected at random to carry out further antioxidant assay works.

Principal component analysis (PCA), which is chemometrics method commonly used in quality control of TCM, was utilized to investigate the variation within datasets. By looking into the loading and score plots (Figure 3b,c) on the first two principal components, rutin (peak 2) and hyperoside (peak 3) play significant factors in differentiating between AV and AP.

Table 3. Results calculated from similarity index method.

Sample Information	SI Value	
	Calculated from AV Mean Chromatograms	Calculated from AP Mean Chromatograms
AV1	0.952	0.347
AV2	0.976	0.379
AV3	0.987	0.314
AV4	0.987	0.347
AV5	0.927	0.394
AV6	0.986	0.350
AV7	0.968	0.348
AV8	0.977	0.355
AV9	0.964	0.316
AV10	0.968	0.312
AP1	0.235	0.954
AP2	0.354	0.997
AP3	0.463	0.961
AP4	0.330	0.993
AP5	0.350	0.994
AP6	0.397	0.988

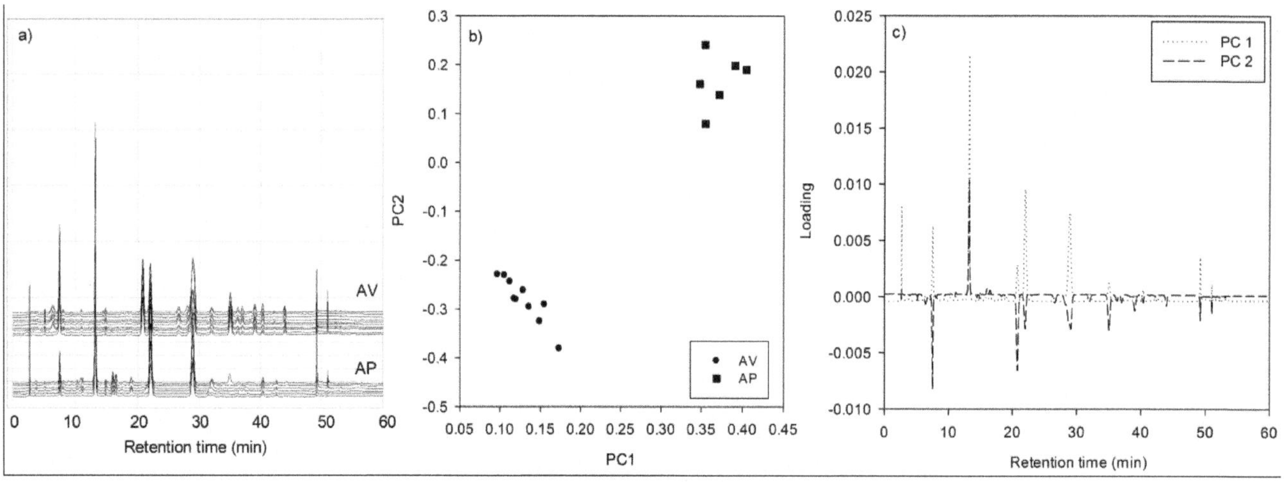

Figure 3. (**a**) HPLC chromatograms of 10 AV samples (upper) and 6 AP samples (lower; (**b**) score plot on the two principal components of HPLC fingerprint data from AV (●) and AP (■) samples; and (**c**) loading plot the PC1 (···) and PC2 (----) for HPLC fingerprint data.

The contents of hyperoside and isoquercetrin in all AV and AP samples were further carried out according to the assay method of Hong Kong Chinese Materia Medica Standard [16]. Validation of the developed HPLC-DAD method on quantification of hyperoside and isoquercetrin was carried out by linearity, precision, accuracy and stability. All the calibration curves of hyperoside and isoquercetrin exhibited good linearity ($r2 > 0.999$ within test ranges). The limit of detection (LOD) and the limit of quantitation (LOQ) of hyperoside and isoquercetrin were 42.5 ng/mL, 212.5 ng/mL and 39.4 ng/mL, 197.0 ng/mL, respectively. Also, the developed assay method of hyperoside and isoquercetrin had good accuracy with 103.26% ± 1.93% and 101.34% ± 2.43%, respectively, in sample ($n = 5$). Stability was tested with both standard and sample extract stored at room temperature within two days. The repeatability of the extracts was less than 5% and two flavonoids were found to be rather stable. A conclusion could be generalized that this assay showed good reproducibility.

The developed HPLC-UV method was applied to analyze two flavonoids in 10 batches of AV and six batches of AP samples and Table 4 summarized the overall results of the contents of hyperoside and isoquercetrin in all the samples collected. It is obvious that the average content of hyperoside in AV was much larger than that in AP. It is believed that hyperoside could serve as a good chemical marker to differentiate AV and AP samples.

Table 4. Content of Hyperoside and Isoquercetrin.

AV Sample	Content of Markers (mg/kg)		AP Sample	Content of Markers (mg/kg)	
	Hyperoside	Isoquercetrin		Hyperoside	Isoquercetrin
AV1	6414	5218	AP1	ND	6289
AV2	6050	5636	AP2	ND	13539
AV3	5542	6015	AP3	ND	17808
AV4	5463	6390	AP4	ND	14987
AV5	5351	5311	AP5	ND	13206
AV6	5894	4406	AP6	ND	14312
AV7	5438	4917			
AV8	5565	4900			
AV9	4232	3483			
AV10	5415	4448			
Mean	5536.4	5072.4	Mean	ND	13356

ND = Not detected.

3.3. Comparison between Free Radical Scavenging Capacities

The free radical scavenging capacities of the AV and AP sample extracts were measured with DPPH free radical scavenging assay (Figure 4). The overall free radical scavenging capacities of AV extract was found to be better than that of AP extract ($n = 3$). The SC_{50} values of AV and AP extracts were determined to be 0.01506 ± 0.00085 and 0.03118 ± 0.00084 mg extract/mL, respectively. The SC_{50} value of AV extract was significantly smaller than that of AP extract ($p < 0.0005$), indicating that AV extract provided higher antioxidant activity than AP extract.

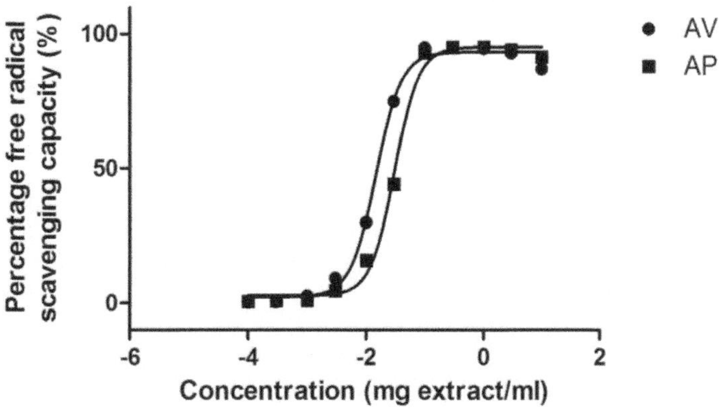

Figure 4. Percentage free radical scavenging capacity (SR %) of AV and AP extracts. Data are expressed as means ± SEM, $n = 3$.

Flavonoid compounds have been reported to possess significant antioxidant power [20,21]. In HPLC analysis, the flavonoid profile among these two samples varied tremendously. Additionally, the overall flavonoid content in AV was shown to be higher than that of the AP. AV extract was found to have a significantly higher free radical scavenging activity than AP extract. It may be contributed to by the difference in their overall flavonoid contents.

3.4. Protective Effect Against H_2O_2-Induced Damage in LO2 Cells

It is agreed that the complicated composition of medicinal herbs makes it impossible to estimate the anti-oxidant potency of individual sample with a single method. Additionally, chemical methods may not be able to completely reflect the actual situation in biological condition [22]. Human embryo liver LO2 cells are considered a normal liver cell line. It is commonly used to study the role of oxidative stress in liver [23]. Any protective effects offered by the herbal extracts in LO2 cells could reflect the beneficial efficacy in a more complex and diverse cell populations such as liver. As shown in Figure 5, H_2O_2 (300 mM) treatment for 24 h reduced the percentage cell viability of LO2 cells to 54.72% ± 3.24%, and 1 mg extract/mL AV extract (but not AP extract) pre-treatment for 2 h protected LO2 cells against H_2O_2-induced damage and increased the percentage cell viability to 68.37% ± 4.41% ($p < 0.05$). Both H_2O_2-induced damage in LO2 cells and DPPH free radical scavenging assay concluded that AV possess higher antioxidant activity that AP.

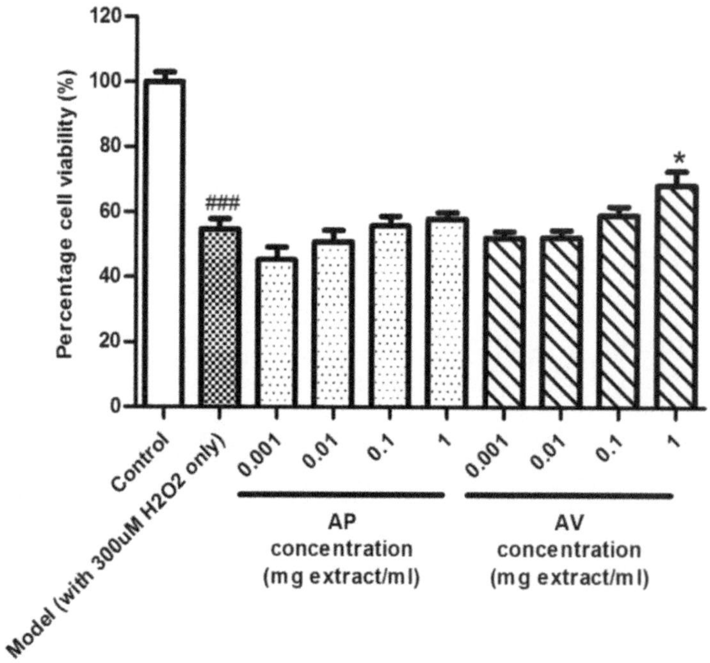

Figure 5. Effect of AV and AP extracts on H_2O_2-induced cytotoxicity in LO2 cells. Data are expressed as means ± SEM; $n = 4$. $^{###}$ $p < 0.001$ represents a significant difference when compared model with the control. * $p < 0.05$ represents a significant difference when compared the AV and AP extracts treated group with model.

4. Conclusions

Our overall results demonstrated that there are significant differences, including chemical constituents and biological activities, between AV and AP. AV is proven to have a better antioxidant activity than AP, both chemically and biologically. In addition, hyperoside could be a suitable chemical marker for discrimination between AV and AP and it is suggested to be used in quality assurance and control of AV.

Acknowledgments

This work was supported by the National Natural Science Foundation of China (No. 81303203), and was part of the work of the Hong Kong Chinese Material Medica Standard Project.

Author Contributions

Si-Bao Chen, Shun-Wan Chan and Daniel Kam-Wah Mok conceived of and designed the experiments. Chi-On Chan performed the experiments and wrote the paper. Ching-Ching Lau participated in HPLC analysis. Li-Jia Xu analyzed the data. Yam-Fung Ng performed the antioxidant assay. All of the authors have read and approved the final manuscript.

Conflicts of Interest

The authors declare no conflict of interest.

References

1. Yi, T.; Zhao, Z.Z.; Yu, Z.L.; Chen, H.B. Comparison of the anti-inflammatory and anti-nociceptive effects of three medicinal plants known as "Snow Lotus" herb in traditional Uighur and Tibetan medicines. *J. Ethnopharmacol.* **2010**, *128*, 405–411.
2. Editorial Committee of the Flora of China of Chinese Academy of Science. *Flora of China*; Science Press: Beijing, China, 1977; p. 157.
3. Editorial Committee of Chinese Pharmacopoeia. *Pharmacopoeia of the People's Republic of China*, 2010th ed.; China Medical Science and Technology Press: Beijing, China, 2010; Volume 1, p. 196.
4. Sandei, N.; Michiko, M.; Chisako, C. Flavonid content in *Apocynum* and *Poacnum* leaves and lipid-peroidation inhibiting efeet of flavonoids. *Nat. Med.* **1994**, *48*, 322–323.
5. Yokozawa, T.; Nakagawa, T. Inhibitory effects of Luobuma tea and its components against glucose-mediated protein damage. *Food Chem. Toxicol.* **2004**, *42*, 975–981.
6. Kagawa, T.; Nakazawa, Y.; Tagashira, E.; Takahashi, Y.; Onizuka, S.; Nishibe, S. Studies on antihypertensive effect of Luobuma (*Apocynum venetum* L.) leaf extract (1). *Nat. Med.* **2004**, *58*, 109–112.
7. Kim, D.W.; Yokozawa, T.; Hattori, M.; Kadota, S.; Namba, T. Effects of aqueous extracts of *Apocynum venetum* leaves on spontaneously hypertensive, renal hypertensive and NaCl-fed-hypertensive rats. *J. Ethnopharmacol.* **2000**, *72*, 53–59.

8. Butterweck, V.; Simbrey, K.; Seo, S.; Sasaki, T.; Nishibe, S. Long-term effects of an *Apocynum venetum* extract on brain monoamine levels and Beta-AR density in Rats. *Pharmacol. Biochem. Behav.* **2003**, *75*, 557–564.
9. Grundmann, O.; Nakajima, J.; Seo, S.; Butterweck, V. Anti-anxiety effects of *Apocynum venetum* L. in the elevated Plus Maze Test. *J. Ethnopharmacol.* **2007**, *110*, 406–411.
10. Kim, D.W.; Yokozawa, T.; Hattori, M.; Kadota, S.; Namba, T. Effects of aqueous extracts of *Apocynum venetum* leaves on hypercholesterolaemic rats. *Phytother. Res.* **1998**, *12*, 46–48.
11. Han, L.W.; Hou, J.J.; Zhao, L.; Liang, T.G.; Li, Q.S. Establishment of HPLC fingerprint and its application in identification of Folium Apocyni Veneti. *Chin. Trad. Herb. Drug* **2008**, *39*, 591–594.
12. An, H.J.; Wang, H.; Lan, Y.X.; Hashi, Y.; Chen, S.Z. Simultaneous qualitative and quantitative analysis of phenolic acids and flavonoids for the quality control of *Apocynum venetum* L. leaves by HPLC-DAD-ESI-IT-TOF-MS and HPLC-DAD. *J. Pharm. Biomed. Anal.* **2013**, *85*, 295–304.
13. Xie, W.; Zhang, X.; Wang, T.; Hu, J. Botany, traditional uses, phytochemistry and pharmacology of *Apocynum venetum* L. (Luobuma): A review. *J. Ethnopharmacol.* **2012**, *141*, 1–8.
14. Yokozawa, T.; Kashiwada, Y.; Hattori, M.; Chung, H.Y. Study on the components of Luobuma with peroxynitrite-scavenging activity. *Biol. Pharm. Bull.* **2002**, *25*, 748–752.
15. Wang, D.Q.; Li, G.Q.; Cheng, Z. The research progress and future respective on Luobuma. *Jiangsu Agric. Sci.* **2011**, *39*, 310–313.
16. Chinese Medicine Division, Department of Health, Government of the Hong Kong Special Administrative Region. *Hong Kong Chinese Medica Materia Standards*; Chinese Medicine Division, Department of Health, Government of the Hong Kong Special Administrative Region: Hong Kong, China, 2013; Volume 5.
17. Liu, X.H.; Zhang, Y.C.; Li, J.S.; Wang, M.; Wang, L.J. Simultaneous Determination of Four Flavonoids in Folium Apocyni Veneti by HPCE-DAD. *Chin. Pharm. J.* **2010**, *45*, 464–467.
18. Zhang, Y.; Liu, C.; Zhang, Z.; Wang, J.; Wu, G.; Li, S. Comprehensive separation and identification of chemical constituents from *Apocynum venetum* leaves by high-performance counter-current chromatography and high performance liquid chromatography coupled with mass spectrometry. *J. Chromatogr. B* **2010**, *878*, 3149–3155.
19. Li, B.Y.; Hu, Y.; Liang, Y.Z.; Huang, L.F.; Xu, C.J.; Xie, P.S. Spectral correlative chromatography and its application to analysis of chromatographic fingerprints of herbal medicines. *J. Sep. Sci.* **2004**, *27*, 581–588.
20. Chan, S.W.; Li, S.; Kwok, C.Y.; Benzie, I.F.F.; Szeto, Y.T.; Guo, D.J.; He, X.P.; Yu, P.H.F. Antioxidant activity of Chinese medicinal herbs. *Pharm. Biol.* **2008**, *46*, 587–595.
21. Coletta, A.; Berto, S.; Crupi, P.; Cravero, M.C.; Tamborra, P.; Antonacci, D.; Daniele, P.G.; Prenesti, E. Effect of viticulture practices on concentration of polyphenolic compounds and total antioxidant capacity of Southern Italy red wines. *Food Chem.* **2014**, *152*, 467–474.
22. Chan, E.; Wong, C.Y.; Wan, C.W.; Kwok, C.Y.; Wu, J.H.; Ng, K.M.; So, C.H.; Au, A.L.; Poon, C.C.; Seto, S.W.; *et al*. Evaluation of anti-oxidant capacity of root of *Scutellaria baicalensis* Georgi, in comparison with roots of *Polygonum multiflorum* Thunb and *Panax ginseng* CA Meyer. *Am. J. Chin. Med.* **2010**, *38*, 815–827.

23. Ye, L.; Shi, Z.; Liu, H.; Yang, X.; Wang, K. Gadolinium induced apoptosis of human embryo liver LO2 cell line by ROS-mediated AIF pathway. *J. Rare Earth* **2011**, *29*, 178–184.

Permissions

All chapters in this book were first published in Antioxidants, by MDPI; hereby published with permission under the Creative Commons Attribution License or equivalent. Every chapter published in this book has been scrutinized by our experts. Their significance has been extensively debated. The topics covered herein carry significant findings which will fuel the growth of the discipline. They may even be implemented as practical applications or may be referred to as a beginning point for another development.

The contributors of this book come from diverse backgrounds, making this book a truly international effort. This book will bring forth new frontiers with its revolutionizing research information and detailed analysis of the nascent developments around the world.

We would like to thank all the contributing authors for lending their expertise to make the book truly unique. They have played a crucial role in the development of this book. Without their invaluable contributions this book wouldn't have been possible. They have made vital efforts to compile up to date information on the varied aspects of this subject to make this book a valuable addition to the collection of many professionals and students.

This book was conceptualized with the vision of imparting up-to-date information and advanced data in this field. To ensure the same, a matchless editorial board was set up. Every individual on the board went through rigorous rounds of assessment to prove their worth. After which they invested a large part of their time researching and compiling the most relevant data for our readers.

The editorial board has been involved in producing this book since its inception. They have spent rigorous hours researching and exploring the diverse topics which have resulted in the successful publishing of this book. They have passed on their knowledge of decades through this book. To expedite this challenging task, the publisher supported the team at every step. A small team of assistant editors was also appointed to further simplify the editing procedure and attain best results for the readers.

Apart from the editorial board, the designing team has also invested a significant amount of their time in understanding the subject and creating the most relevant covers. They scrutinized every image to scout for the most suitable representation of the subject and create an appropriate cover for the book.

The publishing team has been an ardent support to the editorial, designing and production team. Their endless efforts to recruit the best for this project, has resulted in the accomplishment of this book. They are a veteran in the field of academics and their pool of knowledge is as vast as their experience in printing. Their expertise and guidance has proved useful at every step. Their uncompromising quality standards have made this book an exceptional effort. Their encouragement from time to time has been an inspiration for everyone.

The publisher and the editorial board hope that this book will prove to be a valuable piece of knowledge for researchers, students, practitioners and scholars across the globe.

List of Contributors

Mohd Fadzelly Abu Bakar
Institute for Tropical Biology and Conservation, Universiti Malaysia Sabah, Jalan UMS, Kota Kinabalu, Sabah 88400, Malaysia
Faculty of Science, Technology and Human Development, Universiti Tun Hussein Onn Malaysia (UTHM), Parit Raja, Batu Pahat, Johor 86400, Malaysia

Nor Ezani Ahmad
Institute for Tropical Biology and Conservation, Universiti Malaysia Sabah, Jalan UMS, Kota Kinabalu, Sabah 88400, Malaysia

Fifilyana Abdul Karim
Institute for Tropical Biology and Conservation, Universiti Malaysia Sabah, Jalan UMS, Kota Kinabalu, Sabah 88400, Malaysia

Syazlina Saib
Institute for Tropical Biology and Conservation, Universiti Malaysia Sabah, Jalan UMS, Kota Kinabalu, Sabah 88400, Malaysia

Lihua Wang
Department of Chemistry and Biochemistry, Kettering University, 1700 University Ave, Flint, MI 48504, USA

Elizabeth Santos
Department of Chemistry and Biochemistry, Kettering University, 1700 University Ave, Flint, MI 48504, USA

Desiree Schenk
Department of Chemistry and Biochemistry, Kettering University, 1700 University Ave, Flint, MI 48504, USA

Montserrat Rabago-Smith
Department of Chemistry and Biochemistry, Kettering University, 1700 University Ave, Flint, MI 48504, USA

Constant Anatole Pieme
Laboratory of Biochemistry, Department of Biochemistry and Physiological Sciences; Faculty of Medicine and Biomedical Sciences, University of Yaounde I, P.O. Box 1364 Yaounde, Cameroon

Joseph Ngoupayo
Department of pharmacy, Faculty of Medicine and Biomedical Sciences, University of Yaounde I, P.O. Box 1364 Yaounde, Cameroon

Claude Herve Khou-Kouz Nkoulou
Department of pharmacy, Faculty of Medicine and Biomedical Sciences, University of Yaounde I, P.O. Box 1364 Yaounde, Cameroon

Bruno Moukette Moukette
Laboratory of Biochemistry, Department of Biochemistry and Physiological Sciences; Faculty of Medicine and Biomedical Sciences, University of Yaounde I, P.O. Box 1364 Yaounde, Cameroon

Borgia Legrand Njinkio Nono
Department of Biochemistry, Faculty of Sciences; University of Yaounde I, P.O. Box 812 Yaounde, Cameroon

Vicky Jocelyne Ama Moor
Laboratory of Biochemistry, Department of Biochemistry and Physiological Sciences; Faculty of Medicine and Biomedical Sciences, University of Yaounde I, P.O. Box 1364 Yaounde, Cameroon

Jacqueline Ze Minkande
Department of Anesthesia and Reanimation, Faculty of Medicine and Biomedical Sciences, University of Yaounde I, P.O. Box 1364 Yaounde, Cameroon

Jeanne Yonkeu Ngogang
Laboratory of Biochemistry, Department of Biochemistry and Physiological Sciences; Faculty of Medicine and Biomedical Sciences, University of Yaounde I, P.O. Box 1364 Yaounde, Cameroon

Francisco J. Pérez-Cano
Departament de Fisiologia, Facultat de Farmàcia, Universitat de Barcelona; Av. Joan XXIII s/n, Edifici B, 3[a] planta, E-08028 Barcelona, Spain
Institut de Recerca en Nutrició i Seguretat Alimentària (INSA-UB), E-08028 Barcelona, Spain

Malen Massot-Cladera
Departament de Fisiologia, Facultat de Farmàcia, Universitat de Barcelona; Av. Joan XXIII s/n, Edifici B, 3[a] planta, E-08028 Barcelona, Spain
Institut de Recerca en Nutrició i Seguretat Alimentària (INSA-UB), E-08028 Barcelona, Spain

Maria J. Rodríguez-Lagunas
Institut de Recerca en Nutrició i Seguretat Alimentària (INSA-UB), E-08028 Barcelona, Spain
Departament de Ciències Fisiològiques II, Facultat de Medicina, Universitat de Barcelona, Feixa Llarga s/n, L'Hospitalet de Llobregat, E-08907 Barcelona, Spain

Margarida Castell
Departament de Fisiologia, Facultat de Farmàcia, Universitat de Barcelona; Av. Joan XXIII s/n, Edifici B, 3[a] planta, E-08028 Barcelona, Spain

Institut de Recerca en Nutrició i Seguretat Alimentària (INSA-UB), E-08028 Barcelona, Spain

Dildar Ahmed
Department of Chemistry, Forman Christian College (A Chartered University), Lahore-54600, Pakistan

Khaizran Fatima
Department of Chemistry, Forman Christian College (A Chartered University), Lahore-54600, Pakistan

Ramsha Saeed
Department of Chemistry, Forman Christian College (A Chartered University), Lahore-54600, Pakistan

Celina Maria P. Guerra Dore
Laboratory of Glycobiology, Department of Biochemistry, Federal University of Rio Grande do Norte, Av. Salgado Filho, 3000, Bairro L. Nova, CEP 59072-940, Natal, Brazil

Monique Gabriela das Chagas F. Alves
Laboratory of Glycobiology, Department of Biochemistry, Federal University of Rio Grande do Norte, Av. Salgado Filho, 3000, Bairro L. Nova, CEP 59072-940, Natal, Brazil

Maria da Glória L. Santos
Laboratory of Glycobiology, Department of Biochemistry, Federal University of Rio Grande do Norte, Av. Salgado Filho, 3000, Bairro L. Nova, CEP 59072-940, Natal, Brazil

Leonardo Augusto R. de Souza
Laboratory of Glycobiology, Department of Biochemistry, Federal University of Rio Grande do Norte, Av. Salgado Filho, 3000, Bairro L. Nova, CEP 59072-940, Natal, Brazil

Iuri Goulart Baseia
Laboratory of Mycology, Department of Botany, Ecology and Zoology, Federal University of Rio Grande do Norte, Av. Salgado Filho, 3000, Bairro L. Nova 59078-970, Natal, Brazil

Edda Lisboa Leite
Laboratory of Glycobiology, Department of Biochemistry, Federal University of Rio Grande do Norte, Av. Salgado Filho, 3000, Bairro L. Nova, CEP 59072-940, Natal, Brazil

Anelise Samara Nazari Formagio
Faculty of Agricultural Sciences, University Federal of Grande Dourados (UFGD), 79804-970/533 Dourados-MS, Brazil

Carla Roberta Ferreira Volobuff
Biological and Environmental Sciences, University Federal of Grande Dourados (UFGD), 79804-970/533 Dourados-MS, Brazil

Matheus Santiago
Biological and Environmental Sciences, University Federal of Grande Dourados (UFGD), 79804-970/533 Dourados-MS, Brazil

Claudia Andrea Lima Cardoso
Chemistry Program, State University of Mato Grosso do Sul (UEMS), Mato Grosso do Sul, 79804-970/351 Dourados-MS, Brazil

Maria do Carmo Vieira
Faculty of Agricultural Sciences, University Federal of Grande Dourados (UFGD), 79804-970/533 Dourados-MS, Brazil

Zefa Valdevina Pereira
Biological and Environmental Sciences, University Federal of Grande Dourados (UFGD), 79804-970/533 Dourados-MS, Brazil

Joash Ban Lee Tan
School of Science, Monash University Sunway Campus, Bandar Sunway 46150, Malaysia

Wei Jin Yap
School of Science, Monash University Sunway Campus, Bandar Sunway 46150, Malaysia

Shen Yeng Tan
School of Science, Monash University Sunway Campus, Bandar Sunway 46150, Malaysia

Yau Yan Lim
School of Science, Monash University Sunway Campus, Bandar Sunway 46150, Malaysia

Sui Mae Lee
School of Science, Monash University Sunway Campus, Bandar Sunway 46150, Malaysia

Sing Pei Tan
School of Environmental and Life Sciences, University of Newcastle, Ourimbah, NSW 2258, Australia

Costas Stathopoulos
Faculty of Bioscience Engineering, Ghent University Global Campus, Incheon 406-840, Korea

Sophie Parks
School of Environmental and Life Sciences, University of Newcastle, Ourimbah, NSW 2258, Australia
Central Coast Primary Industries Centre, NSW Department of Primary Industries, Ourimbah, NSW 2258, Australia

Paul Roach
School of Environmental and Life Sciences, University of Newcastle, Ourimbah, NSW 2258, Australia

Talambedu Usha
Department of Biotechnology, Maharani Lakshmi Ammanni College for Women, Bangalore-560012, India

Sushil Kumar Middha
Department of Biotechnology, Maharani Lakshmi Ammanni College for Women, Bangalore-560012, India

Malay Bhattacharya
Department of Botany, Kalimpong College, Kalimpong-734301, India

Prakash Lokesh
Padmashree Institute of Management and Science, Bangalore-560060, India

Arvind Kumar Goyal
Bamboo Technology, Department of Biotechnology, Bodoland University, Kokrajhar-783370, India

Sakina Razack
Biochemistry and Nanosciences Division, Defence Food Research Laboratory, Mysore-570011, India

Kandikattu Hemanth Kumar
Biochemistry and Nanosciences Division, Defence Food Research Laboratory, Mysore-570011, India

Ilaiyaraja Nallamuthu
Biochemistry and Nanosciences Division, Defence Food Research Laboratory, Mysore-570011, India

Mahadeva Naika
Applied Nutrition Division, Defence Food Research Laboratory, Mysore-570011, India

Farhath Khanum
Biochemistry and Nanosciences Division, Defence Food Research Laboratory, Mysore-570011, India

Peter F. Surai
Department of Microbiology and Biochemistry, Faculty of Veterinary Medicine, Trakia University, Stara Zagora 6000, Bulgaria
Department of Animal Nutrition, Faculty of Agricultural and Environmental Sciences, Szent Istvan University, Gödöllo H-2103, Hungary
Department of Veterinary Expertise and Microbiology, Faculty of Veterinary Medicine, Sumy National Agrarian University, Sumy 40021, Ukraine
Odessa National Academy of Food Technology, Odessa 65039, Ukraine

Ricardo Bosch
Department of Dermatology, Virgen de la Victoria University Hospital, Málaga 29010, Spain
Dermatology and Medicine Department, University of Málaga, Málag 29071, Spain

Neena Philips
School of Natural Sciences, Fairleigh Dickinson University, 1000 River Road, Teaneck, NJ 07666, USA

Jorge A. Suárez-Pérez
Department of Dermatology, Virgen de la Victoria University Hospital, Málaga 29010, Spain
Dermatology and Medicine Department, University of Málaga, Málag 29071, Spain

Angeles Juarranz
Biology Department, Universidad Autónoma de Madrid, Madrid 28903, Spain

Avani Devmurari
School of Natural Sciences, Fairleigh Dickinson University, 1000 River Road, Teaneck, NJ 07666, USA

Jovinna Chalensouk-Khaosaat
School of Natural Sciences, Fairleigh Dickinson University, 1000 River Road, Teaneck, NJ 07666, USA

Salvador González
Dermatology Service, Memorial Sloan-Kettering Cancer Center, New York, NY 10022, USA
Ramon y Cajal Hospital, Alcala University, Madrid 28034, Spain

Chi-On Chan
State Key Laboratory of Chinese Medicine and Molecular Pharmacology, Department of Applied Biology and Chemical Technology, The Hong Kong Polytechnic University, Shenzhen 518057, China
Department of Applied Biology and Chemical Technology, The Hong Kong Polytechnic University, Hong Kong, China

Ching-Ching Lau
Department of Applied Biology and Chemical Technology, The Hong Kong Polytechnic University, Hong Kong, China

Yam-Fung Ng
Department of Applied Biology and Chemical Technology, The Hong Kong Polytechnic University, Hong Kong, China

Li-Jia Xu
The Institute of Medicinal Plant Development, Chinese Academy of Medical Sciences and Peking Union Medical College, Beijing, 100193, China

Si-Bao Chen
State Key Laboratory of Chinese Medicine and Molecular Pharmacology, Department of Applied Biology and Chemical Technology, The Hong Kong Polytechnic University, Shenzhen 518057, China

Department of Applied Biology and Chemical Technology, The Hong Kong Polytechnic University, Hong Kong, China

Shun-Wan Chan
State Key Laboratory of Chinese Medicine and Molecular Pharmacology, Department of Applied Biology and Chemical Technology, The Hong Kong Polytechnic University, Shenzhen 518057, China
Department of Applied Biology and Chemical Technology, The Hong Kong Polytechnic University, Hong Kong, China

Daniel Kam-Wah Mok
State Key Laboratory of Chinese Medicine and Molecular Pharmacology, Department of Applied Biology and Chemical Technology, The Hong Kong Polytechnic University, Shenzhen 518057, China
Department of Applied Biology and Chemical Technology, The Hong Kong Polytechnic University, Hong Kong, China
The Institute of Medicinal Plant Development, Chinese Academy of Medical Sciences and Peking Union Medical College, Beijing, 100193, China